Windows Server 2022 Administration Fundamentals

Third Edition

A beginner's guide to managing and administering Windows Server environments

Bekim Dauti

BIRMINGHAM—MUMBAI

Windows Server 2022 Administration Fundamentals
Third Edition

Copyright © 2022 Packt Publishing

All rights reserved. No part of this book may be reproduced, stored in a retrieval system, or transmitted in any form or by any means, without the publisher's prior written permission, except for brief quotations embedded in critical articles or reviews.

Every effort has been made to prepare this book to ensure the accuracy of the information presented. However, the information contained in this book is sold without warranty, either express or implied. Neither the author nor Packt Publishing or its dealers and distributors will be held liable for any damages caused or alleged to have been caused directly or indirectly by this book.

Packt Publishing has endeavored to provide trademark information about all of the companies and products mentioned in this book by the appropriate use of capital. However, Packt Publishing cannot guarantee the accuracy of this information.

Group Product Manager: Vijin Boricha
Publishing Product Manager: Mohd Riyan Khan
Senior Editor: Tanya D'cruz
Technical Editor: Rajat Sharma
Copy Editor: Safis Editing
Project Coordinator: Shagun Saini and Deeksha Thakkar
Proofreader: Safis Editing
Indexer: Pratik Shirodkar
Production Designer: Aparna Bhagat
Marketing Coordinator: Hemangi Lotlikar

First published: December 2017
Second edition: October 2019
Third edition: September 2022

Production reference: 1180822

Published by Packt Publishing Ltd.
Livery Place
35 Livery Street
Birmingham
B3 2PB, UK.

ISBN 978-1-80323-215-7

www.packt.com

It is estimated that there are 5 billion (source: Statista) internet users worldwide, including politicians. Will that then be enough to contribute to stopping the wars and bringing PEACE to everyone globally?

– Bekim Dauti

Contributors

About the author

Bekim Dauti works in the administration of servers and computer networks, and training in Cisco, CompTIA and Microsoft.

He has a bachelor's degree from the University of Tirana and a master's degree from UMGC Europe, both in IT. Additionally, he has nearly 20 years of experience as a **Cisco Certified Academy Instructor** (**CCAI**) and more than 15 years of experience as a **Microsoft Certified Trainer** (**MCT**). Bekim holds several certifications from vendors such as ECDL, Certiport, CompTIA, Cisco, Microsoft, and Sun Microsystems. Bekim has contributed to nearly 20 books and dozens of articles for *PC World Albanian* and *CIO Albanian*. He founded Dautti LLC.

These days, he blogs on *Bekim Dauti's Blog*. In addition, he works as a Microsoft Certified Trainer.

> *"I thank God for giving me life, health, and the opportunity to contribute through knowledge sharing. May God Almighty reward my family, friends, the folks at Packt Publishing, my colleagues at elev8 LLC., and everyone who supported me in writing this book. Last but not least, peace and blessings to every reader."*

About the reviewer

Premnath Sambasivam is a server engineer with 10 years of experience in Windows, Azure, VMware, and SCCM administration. He is an MCSE Cloud Platform and Infrastructure professional and a Microsoft-certified Azure architect. He has developed and deployed Microsoft System Center Configuration Manager solutions to manage over 6,000 assets in his client's environment and various VMware solutions. Premnath is a technology enthusiast who loves learning and exploring new technologies. He is currently a senior cloud engineer for one of the major retail brands in the USA. He has also reviewed Packt Publishing's books like *Mastering Windows Server 2019* and *Mastering Windows Security and Hardening*.

> *"I want to thank my wife and son for encouraging me to spend time learning. Reviewing books also refreshes our memory and lets us learn about the latest technologies and software improvements. Special thanks to my mom and dad for always being supportive."*

Table of Contents

Preface   xv

Part 1: Introducing Windows Server and Installing Windows Server 2022

1

Getting Started with Windows Server   3

Technical requirements	4
Getting to know computer networks	4
What is a computer network?	4
The types of computer networks	5
Exploring computer network components	**9**
Clients and servers	9
Hosts and nodes	10
Investigating computer network architectures	**11**
P2P network architecture	11
Client/server network architecture	12
Getting to know IP addressing and subnetting	**13**
IPv4 network addresses	13
IPv6 network addresses	14
IPv4 subnetting	14
Exploring servers	**15**
Server hardware and software	15
Server sizes, form factors, and shapes	16
Understanding a NOS	**17**
Windows Server overview	18
Linux Server overview	19
macOS Server overview	19
Understanding Windows Server	**20**
The Windows Server timeline	21
Chapter exercise – downloading Windows Server 2022	**21**
Summary	23
Questions	23
Further reading	24

2
Introducing Windows Server 2022

Technical requirements	25
An overview of Windows Server 2022	25
Windows Server 2022 editions	27
Windows Server 2022 versus Windows Server 2019	28
The minimum and recommended system requirements	29
What's new in Windows Server 2022?	29
Microsoft Edge Chromium	30
Azure hybrid center	30
Storage Migration Service	31
Storage Replica	32
Secured-core server	33
Azure Kubernetes Service	34
Containers	36
Windows Admin Center	37
Chapter exercise – downloading Windows Admin Center	37
Summary	38
Questions	39
Further reading	40

3
Installing Windows Server 2022

Technical requirements	41
Understanding the installation of Windows Server 2022	42
Getting to know the partition schemes	42
Getting to know the boot options	43
Accessing the advanced startup options	43
Various Windows Server 2022 installation methods	45
Choosing Desktop Experience, Server Core, or Nano Server installation	46
A clean installation	46
Installing over a network	51
Unattended installation	53
An in-place upgrade	58
Migrating network services	61
Trying Windows Server 2022 in Azure	64
Chapter exercise – setting up WDS	67
Installing WDS	67
Setting up WDS	69
Summary	74
Questions	75
Further reading	75

4
Post-Installation Tasks in Windows Server 2022

Technical requirements	77
Understanding devices and device drivers	78
Getting to know computer devices and device drivers	78
Working with devices and device drivers	80

Getting to know PnP, IRQ, DMA, and driver signing	87	Using Server Configuration in Server Core	106

Understanding the Windows Registry and its services — 90

Windows Server Registry	90
Windows Server services	90
Working with Windows Registry and its services	91

Understanding Windows Server initial configuration — 105

Using Server Manager in Desktop Experience — 105

Chapter exercise – performing an initial Windows Server configuration — 107

Performing Windows Server initial configuration using Server Manager	107
Performing Windows Server initial configuration using Server Configuration	114
Summary	**120**
Questions	**120**
Further reading	**121**

Part 2: Setting Up Windows Server 2022

5

Directory Services in Windows Server 2022 — 125

Technical requirements	**126**	The hosts and lmhosts files	139
Understanding the Active Directory infrastructure	**126**	Hostnames	140
		DNS zones	141
DC	128	WINS	142
Domains	128	UNC	143
Tree domain	129	**Understanding OUs and containers**	**143**
Forest	130	What are OUs?	143
Child domain	131	Default containers	144
Operations master roles	132	Hidden default containers	145
Comparing a domain with a workgroup	133	Uses of default containers	145
Trust relationship	133	Delegating control to an OU	146
Functional levels	133	**Understanding accounts and groups**	**146**
Namespaces	135	Domain accounts	147
Sites	136	Local accounts	148
Replication	136	User profiles	149
Schema	136	Computer accounts	150
Microsoft Passport	136	Group types	150
Understanding DNS	**137**	Default groups	151
Adding the DNS role	137		

Group scopes	152	server to a DC	154
Group nesting	153	Summary	157
Chapter exercise – installing the AD DS and DNS roles and promoting the		Questions	157
		Further reading	158

6

Adding Roles to Windows Server 2022 — 161

Technical requirements	162	Understanding Remote Access	178
Understanding server roles and features	162	Remote Assistance	179
		RSAT	180
Server roles	162	RDS	181
Role services	162	RDS Licensing	182
Server features	162	RDG	183
Server Manager	163	VPN	184
Understanding application servers	164	App-V	185
Email server	164	Multiple ports	186
Database server	166	Understanding file and print services	186
Collaboration server	166	File Services role	186
Monitoring server	167	PDS role	187
Data protection server	167	Understanding user rights, NTFS permissions, and share permissions	193
Understanding web services	168		
IIS	168	Understanding file server auditing	196
WWW	170	Chapter exercise – installing the Web Server (IIS) and PDS roles	197
FTP	171		
Separate worker processes	172	Installing the Web Server (IIS) role	197
Adding components to the IIS	173	Installing a PDS role	198
Sites	174		
Ports	175	Summary	200
SSL	176	Questions	200
Certificates	177	Further reading	201

Part 3: Configuring Windows Server 2022

7

Group Policy in Windows Server 2022 205

Technical requirements	205	Chapter exercise – examples of GPOs for system administrators	216
Understanding GP	206	Renaming the administrator account	216
Managing GPOs	207	Renaming the guest account	217
GPO configuration settings	209	Blocking the Microsoft accounts	217
Processing GPOs	210	Prohibiting access to the Control Panel and PC settings	218
Types of GP editors	211	Denying access to all removable storage classes	218
Local Group Policy Editor	211		
Updating local GPOs	212	Summary	218
GPO configuration settings	213	Questions	219
		Further reading	220

8

Virtualization with Windows Server 2022 221

Technical requirements	221	Understanding checkpoints	232
Understanding server virtualization	222	VHD and VHDX formats	235
Virtualization modes	222	P2V conversion	236
The Hyper-V architecture	224	V2P conversion	237
Hyper-V's installation requirements	225	Configuring VM settings	237
Nested virtualization	225	Managing VMs	239
Getting to know Hyper-V Manager	226	Chapter exercise – installing Hyper-V on Windows Server 2022	240
Configuration settings in Hyper-V	227		
Creating and configuring VHDs	228	Summary	242
Managing a VM's virtual memory	229	Questions	243
Setting up a virtual network	231	Further reading	244

9

Storing Data in Windows Server 2022 — 245

Technical requirements	246	Resiliency using S2D	259
Understanding storage technologies	246	High availability	259
Exploring different storage types	246	**Understanding disks**	260
Adapter and controller types	251	HDDs	260
Serial bus technologies	251	SSDs	261
Storage protocols	252	ODDs	262
File-sharing protocols	252	Basic disk	263
The HBA and FC switches	253	Dynamic disk	263
The iSCSI hardware	253	Mount points	265
S2D	253	Filesystems	265
Dedup	254	Mounting a VHD	266
Storage tiering	255	DFS	267
Managing storage	256	**Chapter exercise – enabling Dedup on Windows Server 2022**	268
Understanding RAID	257	**Summary**	269
Types of RAID	258	**Questions**	269
Hardware versus software RAID	258	**Further reading**	270
SDS	259		

Part 4: Keeping Windows Server 2022 Up and Running

10

Tuning and Maintaining Windows Server 2022 — 273

Technical requirements	273	Network interface	277
Understanding server hardware components	274	32-bit and 64-bit architectures	278
		Removable drives	278
Processor	274	Graphics cards	279
Memory	275	Cooling	280
Disk	277	Power supply	280
		Physical ports	281

Understanding performance monitoring	**282**	**Chapter exercise – working with the Performance Logs & Alerts service**	**290**
Performance monitoring methodology	282	Starting the Performance Logs & Alerts service	290
Performance monitoring procedures	283	Accessing the Performance Monitor logs folder	291
Server baselines	283		
Performance Monitor	284	Creating performance data logs	292
Resource Monitor	285	Setting up performance counter alerts	293
Task Manager	287	**Summary**	**294**
Performance counters	288	**Questions**	**295**
Understanding logs and alerts	**289**	**Further reading**	**296**

11

Updating and Troubleshooting Windows Server 2022   297

Technical requirements	**298**	Bootloader	316
Understanding updates	**298**	Boot sector	316
Understanding Windows Update	298	Boot menu	316
Updating Microsoft programs	301	Safe Mode	317
Updating third-party programs	301	**Understanding business continuity**	**318**
Updating the device drivers	303	DRP	319
Getting to know WSUS	304	Data redundancy	319
Understanding the troubleshooting methodology	**306**	Clustering	319
		Folder redirection	320
Best practices, guidelines, and procedures	306	Backup and restore	321
Troubleshooting process	306	Active Directory (AD) restore	323
The systematic versus the specific approach	307	Power redundancy	325
Troubleshooting procedures	307	**Chapter exercise – using Event Viewer to monitor and manage logs**	**325**
ITIL	308		
Event Viewer	309	Setting up centralized monitoring	325
Understanding the startup process	**310**	Filtering Event Viewer logs	327
BIOS	310	Changing the default logs location	327
UEFI	311	**Summary**	**329**
TPM	312	**Questions**	**329**
POST	313	**Further reading**	**330**
MBR	313		
BCD	314		

Part 5: Studying and Preparing for Microsoft Certification Exams

12

Preparing for Microsoft Certifications — 333

What is Microsoft Certification?	334
What is Microsoft role-based certification?	334
Who should take a Microsoft Certification exam?	335
Which skills are measured by Microsoft Certification exams?	336
Deploy and manage Active Directory Domain Services (AD DS) in on-premises and cloud environments (30–35%)	337
Manage Windows Servers and workloads in a hybrid environment (10–15%)	338
Manage virtual machines and containers (15–20%)	339
Implement and manage an on-premises and hybrid networking infrastructure (15–20%)	341
Manage storage and file services (15–20%)	342
What should you expect in a Microsoft Certification exam?	343
How should you prepare for a Microsoft Certification exam?	343
How do you register for a Microsoft Certification exam?	344
On the day of the Microsoft Certification exam	344
New Microsoft Certification validity period and renewal format	345
Summary	346
Further reading	346

Assessments — 347

Index — 355

Other Books You May Enjoy — 374

Preface

Windows Server 2022 is Microsoft's latest operating system for servers as part of the Windows NT family of operating systems, based on the Windows 10 platform.

With Windows Server 2022, Microsoft is continuing to enhance and advance its OS for servers by making it robust, high-performing, secure, and cloud-enabled. The vision was to bring the cloud to everyone by providing platforms and tools to help build IT solutions that drive success. In addition, with Windows Server 2022, Microsoft has consolidated its status in the world of cloud service providers by competing head to head with the **Amazon Web Services cloud** (**AWS cloud**). For this reason, there has been no better time to become a cloud-focused system administrator.

This book begins with the computer network essentials and then moves on to the world of Windows Server 2022. It covers all aspects of the administration-level tasks and activities required to become an expert in Microsoft Windows Server 2022. It begins by introducing Windows Server and Windows Server 2022 and then gradually builds up its content with the installation and deployment of Windows Server 2022 in *Chapter 3, Installing Windows Server 2022*. After becoming familiar with Windows Server's 2022 post-installation tasks in *Chapter 4, Post-Installation Tasks in Windows Server 2022*, you will start functionalizing Windows Server 2022 by adding roles. Doing so, you will find out the following:

- What is a domain controller?
- How to set up a file and print server
- Configuring a web server and hosting a website
- Virtualizing your IT environment
- Automating Windows Server 2022 deployment
- Centrally managing Windows Server 2022 updates

With the help of multiple hands-on exercises, you will gain an immense understanding of Windows Server 2022, which will help you solve complex tasks quickly. At the end of the book, you will be exposed to maintenance and troubleshooting tasks where, with the help of best practices, you will manage Windows Server 2022 with ease.

At its heart, this book aims to teach you the system administrator's craft. Hence, to validate your skills and the knowledge gained from this book, each chapter ends with a concept summary and questionnaire to help you take full advantage of the content. By the end of this book, you will have enough knowledge to administer and manage Windows Server 2022 with ease and be able to get informed and learn about Microsoft certifications just in case you set yourself a challenge such as passing an exam.

Who this book is for

This book is for you if you are an IT professional interested in deploying and configuring Windows Server 2022. This book will also help you get to know about the Microsoft certifications.

What this book covers

Chapter 1, *Getting Started with Windows Server*, introduces Windows Server. At the beginning of this chapter, there is a recap of the most fundamental concepts of computer networks. This chapter is organized into two parts, where each attempts to provide a concise yet complete description of the basic concepts of computer networks. In addition, definitions of key terms such as hosts, nodes, peer-to-peer, and clients/servers are covered in the *Computer network overview* section.

Chapter 2, *Introducing Windows Server 2022*, introduces you to Windows Server 2022. Windows Server 2022 is developed by Microsoft as part of the Windows NT family of operating systems and concurrently with Windows 10 version 21H2. The *Windows Server overview* section uncovers the essentials of Windows Server 2022. In addition, it outlines the various Windows Server 2022 editions and compares Windows Server 2022 to Windows Server 2019 with a focus on what is new in Windows Server 2022.

Chapter 3, *Installing Windows Server 2022*, provides detailed instructions for installing Windows Server 2022. The step-by-step instructions, driven by easy-to-understand graphics, show you how to master the installation of Windows Server 2022. You will quickly learn the installation process without hitting any obstacles. This chapter is an excellent collection of how-to tips and provides information on getting the job done efficiently.

Chapter 4, *Post-Installation Tasks in Windows Server 2022*, explains the steps required during the post-installation stage, including managing devices and device drivers, checking the registry and status of services, and taking care of the initial server configuration. This chapter is divided into three parts. Each topic is accompanied by step-by-step instructions driven by targeted, easy-to-understand graphics.

Chapter 5, *Directory Services in Windows Server 2022*, introduces you to directory services. Now that you have learned how to install Windows Server 2022 and run the initial server configuration, it is time to set up the first services in your organization's IT infrastructure. With that in mind, this chapter explains directory services. Additionally, you will become familiar with **Organizational Units** (**OUs**), default containers, user accounts, and groups to organize your domain's user and computer accounts.

Chapter 6, *Adding Roles to Windows Server 2022*, explains what a role is and the importance of roles in determining the server's function when providing network services. You will also know all the parts and features that Windows Server 2022 supports. Finally, you will learn how to add roles to the server and the requirements after adding functions so that you can set up client/server network services whenever required.

Chapter 7, *Group Policy in Windows Server 2022*, helps you understand **Group Policy** (**GP**) in Windows Server. You will learn about GP processing, become familiar with the GP Management Console, learn about computer and user policies, and learn about local procedures when your server is not part of a domain. At the same time, you will learn the steps involved in configuring computer and user policies in a domain-based network.

Chapter 8, *Virtualization with Windows Server 2022*, teaches you virtualization concepts and familiarizes you with Hyper-V software, enabling the virtualization of Windows-based servers. You will discover the steps to add the Hyper-V role to your server, familiarize yourself with Hyper-V Manager, and learn the steps it takes to create virtual machines. That way, you will understand what virtualization is and how you can enable the Hyper-V role and create virtual machines.

Chapter 9, *Storing Data in Windows Server 2022*, explains storage technologies. As well as understanding storage technologies in general, you will learn about various related topics. These include physical interfaces and disk controllers. We will also explore how data is stored in a medium, the types of storage systems used in network environments, and various storage protocols. Additionally, you will get to know the concepts and types of RAID.

Chapter 10, *Tuning and Maintaining Windows Server 2022*, covers the best practices and considerations for server hardware. By understanding the importance of a server's role in a computer network and learning about each component, we can be vigilant when selecting server hardware. In addition to this, this chapter teaches you server performance monitoring methodologies and procedures. Performance monitoring will help you identify the cause of server performance issues early on.

Chapter 11, *Updating and Troubleshooting Windows Server 2022*, outlines the server startup process; advanced boot options and Safe Mode; backup and restore; the disaster recovery plan; and how to update the operating system, hardware, and software. Event Viewer is also mentioned, which will help you monitor different logs in your system, thus helping you troubleshoot and solve problems. In this way, you will be able to minimize downtime, expressed as money lost from a business point of view.

Chapter 12, *Preparing for Microsoft Certifications*, offers an overview of the Microsoft certifications, including a look at the skills measured in the exam. Additionally, this chapter explains the role-based Microsoft certifications and how to register for the exam. Furthermore, you will find valuable resources to help you gather as much information as possible about the exam in general, discover what it takes to pass it, and, by doing so, launch a successful career.

Assessments provides you with answers to the chapter questions. In addition, many questions accompany each chapter to help you reinforce the concepts and definitions. With this appendix, you can check your answers to those questions.

To get the most out of this book

You must have solid experience working with the Windows 10/11 operating system and have a solid knowledge of computer networks and network operating systems.

Make sure you have a computer with a processor that supports virtualization technology and has 8 GB or 16 GB of RAM.

Download the color images

We also provide a PDF file with color images of the screenshots and diagrams used in this book. You can download it here: https://packt.link/8Cdqe.

Conventions used

There are several text conventions used throughout this book.

`Code in text`: Indicates code words in the text, database table names, folder names, filenames, file extensions, pathnames, dummy URLs, user input, and Twitter handles. Here is an example: "Unlike IPv4, IPv6 is a 128-bit address size of 8 hextets with 16 bits each, divided by a colon for simplicity of interpretation (for example, `2001:0DB8:85A3:0000:0000:8A2E:0370:7334`)."

A block of code is set as follows:

```
html, body, #map {
  height: 100%;
  margin: 0;
  padding: 0
}
```

When we wish to draw your attention to a particular part of a code block, the relevant lines or items are set in bold:

```
[default]
exten => s,1,Dial(Zap/1|30)
exten => s,2,Voicemail(u100)
exten => s,102,Voicemail(b100)
exten => i,1,Voicemail(s0)
```

Any command-line input or output is written as follows:

```
$ mkdir css
$ cd css
```

Bold: Indicates a new term, an important word, or words you see onscreen. For instance, words in menus or dialog boxes appear in **bold**. Here is an example: "Take the time to read the license terms. When done, select the **I accept the license terms** checkbox and click **Next**."

> **Tips or important notes**
> Appear like this.

Get in touch

Feedback from our readers is always welcome.

General feedback: If you have questions about any aspect of this book, email us at `customercare@packtpub.com` and mention the book title in the subject of your message.

Errata: Although we have taken every care to ensure our content's accuracy, mistakes happen. If you have found an error in this book, we would be grateful if you would report this to us. Please visit `www.packtpub.com/support/errata` and fill in the form.

Piracy: If you come across any illegal copies of our works on the internet, we would be grateful if you would provide us with the location address or website name. Please contact us at `copyright@packt.com` with a link to the material.

If you are interested in becoming an author: If there is a topic that you have expertise in and are interested in either writing or contributing to a book, please visit `authors.packtpub.com`.

Share Your Thoughts

Once you've read *Windows Server 2022 Administration Fundamentals*, we'd love to hear your thoughts! Scan the QR code below to go straight to the Amazon review page for this book and share your feedback.

https://packt.link/r/1803232153

Your review is important to us and the tech community and will help us make sure we're delivering excellent quality content.

Part 1: Introducing Windows Server and Installing Windows Server 2022

Part 1 covers the Windows Server in general, and Windows Server 2022 in particular. In addition, it also covers the installation of Windows Server 2022. Thus, upon completing this part, you will be knowledgeable about Windows Server in general and Windows Server 2022. Additionally, you will be able to clean, install, upgrade, and migrate Windows Server 2022. You will also be able to complete network and unattended installation.

This part of the book comprises the following chapters:

- *Chapter 1, Getting Started with Windows Server*
- *Chapter 2, Introducing Windows Server 2022*
- *Chapter 3, Installing Windows Server 2022*
- *Chapter 4, Post-Installation Tasks in Windows Server 2022*

1
Getting Started with Windows Server

This chapter introduces you to computer networking, in general, and **Windows Server**, in particular. As such, this chapter contains concepts from these two main topics. It begins with a section on computer networks, which covers the ideas and types. Then, discussions about hosts, nodes, **Peer-to-Peer** (**P2P**) networking, and clients/servers are covered in the *Exploring computer network components* section. Moreover, the `Getting to know IP addressing and subnetting` section covers IP addressing and subnetting. Together, they will help you learn and remember the basic concepts of computer networks because networking is considered part of the Windows Server **Operating System** (**OS**).

In contrast, Windows Server is covered in the second part of this chapter, where basic concepts such as hardware, software, and the **Network Operating System** (**NOS**) are explained. In addition, a brief introduction to Windows Server, Linux Server, and macOS Server are included. Furthermore, the Windows Server timeline is presented, too. Finally, in the chapter exercise, you can download Windows Server 2022 and create installation media.

In this chapter, the following sections will be covered:

- Getting to know computer networks
- Exploring computer network components
- Investigating computer network architectures
- Getting to know IP addressing and subnetting
- Exploring servers
- Understanding a NOS
- Understanding Windows Server
- Chapter exercise – downloading Windows Server 2022

Technical requirements

To complete the exercise in this chapter, you will need a PC with Windows 11 Pro, at least 8 GB of **Random Access Memory** (**RAM**), 500 MB of HDD, and access to the internet.

Getting to know computer networks

When you start reading this section, you will naturally wonder why you should learn about computer networking if you are interested in learning about Windows Server. Initially, your statement would stand. However, the more you delve into Windows Server, the more you will justify the necessity of learning about computer networks. As such, this section has been designed to instill fundamental networking skills in you, which are very much needed for installing and supporting Windows Server 2022. So, let's go back to computer networks.

It all began many years ago when sharing resources became necessary. As time went by and demand increased, the development and advancement of computer network technologies also occurred. Therefore, the need to connect and interconnect computers within computer networks and among more geographic locations created a demand for well-defined terms and concepts to describe computer networks. Because of that, concepts such as computer network types, computer network topologies, computer network architectures, and computer network components were born. As such, a computer network represented one of humanity's most significant inventions in communication. That said, simply think about the internet, and you will immediately understand how great of a benefit a computer network is to society. More computers were connected to computer networks, and geographical distances were diminished in communication. Therefore, it created a need for well-defined terms and concepts to describe computer networking. Because of that, different types of computer networks, network topologies, architectures, and components have emerged.

Let's begin by understanding what a computer network is.

What is a computer network?

Merriam-Webster defines a network as "*a group of people or organizations that are closely linked and that work with each other.*" Furthermore, from the same dictionary, networking is defined as the "*exchange of information or services among individuals, groups or institutions.*" These definitions will serve as a simple, clear, and concrete way to define computer networks next.

From what was mentioned in the preceding paragraph, a computer network is a group of computers connected through networking devices and networking media to share resources. Usually, when talking about resources, they can be data, network services, and peripheral devices. So, anyone with experience with computer networks has seen that sharing files, applications, printers, and other peripheral devices is simple. Yet, people often confuse what a computer network is with what a computer network does. While the former explains what constitutes a computer network, the latter shows the benefits we get out of it. That is best illustrated in *Figure 1.1*, where you can see that a computer network is indeed a group of computers connected to share resources:

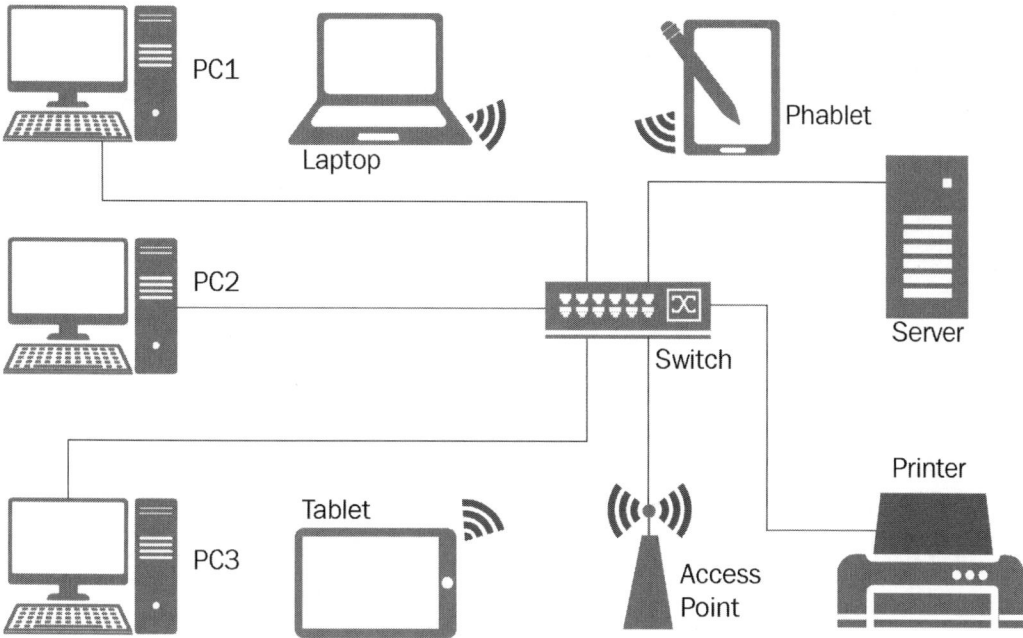

Figure 1.1 – A typical computer network

A computer network is divided into different types. Let's take a look at each of them individually.

The types of computer networks

Indeed, the most exciting thing about a computer network is the process of designing and building one. The design and deployment of computer networks are linked to the definition of networking itself. Therefore, the minimal requirement for building a computer network is that there must be two computers. The number of computers on a particular computer network and how they access the resources from the same network determine the categorization of computer network types, which will be explained in the following section.

In general, the categorization of computer networks consists of the **area** they cover and the **purpose** they serve. The following subsections describe some of the most popular computer networks used today.

Personal area network

A **Personal Area Network** (**PAN**), as shown in *Figure 1.2*, is a computer network that connects and transmits data between devices in a private area. This so-called private area refers to the space that belongs to an individual. For example, at a working desk at home, you can have your laptop, smartphone, printer, and headphones connected to a PAN via Bluetooth. **Bluetooth** and **Wi-Fi** are the most common communication technologies to interconnect devices in a PAN. Often, a PAN is also known as a **Home Area Network** (**HAN**):

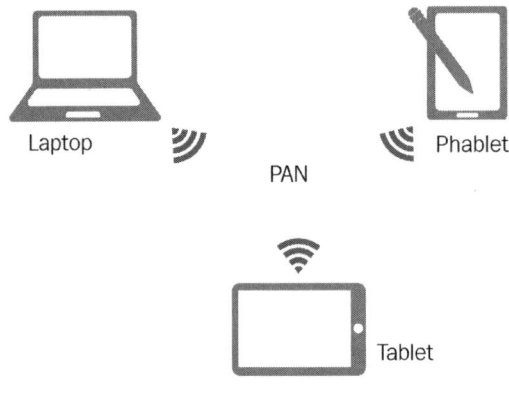

Figure 1.2 – A PAN

Another type of network is the **Local Area Network** (**LAN**). However, its coverage is far greater than a PAN. Let's learn more about it in the next section.

Local area network

A LAN, as shown in *Figure 1.3*, is a computer network that connects two or more computers within a local area. Imagine a local area as one single room, a floor, several floors, a building, or several buildings adjacent to each other at a distance that Ethernet communication technology IEEE 802.3 permits. Usually, a LAN utilizes a central device that uses twisted pair, coaxial, or fiber optic cables as a networking media to interconnect computers.

Now that you have learned about PAN, it will be easier to understand what LAN is. Next, let's compare it with the PAN. A PAN is dominated by portable devices (smartphones), while a LAN mainly consists of fixed appliances. Both computer networks cover the local area. However, the LAN has more extensive coverage than the PAN. For example, a LAN can cover a single floor of the building, several floors of the building, an entire building, or even a few buildings close to one another. Furthermore, while a PAN is primarily organized around a person, a LAN is organized around a site:

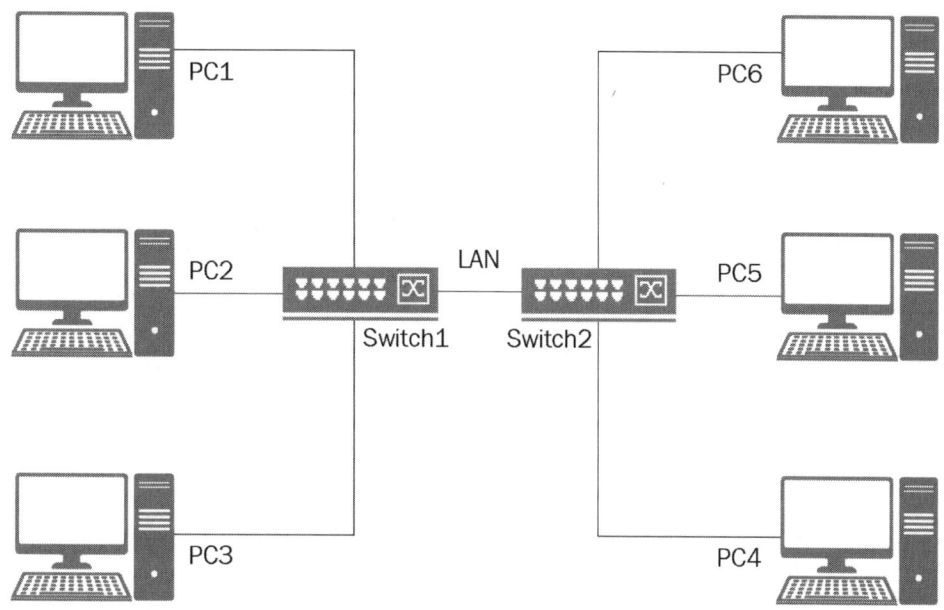

Figure 1.3 – A LAN

The next type of network we will look at is the **Metropolitan Area Network** (**MAN**). Its coverage is even more significant than a LAN.

Metropolitan area network

In contrast to a LAN, a MAN, as shown in *Figure 1.4*, represents a group of LANs interconnected within the geographical boundary of a town or city. As was the case with the PAN and the LAN, the MAN's existence is the need for sharing and accessing the resources inside the city or metro. As a result, a MAN is larger than a LAN in terms of coverage and smaller than a **Wide Area Network** (**WAN**). At the same time, the MAN is faster than the LAN and the WAN in data transmission speeds. Nowadays, fiber optics and gigabit Layer 3 switches interconnect LANs and route the traffic, thus enabling MAN's high speeds:

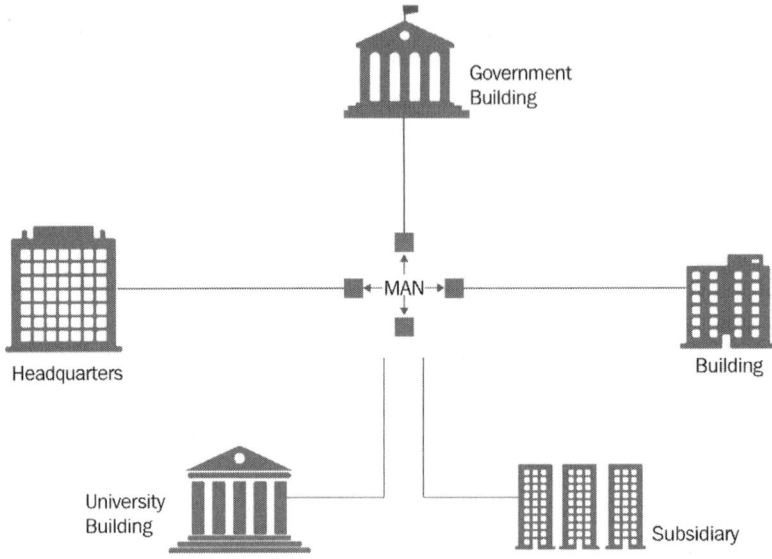

Figure 1.4 – A MAN

Finally, we will understand a WAN with the most significant coverage.

Wide area network

A WAN covers areas that a LAN or a MAN does not cover. Therefore, unlike a MAN, a WAN, as shown in *Figure 1.5*, is a computer network covering a wide geographic area using dedicated telecommunication lines such as telephone lines, leased lines, or satellites. As such, WANs do not have geographic restrictions. The internet is the best example of a WAN:

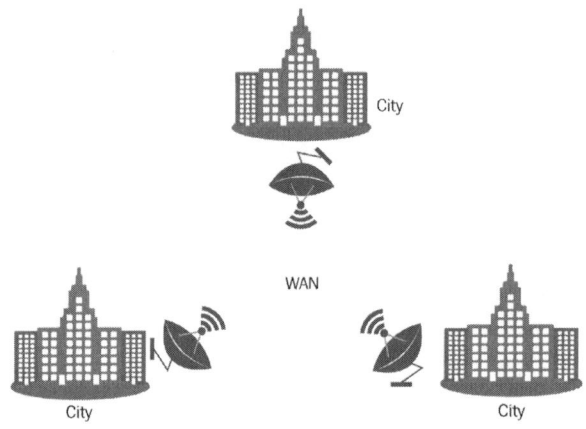

Figure 1.5 – A WAN

> **Important note**
>
> You can learn more about the types of computer networks at https://www.lifewire.com/lans-wans-and-other-area-networks-817376.

Now that we've understood the different types of computer networks, let's look at the underlying components that make up those networks.

Exploring computer network components

Just as PCs have *components*, computer networks have their *components*, too. Usually, while PCs and peripheral devices are known to most people, IT professionals mostly understand components such as networking devices, networking media, and NOSes.

First, let's understand what clients and servers in a computer network are.

Clients and servers

Let's assume that the network resource is the reference point for clients and servers. Then, in a computer network, **clients** usually request access to resources. On the other hand, **servers** are responsible for providing resources and managing access to those resources. Both clients and servers play an active role in the computer network. For example, in *Figure 1.6*, a server with a directly connected printer provides print services to PCs in the role of print requesters:

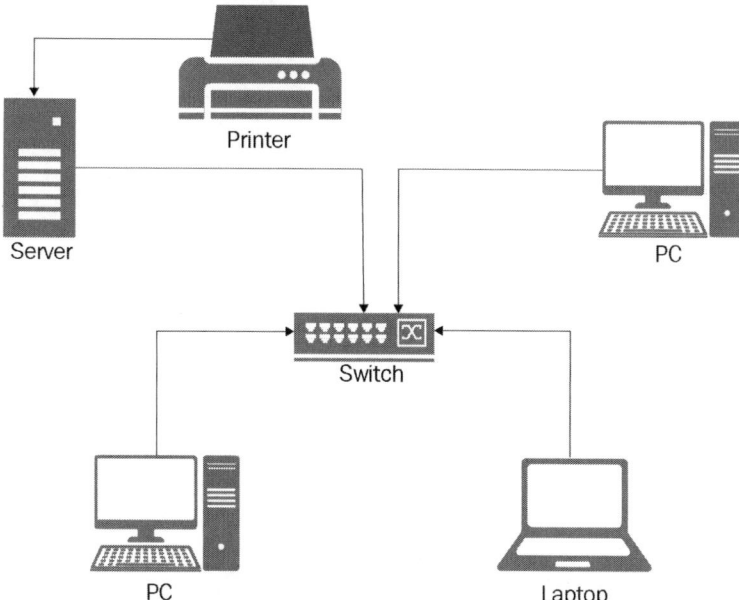

Figure 1.6 – The client and server in a computer network

> **Important note**
>
> Interestingly, the origin of the word *servers* comes from the word *serve*. If you search for the word `serve` in the Merriam-Webster dictionary, among the results, you will find one that says "*to provide services that benefit or help*". Therefore, we can think of a server in a computer network as the computer that provides services to clients. In conclusion, the server serves the clients.

Although clients and servers are the most critical components of a computer network, they take a different naming approach in computer network terminology. So, let's see how that fits into this structure.

Hosts and nodes

Have you heard about **hosts** and **nodes** and wondered what they are? Our first impressions might make us think that hosts and nodes are the same, but they are not! While all hosts can be nodes, not every node can be a host. Hence, a host represents any device with an IP address assigned to its network interface that requests or provides networking services. Usually, clients, servers, and routers act as hosts.

> **Important note**
>
> An **Internet Protocol** (**IP**) address is a logical element of decimal numbers separated by a dot. It is assigned to the host's network interface to identify it in a computer network.

However, a node is any device that can receive and transmit the network traffic but has no interface with an IP address assigned to it. However, nodes have a network interface that is used for their management. For example, in *Figure 1.7*, the PCs and the file server act as hosts, while switches act as nodes:

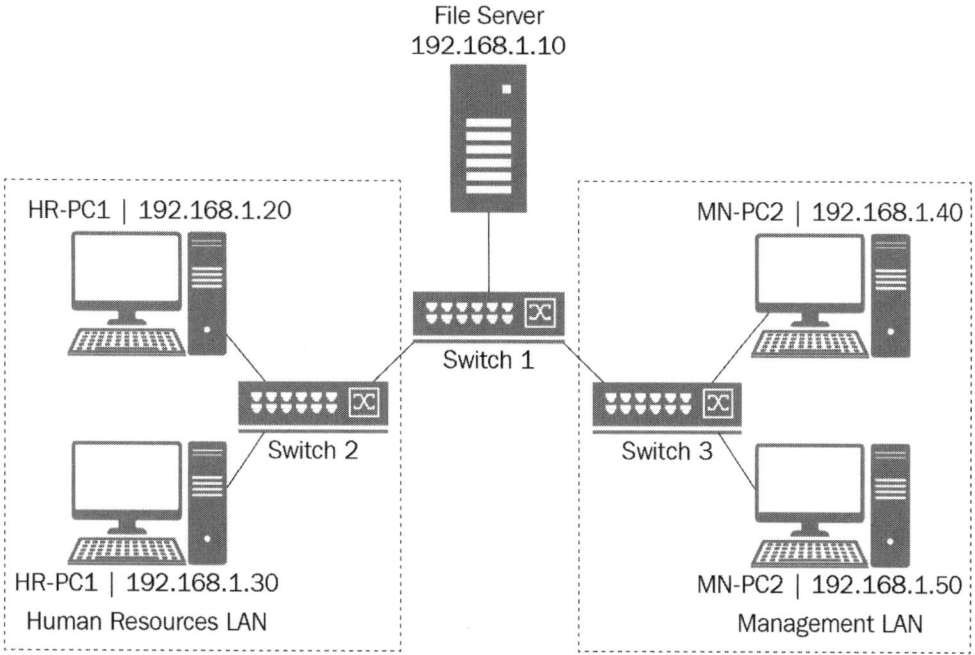

Figure 1.7 – Hosts and nodes within a computer network

Now that we have learned what a network is and its components, we can understand its architecture.

Investigating computer network architectures

Talking about computer networks usually involves discussions about the essential and broader concepts, for example, the elements that make up a computer network. In this debate, the computer network types deal with the area coverage, whereas physical and logical topologies deal with the computer network's physical arrangement and logical structure. Computer network architecture represents a framework incorporating many aspects: physical and logical topology, network components, communication protocols, and operational principles and procedures. Moreover, the computer network architecture is a design that enables computers to communicate based on the request and response paradigm. The most popular network architectures are P2P and client/server.

First, let's understand the P2P network architecture.

P2P network architecture

P2P, often known as a workgroup, is a computer network (see *Figure 1.8*) where hosts have no predefined roles. Instead, they switch the roles from client to server, and vice versa, based on their actual activities on the network. For example, if *PC1* requests services from *PC2*, *PC1* acts as the client, and *PC2* acts as the server. Likewise, if *PC2* requests services from *PC1*, *PC2* acts as a client, and *PC1* acts as the server. Usually, PANs are the best example of a P2P network:

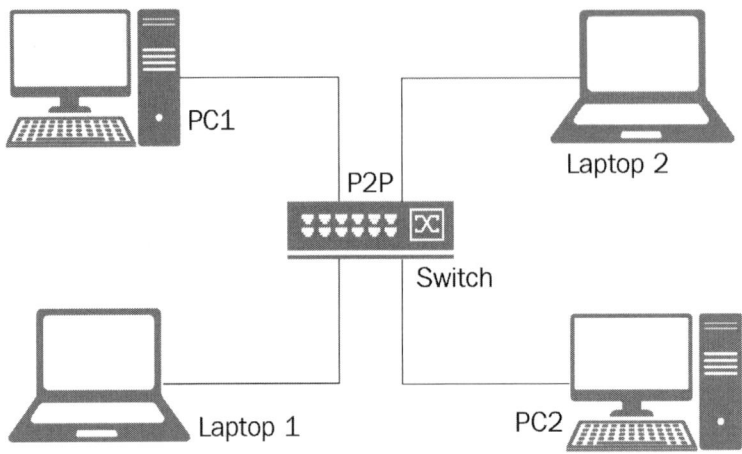

Figure 1.8 – A P2P computer network

> **Important note**
> P2P network architecture refers to a network model where hosts or computers are equally privileged in network participation. Each host may act as a client or server depending on the request and response paradigm. But, of course, that depends on whether it requests or provides services in that network.

The next type of network architecture is the client/server architecture.

Client/server network architecture

A **client/server** network architecture, or a domain-based network, is a computer network (see *Figure 1.7*) where hosts have a predefined role. In such networks, hosts that request services are called clients, whereas hosts that provide services are called servers. The client/server network architecture has dedicated clients and servers.

We now have a greater understanding of how a network operates. However, for a computer to communicate in a network, it requires an IP address. In more detail, we will learn about this in the next section.

Getting to know IP addressing and subnetting

For a computer to communicate within a computer network, it must have an **IP address**. As explained earlier, the IP address identifies the computer on that network. In addition, we encounter the term "subnet" in complex networks, which helps determine the specific network within the overall network. So far, the world of networks recognizes two IP-addressing technologies: **Internet Protocol version 4 (IPv4)** and **Internet Protocol version 6 (IPv6)**. Nevertheless, even though IPv6-addressing technology is becoming increasingly plausible, it still prefers the role of spectator in the great arena of the internet, in which IPv4-addressing technology continues to be the norm.

First, let's take a look at IPv4 network addresses.

IPv4 network addresses

A computer must have an IPv4 address assigned to its network interface to communicate. Therefore, an **IPv4** addressing technology is often referred to as just an IP address in its most straightforward format. The v4 label represents the fourth version of IP addressing specified in the IETF publication, RCF 791. It is a logical element in a network that consists of 32 bits organized into four octets with 8 bits each, divided by a decimal point for simplicity of interpretation (for example, 192.168.1.1). Additionally, IETF's RFC 791 document organizes IP addresses into 8-bit, 16-bit, or 24- bit prefixes. That introduces the **classful addressing** that enables IP addresses to be classified as A, B, C, D, and E. The classful addressing organizes the IPv4 addresses into the bits used for the network and the host portions for a given class. From what has been said about IPv4, if you think that internet traffic is realized mainly through IPv4 addresses, I will tell you that you are not wrong:

IPv4 classes	IPv4 range of the first octet
A	1–127
B	128–191
C	192–223
D	224–239
E	240–255

Table 1.1 – IPv4 classes and their corresponding ranges

Now, let's look at the IPv6 addressing technology introduced to overcome the IPv4 address exhaustion of IPv4 network addresses.

IPv6 network addresses

Today, internet traffic is supposed to be IPv6-driven, but it is not. However, at the time of writing, according to Google IPv6 statistics, 32.57% of internet traffic is managed by IPv6 addressing technology, which is not bad! Therefore, an **IPv6** addressing technology is another logical element in identifying a device on a computer network. The label, v6, represents the sixth version of IP addressing, as specified in the IETF publication, RFC 2460. Unlike IPv4, IPv6 is a `128`-bit address size of 8 hextets with `16` bits each, divided by a colon for simplicity of interpretation (for example, `2001:0DB8:85A3:0000:0000:8A2E:0370:7334`). The fact that IPv6 uses 128 bits makes it possible to use 2,128 IPv6 addresses, which gives an approximate number of 340 undecillion IPv6 addresses. Undoubtedly, that represents a vast number of available IPv6 addresses.

Next, we look at IPv4 subnetting, which is vital in identifying the network addresses.

IPv4 subnetting

Subnetting represents a logical division of one extensive network into multiple smaller networks. A subnet mask plays an essential role in identifying the network and determining the size of the subnet. Additionally, subnetting enables you to specify a given network's network address, host addresses, and broadcast address. A subnet mask is a 32-bit address combined with an IPv4 address to indicate a network and its hosts.

The default subnet masks, otherwise known as **classful networks**, for each class of IPv4 addresses are shown in *Table 1.2*:

IPv4 class	Default subnet mask
A	255.0.0.0
B	255.255.0.0
C	255.255.255.0

Table 1.2 – The IPv4 classful networks

> **Important note**
>
> You can learn more about IPv4-addressing technology, address space exhaustion, and classful networks at `https://blogs.igalia.com/dpino/2017/05/25/ipv4-exhaustion/`.

So far, we have understood what a computer network is and the various types, components, and architectures available. The following section will introduce Windows Server and its related concepts.

Exploring servers

Since we have already provided a basic definition for the server, we will introduce Windows Server in this section. Throughout its history, Windows Server has evolved from a simple file server to an OS capable of handling network services in complex environments such as corporate networks. Therefore, Windows Server can provide network services such as domain controllers, web servers, print servers, and file servers. In addition, it often acts as a separate platform in which enterprise applications such as **Exchange Server**, **SQL Server**, **SharePoint Server**, and others are executed. With its robust performance and advanced security, nowadays, Windows Server is shaping cloud computing.

Server hardware and software

As you might recall, computer hardware and software represent physical and logical components. Therefore, since the server's primary role is to provide network services to the clients, a server requires powerful hardware. That is because software such as Windows Server is designed to provide advanced network services. Therefore, its hardware must be durable and high-quality materials to deliver services and support network-based operations continually. Aside from distinguishing itself from the ordinary computer, a server is also specific in the types of services it provides. For example, a database server requires more memory capacity and storage space.

The **Central Processing Unit** (**CPU**), memory, disk, and network are *key hardware components*. As such, these components affect the overall performance of servers. Therefore, it is recommended that the actual performance of system components is continuously monitored to maintain the optimal performance of servers for both regular and heavy workloads.

First, let's understand what a CPU is.

CPU

A CPU, or processor, is a chip on a server's motherboard. In literature, you often encounter the term *computer's brain*. That is a component that does all the processing and calculations. *Intel* and *AMD* are the biggest CPU manufacturers for PCs and servers. Their newest CPUs on the market are based on 64-bit architecture, which differs from 32-bit architecture-based processors. In 64-bit architecture, 64 bits of data are exchanged between the CPU and RAM in each communication session. On the other hand, in 32-bit architecture, only 32 bits of data are exchanged per communication session between RAM and the CPU. That is 50% less data being communicated via a 32-bit architecture compared to a 64-bit architecture.

To give out performance, the CPU depends on RAM. Let's learn about that in the next section.

Memory

RAM represents the server's working memory used by Windows Server 2022 and the server's applications. Therefore, the more RAM there is on the server, the more multitasking can be performed, which can be interpreted as more applications running simultaneously. You can learn more about RAM in the *Memory* section of *Chapter 10, Tuning and Maintaining Windows Server 2022*.

Now, let's understand what a disk is in a server.

Disks

As you know, data is usually stored on a disk. In the case of servers, they mostly have more than one disk, referred to as the server's disk subsystem. As for disk performance, read/write speed is an element that must be considered because the faster the disk's throughput, the higher the performance of your disk subsystem. In terms of disk technology used in the server, usually, we encounter types such as **Solid State Drive (SSD)** and **Hard Disk Drive (HDD)**. SSD has no moving parts and contains high read and writing speeds. In contrast, HDD has moving parts and durability and contains high-capacity storage spaces.

Now, let's understand what a network interface is.

Network interface

A **network interface** enables the server to connect to an organization's LAN and the internet. Usually, servers have more than one network interface. That is because the faster the server's network connection speed is, the more data the server can send and receive to and from the network.

Now that we have understood a server, let's look at the various server sizes, form factors, and shapes.

Server sizes, form factors, and shapes

Essentially, the server is a computer, so everything that applies to the form factor of the laptop also applies to the form factor of the server. So, the question that arises is, what is a form factor? A form factor is a hardware design that defines and describes an electronic device's size, shape, and technical specifications. So, in terms of size, shape, and tech specs, today's servers are presented in the following three form factors:

- As their name suggests, **rack-mountable servers** are usually built to be mounted inside a rack. These servers are considered general-purpose computers and can support various applications and network services. In addition, these servers usually populate on-premises server rooms or data centers. And because of their weight, these servers are fixed to the rack, as shown in *Figure 1.9*:

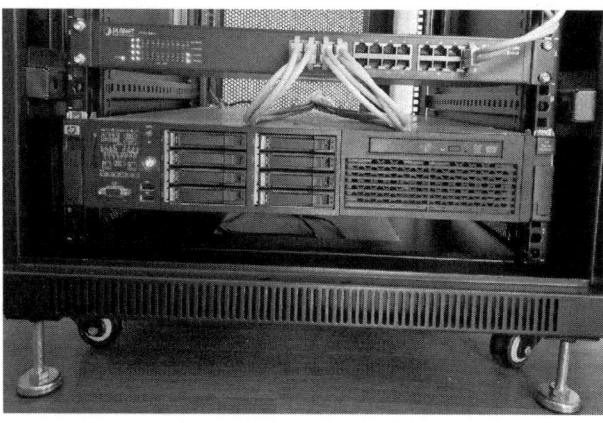

Figure 1.9 – An HP server in a rack

- **Blade servers** are modular servers that allow multiple servers to be deployed within a smaller area. They are thin in design and contain mainly the CPU, memory, network interface, and storage disks. Put simply, blade servers usually populate data centers or supercomputer facilities. That is because they can fit multiple servers on a single shelf, providing high processing power.
- **Tower servers** refer to a type of server that looks identical to a PC's vertical case. However, they contain very advanced hardware and, as a result, offer higher processing power when compared to ordinary PCs. Usually, these servers are used for testing or local services in a **Small Office/Home Office (SOHO)**.

> **Important note**
> A 64-bit Windows server installed on a 64-bit hardware server can process double the amount of data compared to a 32-bit Windows server installed on a 32-bit hardware server.

The server has an OS that enables network services such as a computer. Let's learn more about it next.

Understanding a NOS

A NOS is software capable of managing, maintaining, and providing services within a network. Additionally, a NOS can share files and applications, provide web services, provide authentication and authorization, control access to resources, administer users and computers, provide tools for configuration, maintain and provide resources, and perform other functions related to network resources. With that in mind, a NOS is crucial for managing computer network resources.

These days, Windows Server, Linux Server, and macOS Server versions are all considered NOSes because they can provide network services. So, let's understand each one of them individually.

Windows Server overview

As you know, in general, Windows OS is a Microsoft product. The same applies to Windows Server. Therefore, its server line began with Windows NT 3.5 in the early 1990s, followed by other Windows Server versions, starting with Windows 2000 Server. Windows Server has a **Graphical User Interface** (**GUI**)-based OS at its core. However, as of Windows Server 2008, a Server Core edition has been introduced, a **Command-Line Interface** (**CLI**)-based OS. From Windows Server 2003 to Windows Server 2008, the architecture was 32-bit and 64-bit; however, since Windows Server 2012, it's only 64-bit. The **New Technology File System** (**NTFS**) remains its native filesystem. However, with Windows Server 2012, the **Resilient File System** (**ReFS**) was introduced to replace NTFS. Regardless, in Windows Server 2022 (see *Figure 1.10*), NTFS is a native filesystem, whereas ReFS is used in database applications. Nowadays, Windows Server powers both on-premises and cloud network services:

Figure 1.10 – The NTFS continues to be used by Windows Server 2022

An overview of ReFS can be found at `https://docs.microsoft.com/en-us/windows-server/storage/refs/refs-overview`.

Linux Server overview

If something is interesting to talk about in the world of OSs, the Linux OS is unequivocal. That is because the world of technology does not recognize any innovative initiatives as having gathered more volunteers than Linux has. Everything started as a desire to improve functionality in an existing OS such as MINIX. Instead of an enhanced MINIX, it turned out that, in the early 1990s, Linus Torvalds had developed a new OS called Linux. So, the GNU GPL project took over the licensing of Linux, and a penguin became the Linux mascot. The first Linux booklet published was called *Linux Installation and Getting Started*, and the first Linux virus was *Bliss*.

Journal and Linux Weekly News marked the first-release Linux magazines. And just like that, many other global activities followed that would form the so-called Linux community, which turned out to be one of the world's largest volunteer communities, contributing globally to the further development of Linux. Nowadays, Linux servers (see *Figure 1.11*) power most web servers and supercomputers due to their security and open source nature, both on-premises and on cloud:

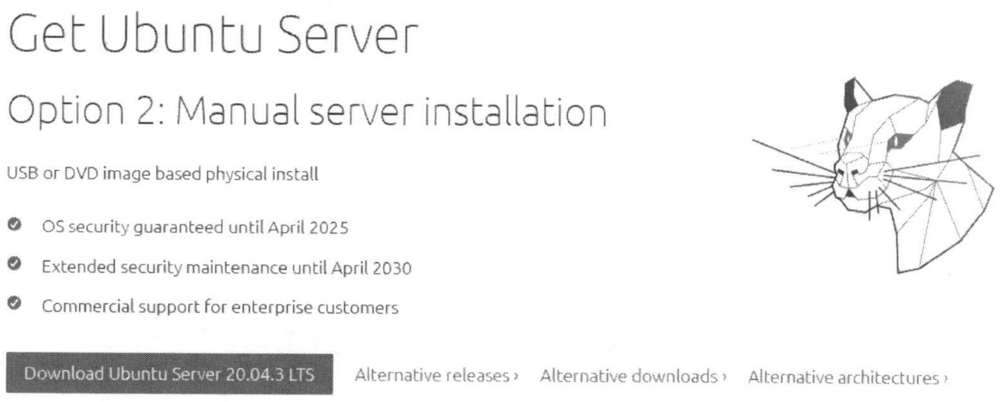

Get Ubuntu Server

Option 2: Manual server installation

USB or DVD image based physical install

- OS security guaranteed until April 2025
- Extended security maintenance until April 2030
- Commercial support for enterprise customers

Download Ubuntu Server 20.04.3 LTS   Alternative releases ›   Alternative downloads ›   Alternative architectures ›

Figure 1.11 – Downloading Ubuntu Server from ubuntu.com

You can find out how to run the Linux subsystem on Windows Server 2022 at https://docs.microsoft.com/en-us/windows/wsl/install-on-server.

macOS Server overview

Although macOS Server might have a smaller percentage of use than the Windows Server and Linux Server OSs, its most positive characteristic is its reliability. At its core, macOS Server is a modified Unix OS that already conforms to the familiar Apple GUI for Mac computers. Like Windows and Linux, macOS Server is also offered on 32-bit and 64-bit platforms. However, since Apple was designated to use Intel processors for their computers and servers, macOS Server is distributed only on 64-bit. Although we cannot tell the exact number of servers powered by macOS Server, Apple continues to release new versions of its macOS Server (see *Figure 1.12*) and support it:

Figure 1.12 – Downloading macOS Server from the Mac App Store

You can learn more about macOS servers at https://www.apple.com/macos/server/.

In this section, we have understood what a server is and learned about server hardware such as CPUs, memory, disks, and network interfaces. Additionally, we have understood server sizes, form factors, shapes, and NOSs. Moreover, we have become acquainted with Windows Server, Linux Server, and macOS Server. In the next section, we will extend our learning of Windows Server.

Understanding Windows Server

What would your answer be if someone asked you what Windows Server is? I guess your answer would be more or less like the following: *Windows Server is the server OS developed by Microsoft as part of the Windows NT family of OSs.* Whether a server is based on a Windows Server, Linux Server, or macOS Server OS, it does not make any difference if the version used continues to provide adequate services within an organization's network. However, many differences are evident when looking at them from a deployment perspective, a user-interface perspective, a managing resources perspective, and maintaining a server perspective.

Let's look at the Windows Server timeline to understand how it has evolved over the years.

The Windows Server timeline

So far, in the 26-year history of Windows Server, including Windows NT, I think Microsoft has been quite intuitive in adopting new requirements in the server world. As a result, the Windows Server timeline looks exciting, and I want to share it with you. In particular, notice the transition of the Windows Server technology over time. Simply put, it's impressive. The Windows Server timeline is shown in the following table:

Server for the masses era, 1996–2000	Enterprise era, 2000–2008	Data center era, 2009–2013	Cloud for the masses era, 2016–present
Windows NT Server 3.5	Windows 2000 Server	Windows Server 2008	Windows Server 2016
Windows NT Server 4.0	Windows Server 2003	Windows Server 2012	Windows Server 2019
			Windows Server 2022

Table 1.3 – The Windows Server timeline

In this section, you have learned about Windows Server and become acquainted with its timeline. The following section will take you through the steps for downloading Windows Server 2022.

Chapter exercise – downloading Windows Server 2022

To download Windows Server 2022 onto your Windows 11 computer, complete the following steps:

1. Press *WinKey + R* to open **Run**.
2. Enter **Microsoft-edge** and press *Enter*.
3. In **Microsoft Edge**, click on the address bar and press *Enter*.
4. Type in https://www.microsoft.com/en-us/evalcenter/, and then press *Enter*.
5. On the **Evaluation Center** page, click on the *search* icon in the upper-right corner, enter Windows Server 2022, and press *Enter*.
6. From the search results, select **Windows Server 2022**.

7. Select your evaluation file type (notice that you have the option to try out Windows Server 2022 in Azure), and then click on **Continue**.

Complete the form, as shown in *Figure 1.13*, and then click on **Continue**:

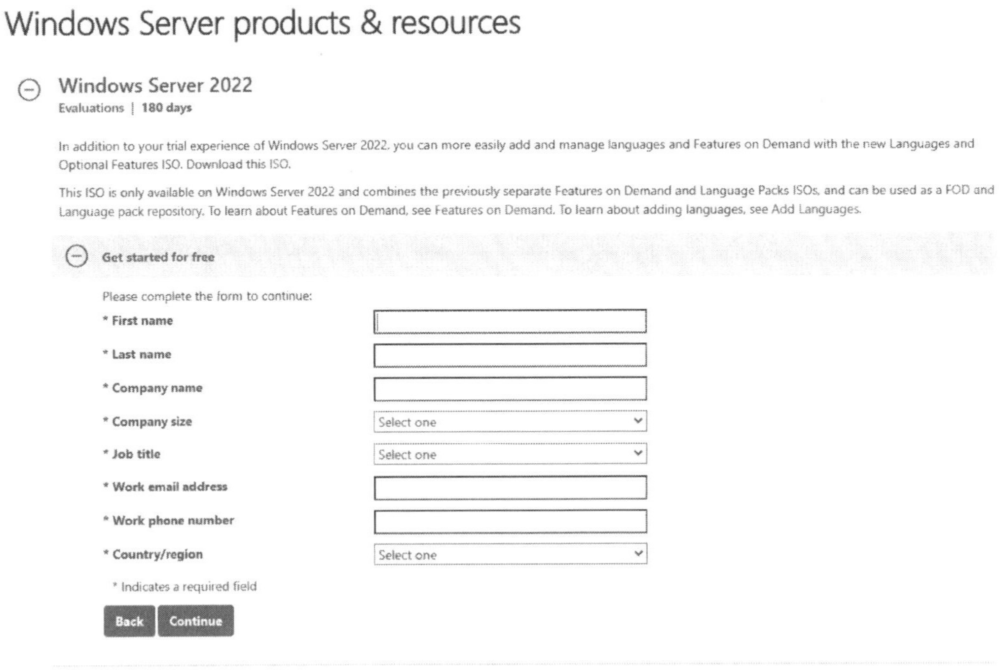

Figure 1.13 – Downloading the Windows Server 2022 evaluation

1. Select your desired **language**, and then click on **Download**.
2. Shortly after, the Windows Server 2022 download will begin. If not, you might want to click on the **Download** button.

> **Important note**
>
> Once the Windows Server 2022 download completes, you should burn the ISO file to a USB flash drive. Information about creating a bootable USB can be found at https://www.lifewire.com/how-to-burn-an-iso-file-to-a-USB-drive-2619270. Once completed, you are all set to move on with installing the Windows Server 2022 evaluation version.

Summary

In this chapter, you learned the basic concepts of computer networks and had a chance to get to know Windows Server. Specifically, you learned a computer network and were introduced to different computer networks, components, network architectures, IP addressing, and subnetting. Furthermore, you learned about server hardware and software, server sizes, form factors, shapes, and NOSs. Finally, you had a chance to view the Windows Server timeline.

The chapter included a chapter exercise that provided instructions on downloading Windows Server 2022 from the Technet Evaluation Center portal to make things more lab-oriented. With the things you have learned in this chapter, you will now understand what a computer network is and be able to identify network architectures, IP addressing, and subnetting. You'll also be able to identify key hardware components and understand a NOS, including the Windows Server timeline.

In the following chapter, we will learn about Windows Server 2022 specifically.

Questions

1. The computer network architecture is a design that enables computers to communicate based on the request and response paradigm. [True | False]
2. _____usually request access to resources, and _____are responsible for providing resources and managing access to those resources.
3. Which of the following are considered to be computer networks?

    A. PAN

    B. LAN

    C. MAN

    D. WAN

    E. All of the above

4. Windows Server is Microsoft's server OS as part of the Windows NT family. [True | False]
5. _____can provide network services such as domain controllers, web servers, print servers, and file servers.
6. The subnet helps to identify a specific network within the overall network. [True | False]
7. Which of the following are considered network architectures? (Choose two.)

    - P2P
    - Client/server
    - NOS
    - Network topology

8. The CPU, memory, disk, and network are the critical system components that affect the overall performance of your servers. [True | False]

9. The_____represents the physical component, while the_____represents the logical component of a server.

10. Which of the following are considered to be IP- addressing technologies? (Choose two.)

    i. IPv2

    ii. IPv4

    iii. IPv6

    iv. IPv8

Further reading

To learn more about the topics that were covered in this chapter, take a look at the following resources:

- *What is a computer network*: `https://www.ibm.com/cloud/learn/networking-a-complete-guide`

- *Windows Server 2022 is now generally available—delivers innovation in security, hybrid, and containers*: `https://cloudblogs.microsoft.com/windowsserver/2021/09/01/windows-server-2022-now-generally-available-delivers-innovation-in-security-hybrid-and-containers/`

- *Linux vs. Windows Server: The Ultimate Comparison*: `https://phoenixnap.com/blog/linux-vs-microsoft-windows-servers`

2
Introducing Windows Server 2022

This chapter teaches you about Microsoft's new server operating system, Windows Server 2022. The first part of this chapter will cover various Windows Server 2022 editions, the differences between Windows Server 2022 and Windows Server 2019, and the minimum recommended system requirements.

The second part of this chapter will cover the new features introduced in Windows Server 2022. Microsoft Edge Chromium, a hybrid cloud center, Storage Migration Service, Storage Replica, Secured-core server, and support for Docker containers and Kubernetes are the new features introduced to improve Windows Server 2022. Finally, this chapter concludes with a chapter exercise on downloading Windows Admin Center.

In a nutshell, the following topics will be covered:

- An overview of Windows Server 2022
- What's new in Windows Server 2022?
- Chapter exercise – downloading Windows Admin Center

Technical requirements

To complete the lab for this chapter, you will need the following equipment:

- A PC with Windows 11 Pro, at least 8 GB of RAM, 500 GB of HDD, and access to the internet

An overview of Windows Server 2022

Windows Server 2022 (see *Figure 2.1*) is the latest version of the server operating system from Microsoft, as part of the Windows NT family of OSs. With a preview program rolled in March 2021, Windows Server 2022 was released on August 18, 2021. Furthermore, its general availability was announced on September 1, 2021, followed by a launch event as part of the Windows Server Summit on September 16, 2021. Like Windows Server 2016 and Windows Server 2019, Windows Server 2022 is also based on Windows 10 code. Therefore, it uses the version number of the November 2021 update of Windows 10, 21H2.

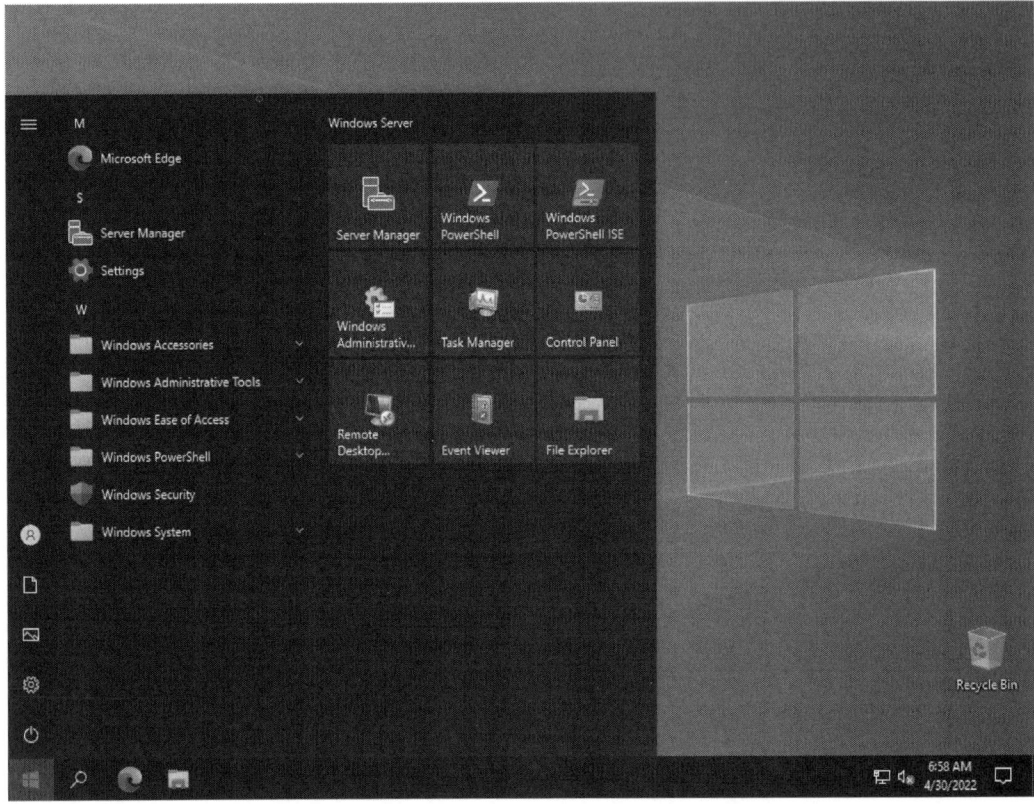

Figure 2.1 – Windows Server 2022 Desktop and Start menu

First and foremost, we live in the cloud era, requiring the server OS to be cloud-capable. With that in mind, remember that when Windows Server 2016 was released, Microsoft named it Windows Server for the cloud. Likewise, that is true for Windows Server 2022, which strains this paradigm even more. Meanwhile, cloud computing has become a more mature platform. Therefore, to comply with the fundamental requirement of having a cloud-capable server OS, Microsoft has introduced Windows Server 2022, which is more secure, flexible, and robust, and supports hybrid deployments through a specially developed edition of Windows Server 2022 Datacenter: Azure. Moreover, Windows Server 2019 introduced new features such as System Insights, hybrid cloud, Storage Migration Service, Storage Replica, the improved Windows Defender, and support for Kubernetes. However, Microsoft with Windows Server 2022 continues to bring significant improvements in security, secure connectivities, Azure hybrid capabilities, application platform, storage, and other vital features.

In addition to the enriched and optimized interface, integration with Microsoft Azure makes Windows Admin Center the modern management platform, enabling businesses to manage, configure, and integrate workloads running in Windows Server 2022 like never before. For more details on this, look at the *What is Windows Admin Center?* and *Downloading Windows Admin Center* sections later in this chapter.

> **Important note**
>
> In contrast to **Remote Server Administration Tools** (**RSAT**), Windows Admin Center is a new server management tool that uses web technology. Its interface is based on Azure's interface, enabling multiple server management features. Once you install Windows Server 2022 and access the desktop, the Windows Admin Center dialog box pops up. It is free of cost and can be downloaded from `https://www.microsoft.com/en-in/evalcenter/evaluate-windows-admin-center`.

First, let's look at the various editions of Windows Server 2022.

Windows Server 2022 editions

Windows Server 2022 seems to follow its predecessors, Windows Server 2016 and Windows Server 2019, regarding available editions. Hence, the editions available in Windows Server 2022 are as follows:

- The Windows Server 2022 Datacenter: Azure edition is a new edition designed to enable enterprises to use the benefits of the cloud through Azure Automanage.
- Windows Server 2022 Datacenter is the complete edition designed for enterprises that own highly virtualized datacenters or act as cloud providers.
- Windows Server 2022 Standard is the full-featured edition designed for medium-sized businesses that own servers on-premises.
- Windows Server 2022 Essentials is designed for small businesses that own a single server in their IT infrastructure.
- Microsoft Hyper-V Server 2022 is designed as a free product that delivers enterprise-class virtualization for data centers and hybrid clouds.

The preceding points give us a clear idea of the different editions within Windows Server 2022, but how different is it compared to the previous version? Let's discuss this next.

Windows Server 2022 versus Windows Server 2019

As is customary these days, every time we hear about a new technological product, instantaneously, the impression is created that the new product differs only in the naming from the previous one. Perhaps at first glance, the same can be said for Windows Server 2022 – that it is just a slight superficial improvement and nothing majorly different than Windows Server 2019. However, once you start digging into the features of Windows Server 2022, you will realize that it is more than that. Therefore, in the following section, an attempt is made to highlight some of the most significant differences between these two OSs regarding improvements or new features:

- **Features related to hybrid cloud capabilities**: Windows Server 2022 improves Azure Arc with a default enabling of both HTTPS and TLS 1.3, and Storage Migration Services with improved deployment and management. Also worth mentioning is Azure Automanage, which offers new capabilities for Windows Server 2022 Datacenter: Azure edition, such as Hotpatch, the **Server Message Block** (**SMB**) over **Quick UDP Internet Connections** (**QUIC**), and an Extended network for Azure.

- **Features related to improved security**: Only Windows Server 2022 supports Hypervisor-based code integrity, Secured-core server, and Hardware-enforced stack protection, whereas both OSs support TLC 1.2, with Windows Server 2022 running TLS version 1.3 by default.

- **Features related to the upgraded Hyper-V Manager**: Only Windows Server 2022 supports the action bar, live storage migration, affinity and anti-affinity rules, VM clones, running workloads between servers, and the new partitioning tool.

- **Features of improved platform flexibility**: Only Windows Server 2022 supports **Dynamic Source Routing** (**DSR**) and **Group Managed Service Accounts** (**gMSA**) domain joining. Both OSs support an uncompressed image size and a virtualized time zone. However, Windows Server 2022 brings improvements, with an uncompressed image size of approximately 2.7 GB and a virtualized time zone configurable within a container.

- **Features of Windows Admin Center**: Only Windows Server 2022 supports automated extension life cycle management, an event workspace to track data, a configurable destination virtual switch, and customizable columns for VM information. While the **Detachable Events Overview** screen can be configured in Windows Server 2019, it comes as a built-in feature in Windows Server 2022.

- **Features of an improved Kubernetes experience**: Only Windows Server 2022 supports HostProcess containers and multiple subnets.

You will quickly determine which version you need by recognizing the differences between Windows Server 2022 and Windows Server 2019. Now, let's look at the minimum and recommended system requirements.

The minimum and recommended system requirements

Before looking at the minimum and recommended system requirements, it is worth mentioning that while the first has more to do with the hardware that generally enables the installation, the latter refers to the hardware that guarantees work comfort. Knowing the hardware requirements helps us to select the proper hardware, depending on what we aim to run and provide. Hence, from the publications released by Microsoft, Windows Server 2022 shares the exact hardware requirements as its predecessors:

- Minimum system requirements:

    - Processor: 1.4 GHz 64-bit processor
    - RAM: 512 MB (2 GB for Server with the Desktop Experience installation option)
    - Disk space: 32 GB
    - Network: An Ethernet adapter capable of at least 1-gigabit throughput
    - Graphics device and monitor: Capable of Super VGA (1024 x 768) or higher resolution
    - Other hardware: A DVD drive (if you intend to install the OS from DVD media), a keyboard, a mouse (or another compatible pointing device), a TPM, and internet access

- Recommended hardware requirements:

    - Processor: 2.0 GHz 64-bit processor or higher
    - RAM: 32 GB or higher
    - Disk space: 256 GB SSD and 1 TB HDD
    - Network: At least 1 gigabit Ethernet **Network Interface Card** (**NIC**)
    - Graphics device and monitor: Capable of Super VGA (1024 x 768) or higher resolution
    - Other hardware: A DVD drive, a keyboard, a mouse (or another compatible pointing device), TPM, and internet access

In this section, you have learned about Windows Server 2022 and its editions, compared Windows Server 2022 with Windows Server 2019, and found the minimum and recommended system requirements. In the following section, you will learn about the new features of Windows Server 2022.

What's new in Windows Server 2022?

There is no doubt that Windows Server 2022 is built on the solid foundation of Windows Server 2019, which makes them share standard features; nonetheless, it brings numerous new elements too. With that in mind, this section will cover some of the new features, among many, introduced in Windows Server 2022.

Microsoft Edge Chromium

Built on Google's Chromium engine and backed by Microsoft security and innovation, Windows Server 2022 includes Microsoft Edge Chromium, replacing the well-known and retired Internet Explorer. In addition, it can be used with the Desktop Experience installation options (see *Figure 2.2*).

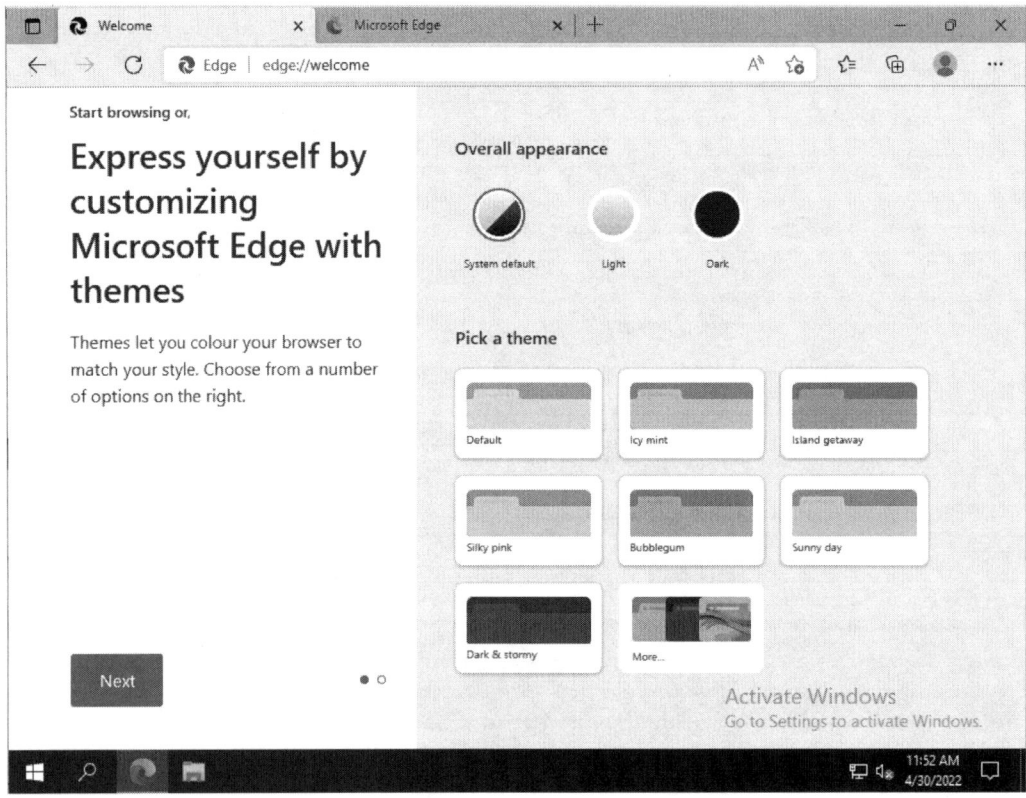

Figure 2.2 – Microsoft Edge Chromium in Windows Server 2022

In addition to Microsoft Edge Chromium, Windows Server 2022 offers hybrid cloud support, another exciting part of Windows Server 2022.

Azure hybrid center

As shown in *Figure 2.3*, the **Azure hybrid center** operates as a centralized hub that enables access to all of Azure's integrated services. Furthermore, it provides improved connections between premise servers and cloud services, including Azure Automanage, Azure Arc, Azure Backup, File Sync, disaster recovery, and other cloud services. For applications running on local servers, take advantage of cloud services.

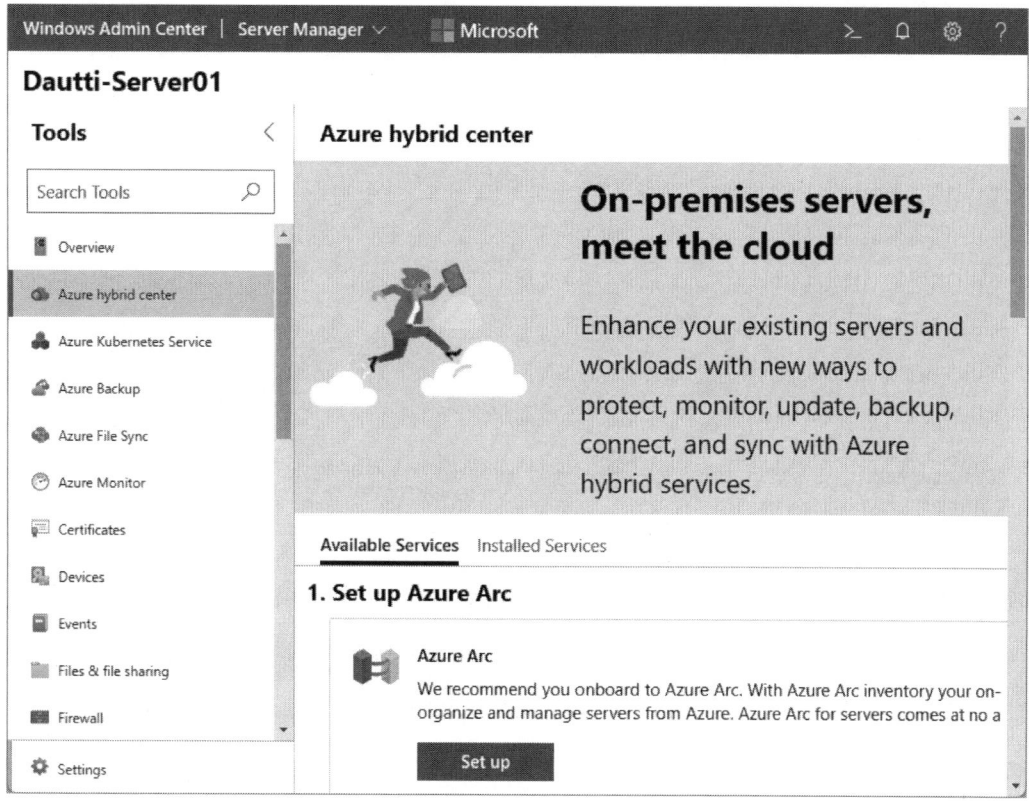

Figure 2.3 – The hybrid cloud center in Windows Server 2022

Next, we will look at the improved feature of the Storage Migration Service in Windows Server 2022.

Storage Migration Service

Storage Migration Service, shown in *Figure 2.4*, is an improved feature in Windows Server 2022 that facilitates migrating servers to a more recent version. It uses a GUI interface and Windows PowerShell to inventory servers' data and transfers the configuration to newer servers. Additionally, it can optionally move the identities of the old servers to the new servers. That way, the apps, and users do not have to change anything.

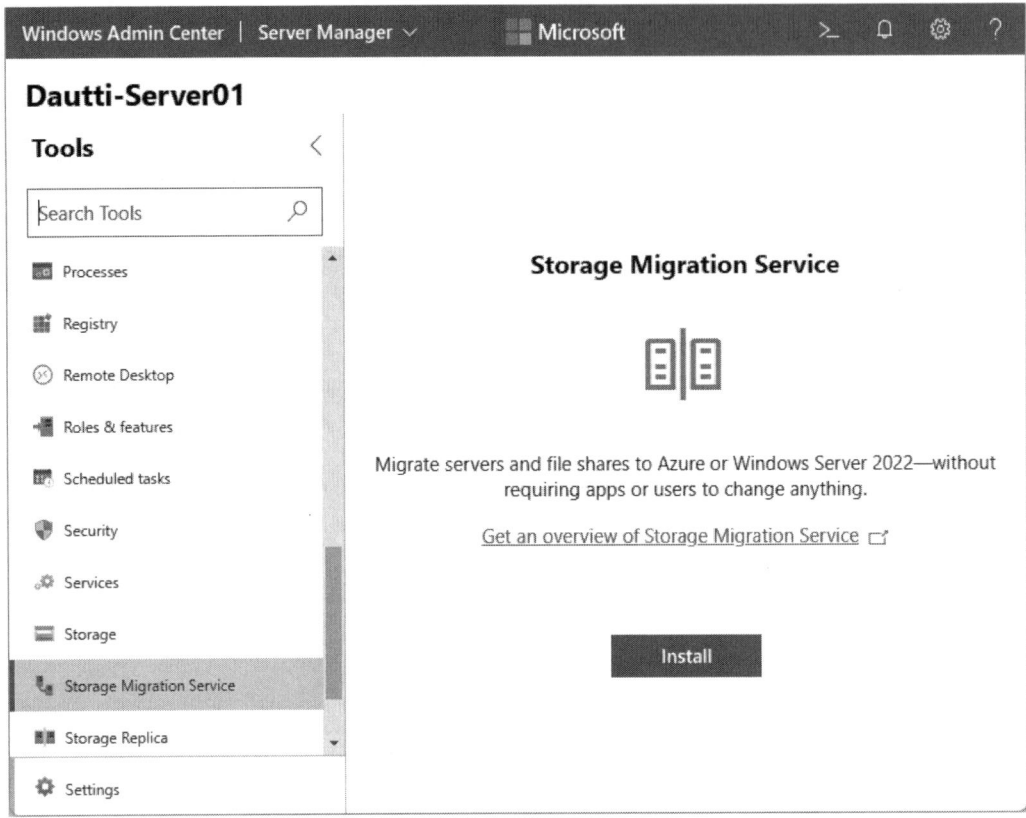

Figure 2.4 – Storage Migration Service in Windows Server 2022

In addition to the new features introduced specifically for Windows Server 2022, there are also those present in earlier versions but improved and enhanced in Windows Server 2022.

The next feature that we will learn about is the Storage Replica feature.

Storage Replica

As shown in *Figure 2.5*, **Storage Replica** is a feature introduced in Windows Server 2016 but improved in Windows Server 2022. It enables the replication of volumes between servers or clusters synchronously and asynchronously. Usually, such replication is used for disaster recovery purposes. Additionally, Storage Replica allows users to create a stretching failover cluster for high availability with nodes spread over two different sites, maintaining synchronization among storage.

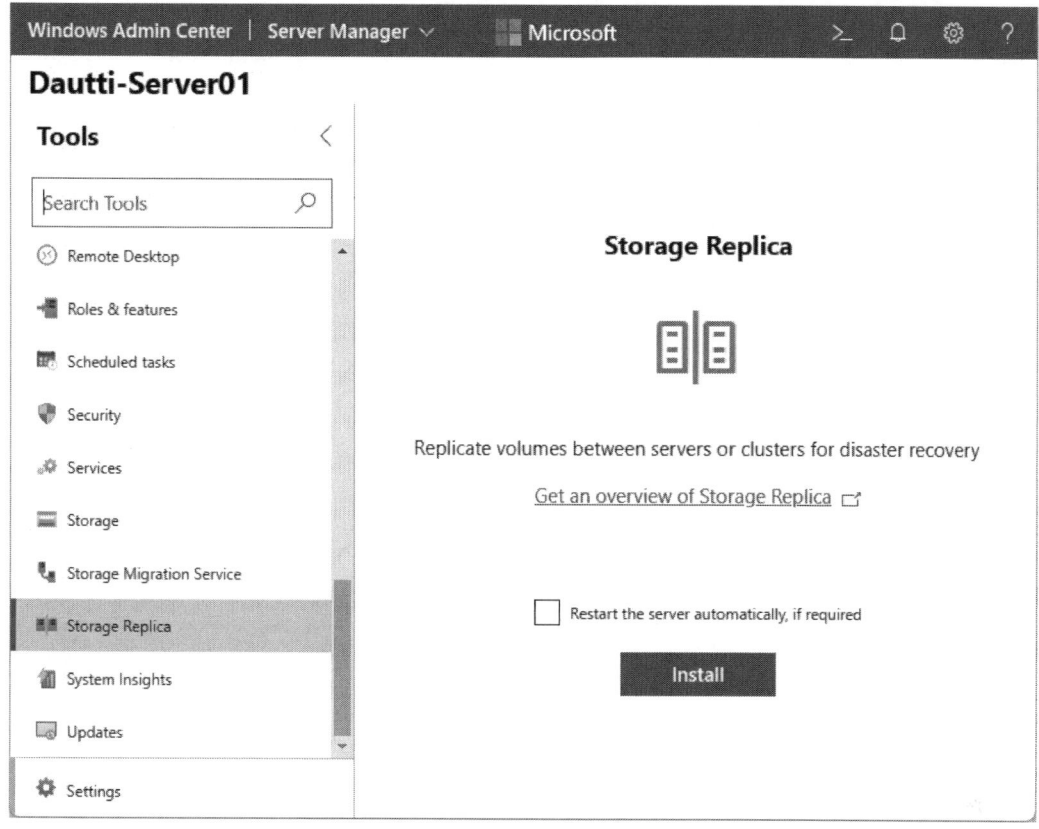

Figure 2.5 – Storage Replica in Windows Server 2022

> **Important note**
> A stretch cluster represents a method of deploying Storage Replica, allowing the configuration of computers and storage in a single cluster. In such a deployment method, servers share their storage synchronously, replicated with the site awareness. In addition to stretching cluster solutions and offering high availability, it also enables disaster recovery.

New features introduced with Windows Server 2022 also cover security. However, first, let's learn about Secured-core server, a new security feature.

Secured-core server

Some of the many new features introduced in Windows Server 2022 are security-related, such as the **Secured-core server** (see *Figure 2.6*). It combines powerful protection against threats to provide multi-layered security across hardware, firmware, and OSs. In addition, it uses **Trusted**

Platform Module (**TPM**) 2.0 and System Guard to boot Windows Server safely and minimize the risks of firmware vulnerabilities. Secured-core servers keep the threat from causing disruptions and launching ransomware attacks, helping organizations improve their overall security posture.

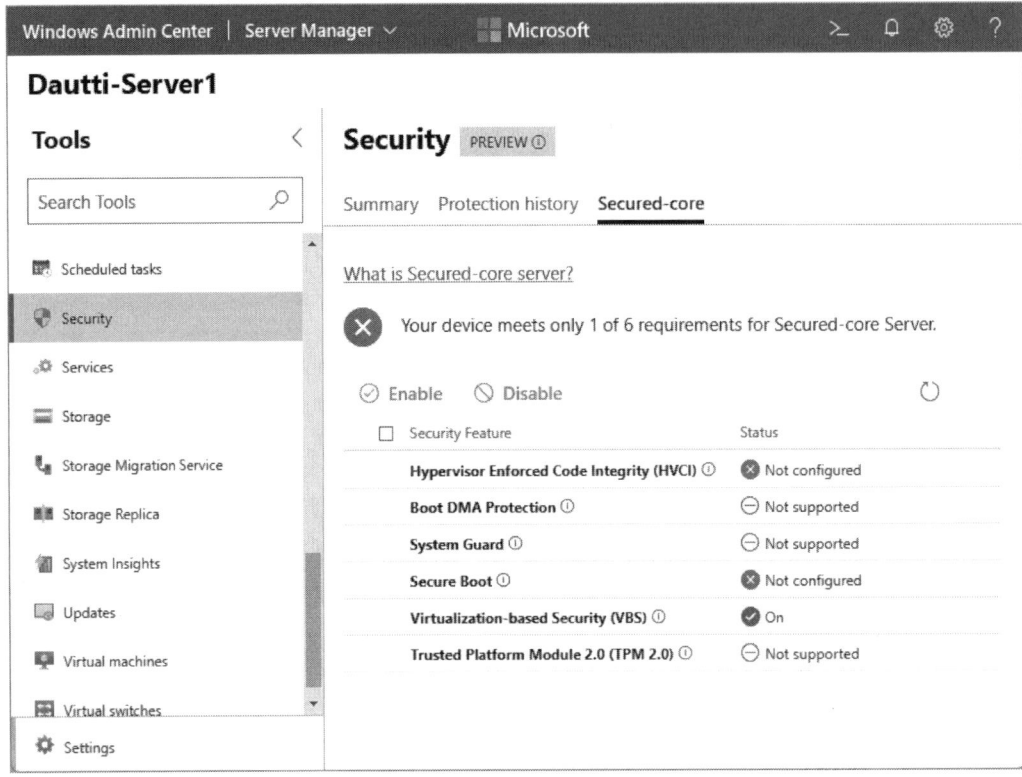

Figure 2.6 – The Secured-core server feature in Windows Server 2022

Now, let's learn about support for Kubernetes, a virtualization platform that, in collaboration with Docker, brings containerization to the Windows world.

Azure Kubernetes Service

Originally designed by Google and maintained by the Cloud Native Computing Foundation, **Kubernetes** is an open-source platform enabling automated deployment, scaling, and operations of application containers in a virtualized environment. Everything began with the introduction of the Docker containerization project launched in 2013. Docker technology enabled the execution of applications in an environment known as **isolated containers**. However, Docker was not capable of managing large and distributed containerized applications. So, Kubernetes came in by providing the platform for supporting applications running in a cluster of containers, which is dramatically easier to work at scale.

Kubernetes consists of **nodes** and **Pods**, where the former can be physical machines or **Virtual Machine (VMs)**, and the latter represents a single instance of an application. Therefore, Kubernetes has become a vital part of Docker's container technology.

Microsoft, through Windows Admin Center, has made it reasonably easy to set up a local Kubernetes group once configured to your Azure Kubernetes Service host – for example, the control panel in the Azure Kubernetes Service tool, as shown in *Figure 2.7*. However, it requires connecting a system with Azure Kubernetes Service.

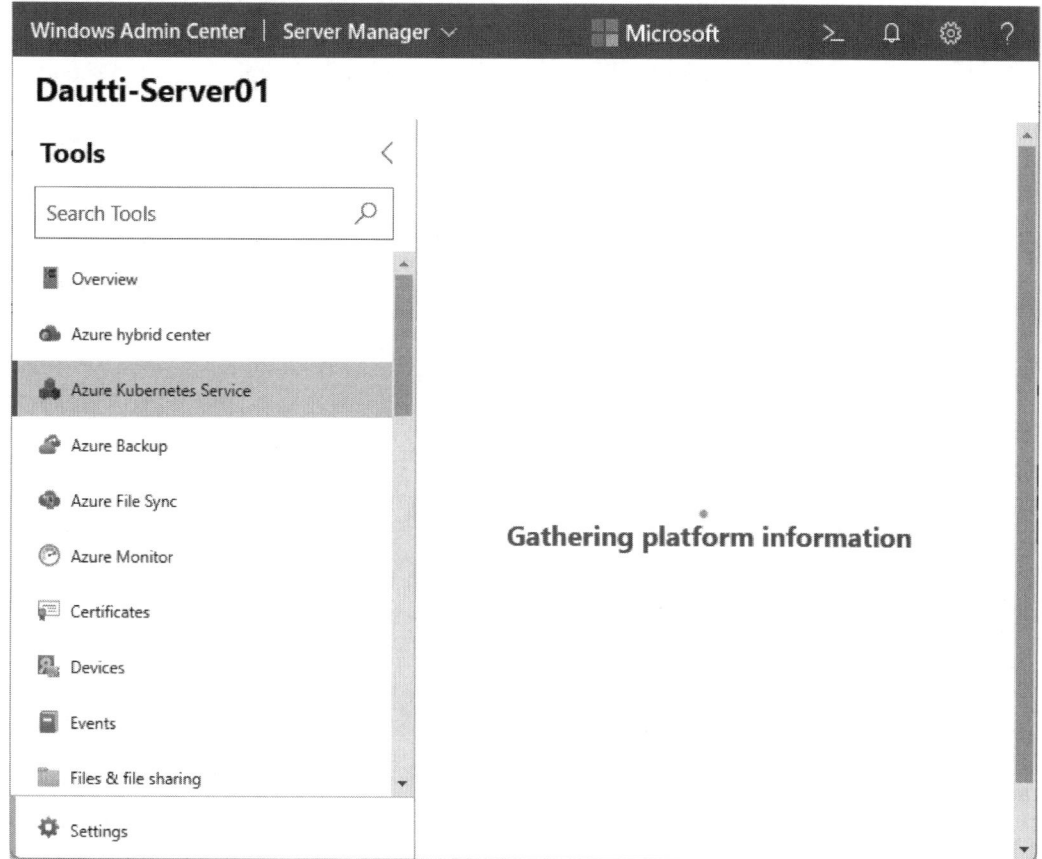

Figure 2.7 – The Secured-core server feature in Windows Server 2022

> **Important note**
> With Kubernetes v1.23, Windows Server 2022 provides OS compatibility for Windows nodes (and Pods).

If Kubernetes enables automated deployment, scaling, and operations of application containers in a virtualized cluster environment, Docker conversely empowers the applications to run their own isolated virtualized environment. Next, let's learn about the support for Docker containers in Windows Server 2022.

Containers

As you may know, Microsoft recommends that every client/server application runs on a dedicated server. Whereas with a **traditional deployment approach**, this would require a single physical server for each application, with a **virtualized deployment approach**, each application requires its VM. By contrast, the **container deployment approach** (see *Figure 2.8*) implies that the OS on the host is shared among the applications. In that way, a container enables the application to run in its isolated environment without knowledge of other applications being executed outside its container.

As with Kubernetes (see the *Getting to Know the Azure Kubernetes service* section), **Docker** has similarly been a game changer in application containerization. Hence, Docker technology enables easy-to-build, easy-to-deploy, and easy-to-run application images. Docker consists of **Docker Engine**, which powers Docker containers. Initially, it was written for Linux, and, after a lot of development work, it is now supported on Windows and macOS:

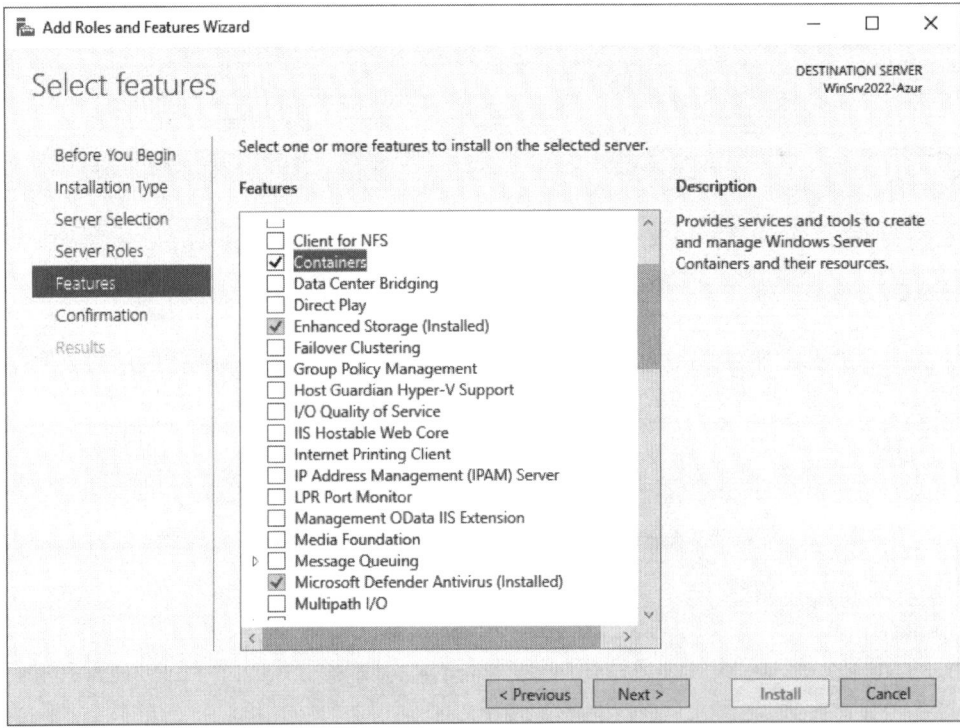

Figure 2.8 – Installing the Containers feature in Windows Server 2022

Now, let's understand what Windows Admin Center is and how it makes working with servers much more manageable. So, instead of logging into the server via Remote Desktop, you bring a vibrant server interface to your computer's desktop with Windows Admin Center.

Windows Admin Center

Formerly known as Project Honolulu, **Windows Admin Center** is considered a modern server management application, introduced in 2018 but enhanced and available in Windows Server 2022. It enables system administrators to work with servers on-premises and in the cloud. In addition, Windows Admin Center uses a browser-based interface to manage Windows Server. This way, Windows Admin Center becomes a replacement platform for older tools such as **Server Manager**, **Computer Management**, and **Remote Server Administrative Tools** (**RSAT**). You can download Windows Admin Center from Microsoft's website (see this chapter's exercises) and install it on Windows Server 2022, Windows 11/10, and earlier versions of Windows client and Windows Server.

This section taught you about Microsoft Edge Chromium, hybrid cloud support, Storage Migration Service, the Storage Replica feature, and Secured-core server. You have also learned about Kubernetes, Docker containers, and Windows Admin Center support. Next, let's learn how to download Windows Admin Center.

Chapter exercise – downloading Windows Admin Center

To download Windows Admin Center for the Windows 11 computer, complete the following steps:

1. Open your browser and go to `https://www.microsoft.com/en-us/cloud-platform/windows-admin-center`.
2. Click the **Download now** button on the Windows Admin Center download site, as shown in *Figure 2.9*.

38 Introducing Windows Server 2022

Figure 2.9 – The Windows Admin Center download site

3. You will be redirected to Microsoft's **Evaluation Center** portal to download Windows Admin Center.
4. Click the **Continue** button, and then fill in the form.
5. Click the **Continue** button, and you will be prompted to save the file you are about to download.

Summary

In this chapter, you have learned about Windows Server 2022, Microsoft's latest version of the **Network Operating System** (**NOS**), part of the Windows NT family of OSs.

This chapter compared the differences between Windows Server 2019 and Windows Server 2022, concepts of authentication and authorization, listed various editions of Windows Server 2022, and provided the minimum and recommended system requirements for installing Windows Server 2022. Furthermore, you have learned about Microsoft Edge Chromium, Hybrid cloud center, Storage Migration Service, Storage Replica, and Secured-core server. Additionally, you have learned about Docker and Kubernetes containerization technologies and got to know the Windows Admin Center.

Finally, the chapter concluded with an exercise that provided instructions on downloading Windows Admin Center, a web-based utility to manage modern Windows Server.

The next chapter will teach you how to install Windows Server 2022.

Questions

1. Windows Admin Center is a new server management app introduced with Windows Server 2022 (True | False).
2. _____ technology has enabled easy-to-build, easy-to-deploy, and easy-to-run application images.
3. Which of the following editions is available in Windows Server 2022? (Choose two.)

    A. Windows Server 2022 Datacenter

    B. Windows Server 2022 Enterprise

    C. Windows Server 2022 Standard

    D. Standard Windows Server 2022 Beginner

4. Microsoft Defender ATP is a unified platform that enables preventative protection, post-breach detection, automated investigation, and response (True | False).
5. _____ is a new feature available in Windows Server 2022 that locally analyzes Windows Server system data by providing an insight into the functionality of servers and helping system administrators to keep everything running smoothly.
6. Which are the minimum CPU requirements for installing Windows Server 2022?

    A. A 1.4 GHz 64-bit processor

    B. A 1.4 GHz 32-bit processor

    C. A 2.4 GHz 64-bit processor

    D. A 2.4 GHz 64-bit processor

7. The hybrid cloud support in Windows Server 2022 improves the connections between the on-premises servers and the cloud services on Amazon Web Services (True | False).

8. _____ is a brand-new feature in Windows Server 2022 that facilitates migrating servers to a more recent version of Windows Server.

9. Which of the following features is new in Windows Server 2022?

    A. Microsoft Defender ATP

    B. Storage Migration Service

    C. The Kubernetes platform

    D. All of the above

10. Windows Server 2022 is Microsoft's penultimate version of the Server OS as part of the Windows NT family of OSs (True | False).

11. _____ consists of nodes and Pods, where the former can be physical machines or VMs, and the latter represents a single instance of an application.

12. What is the new server management app supported with Windows Server 2022?

    A. Windows Administrative Tools

    B. Windows PowerShell

    C. Windows Admin Center

    D. Active Directory Administrative Center

Further reading

- *What's new in Windows Server 2022*: https://docs.microsoft.com/en-us/windows-server/get-started/whats-new-in-windows-server-2022

- *Hello, Windows Admin Center!*: https://docs.microsoft.com/en-us/windows-server/manage/windows-admin-center/overview

- *Storage Migration Service overview*: https://docs.microsoft.com/en-us/windows-server/storage/storage-migration-service/overview

- *Getting started with Docker and Kubernetes on Windows 10*: https://learnk8s.io/installing-docker-kubernetes-windows

3
Installing Windows Server 2022

This chapter is designed to provide you with detailed instructions for installing Windows Server 2022. These step-by-step instructions, illustrated with easy-to-understand graphics, explain and show you how to master the installation of Windows Server 2022. With the guidance provided by this easy-to-follow chapter, you will quickly learn the installation process without any obstacles. It is an excellent resource, with helpful tips on how to get the job done promptly and efficiently.

With that in mind, this chapter covers the following installation types: clean installation, installation over a network using **Windows Deployment Services** (**WDS**), unattended installation using the **Windows Assessment and Deployment Kit** (**Windows ADK**) and the **Microsoft Deployment Toolkit** (**MDT**), an in-place upgrade, migration of network services in a new server, and trying out Windows Server 2022 in Azure.

Finally, this chapter concludes with an exercise in which we will set up WDS.

The following topics will be covered in this chapter:

- Understanding the installation of Windows Server 2022
- Various Windows Server 2022 installation methods
- A chapter exercise – setting up WDS

Technical requirements

To complete the lab for this chapter, you will need the following equipment:

- A PC with Windows 11 Pro, at least 16 GB of RAM, 1 TB of HDD, and access to the internet
- A virtual machine with Windows Server 2012 R2 Standard (Desktop Experience), at least 2 GB of RAM, 100 GB of HDD, and access to the internet

- A virtual machine with Windows Server 2019 Standard (Desktop Experience), at least 4 GB of RAM, 100 GB of HDD, and access to the internet
- A virtual machine with Windows Server 2022 Standard (Desktop Experience), at least 4 GB of RAM, 100 GB of HDD, and access to the internet

Understanding the installation of Windows Server 2022

One of the daily tasks performed by a system administrator is installing a new **Operating System (OS)**. However, it is more than an installation, as it includes steps such as preparing for the installation, installing the OS, verifying the installation, and initial server configuration. Simply put, it's the starting point for everything! Although there might be rare situations when servers come with preloaded OSs, in most cases, it is a system administrator's responsibility to make sure that the server has an OS and, with that, get the job done.

Let's begin by understanding partition schemes, as this will help us manage the partitions on the disks.

Getting to know the partition schemes

The disk partition is a disk's logical division so that an OS can manage data. In comparison, the partition scheme represents the technology used to manage the sections on the disks. In general, there are two partition schemes:

- **Master Boot Record (MBR)**: This is an old partition scheme, known today as a legacy boot option. It operates on a 512-byte disk sector with a maximum of four primary partitions, or three primary partitions and one extended partition. An extended partition can have up to 26 logical partitions. The MBR uses **Logical Block Addressing (LBA)** to support disks up to 2 TB. The MBR has always proven to be a beneficial partition scheme for multiboot platforms.
- **GUID Partition Table (GPT)**: This coexists with the MBR and is a new partition scheme that overcomes the limitations of the MBR. The **Globally Unique Identifier (GUID)** in a GPT is a 128-bit number that Microsoft uses to identify resources. In a GPT, block sizes from 512 bytes and up are supported, where the most common default these days is 4,000 or 4,096 bytes, and the size of the partition entry is 128 bytes. The GPT is part of the **Unified Extensible Firmware Interface (UEFI)** standard that replaces the old **basic input/output system (BIOS)** to support modern hardware. The GPT is fault-tolerant and supports up to 18 EB disk storage and 128 partitions on each disk.

Next, let's look at the boot options to further help with the installation.

Getting to know the boot options

Depending on the manufacturer, different keys on a keyboard can be used to access the BIOS. The most frequently used keys are *Delete* and *F2*. Upon entering the BIOS, there are several boot options available:

- **Installation media**: In most cases, there may be a DVD. Before accessing the BIOS, ensure that you insert the bootable DVD into the DVD drive. Specify the DVD as a first boot option, save the changes, and exit the BIOS.
- **USB flash drive**: The capacity of a USB flash drive must be a minimum of 8 GB. Plug in your bootable USB flash drive before you access the BIOS. Specify the USB flash drive as a first boot option, save the changes, and exit the BIOS.
- **Network boot**: This occurs when installing Windows Server 2022 over a network. First, enable booting from the **local area network** (**LAN**) and then specify booting from the network as a first boot option. Finally, save the changes and exit the BIOS.

Regardless of which option you are using, your computer will soon restart and attempt to boot from the specified boot option. For example, *Figure 3.1* shows the boot from a DVD.

```
Press any key to boot from CD or DVD...
```

Figure 3.1 – Booting from a DVD

> **Tip**
> To make a bootable USB flash drive, you can use the Windows 7 USB/DVD download tool, which can be downloaded from https://www.microsoft.com/en-us/download/windows-usb-dvd-download-tool.

Next, let's look at the advanced startup options, which are very helpful once the installation is complete.

Accessing the advanced startup options

In Windows Server 2022, there is no *F8* option. Instead, you can use the **Advanced startup** options to recover the server OS. That said, to access the **Advanced startup** options, complete the following steps:

1. Click the **Start** button.
2. Select **Settings** from the **Start** menu.
3. In **Windows | Settings**, choose **Update & Security**.

4. Choose **Recovery** from the navigation menu on the left side of the screen.
5. Click the **Restart now** button (as shown in *Figure 3.2*) and click on **Continue**.

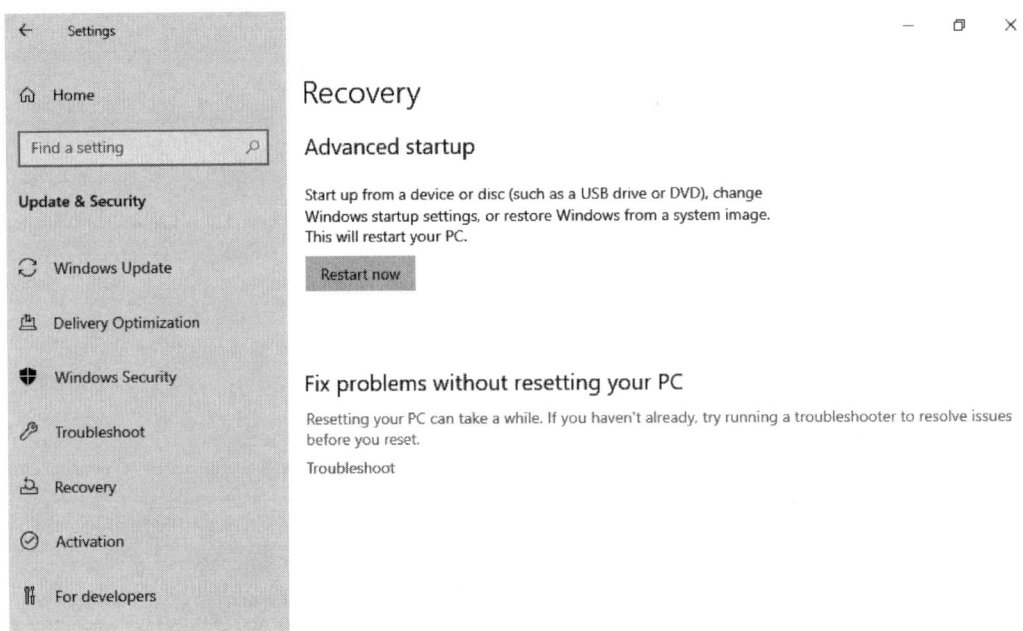

Figure 3.2 – Accessing the advanced startup options in Windows Server 2022

6. After a short time, options such as **Continue**, **Troubleshoot**, and **Turn off your PC** will be displayed.
7. Click **Troubleshoot** to access the advanced options.
8. Select any available options from the **Advanced options** screen, as shown in *Figure 3.3*.

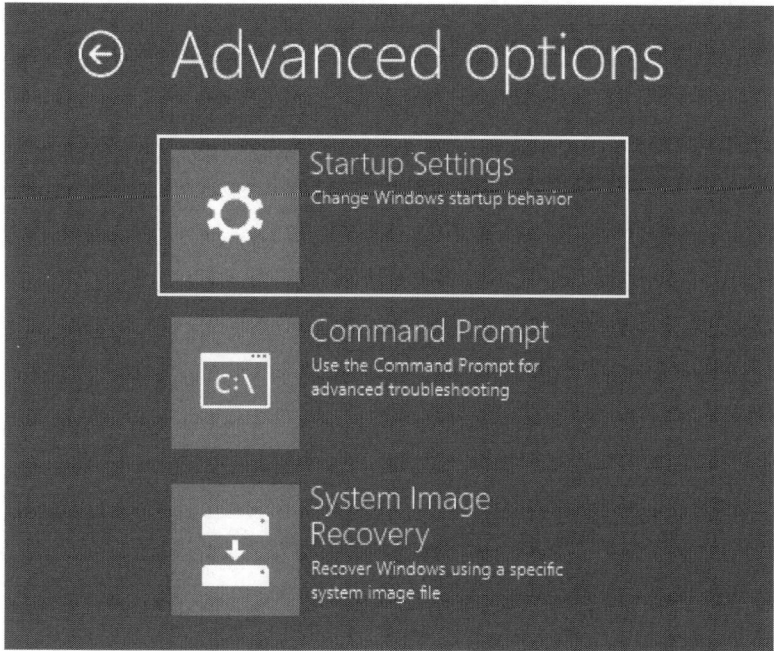

Figure 3.3 – The Advanced startup options in Windows Server 2022

This section taught you the partition schemes, boot options, and advanced startup options. The following section will delve deeper into the server installation options.

Various Windows Server 2022 installation methods

When it comes to installing Windows Server 2022, there are many methods. So, depending on the environment in which you will deploy Windows Server 2022, you can choose from the following:

- A clean installation
- Installation over a network using WDS
- Unattended installation
- In-place upgrade migration

In addition to the aforementioned installation methods, Microsoft Endpoint Configuration Manager (former System Center Configuration Manager) is used to deploy Windows Server 2022 in a corporate network. However, such an installation method is beyond this book's scope because it is at a significantly more advanced level and, as such, is used in an enterprise's IT infrastructure.

Let's now dive into the several installation options you can choose from.

Choosing Desktop Experience, Server Core, or Nano Server installation

Windows Server 2022 offers three installation options. However, the selected installation option affects the availability of roles and features, and therefore, you should consider the options before choosing your desired installation option:

- **Desktop Experience**: This installation option contains everything from Windows Server 2022, so choosing Desktop Experience means you have installed everything on Windows Server 2022. However, your hardware needs to exceed the minimum requirements to benefit from the full-featured **Graphical User Interface** (**GUI**).
- **Server Core**: This is an installation option recommended by Microsoft due to its minimal hardware resource consumption and higher security. The roles and features can be installed locally through Windows PowerShell or remotely through Server Manager.
- **Nano Server**: This replacement for Server Core takes up far fewer hardware resources, has more periodic updates, and supports only 64-bit applications. It is administered remotely, since it has no local login capabilities. This installation option is best understood as *set it and forget it*.

Now that you have been acquainted with the installation options, let's learn how to perform a clean installation.

A clean installation

Whether installing Windows Server 2022 on a new or existing hard disk, a clean installation overwrites the current OS on a hard disk. Be aware that the clean option requires user interactivity, although that might be more limited than the upgrade option.

To perform a clean installation of Windows Server 2022, complete the following steps:

1. Turn on your computer, depending on the selected boot option, and wait for the boot prompt on the screen. The message on the screen requires user confirmation to boot the system from a DVD, USB flash drive, or network boot.
2. The installation files are loaded in memory (i.e., the RAM).
3. Enter the language and other preferences to install, as shown in *Figure 3.4*. Click **Next** to continue.

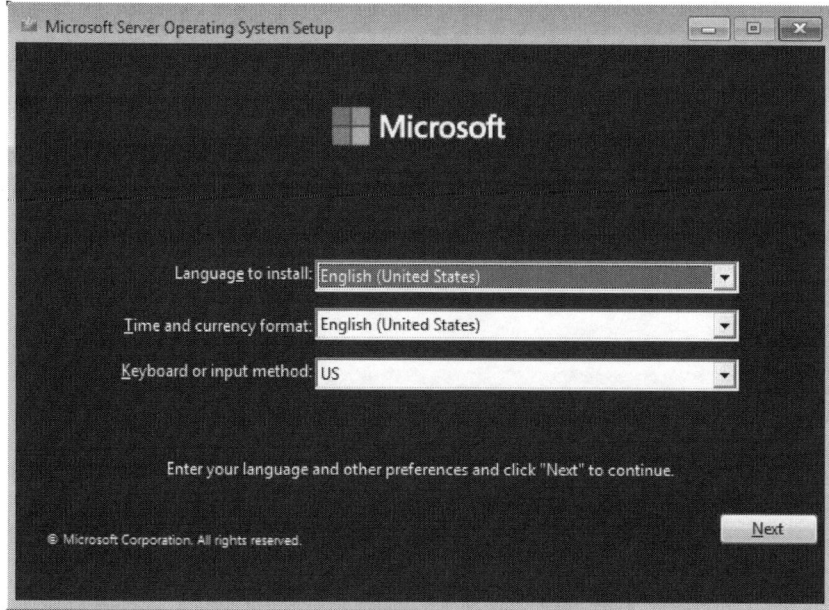

Figure 3.4 – Windows Server 2022 setup

4. Click **Install Now** to install Windows Server 2022, as shown in *Figure 3.5*.

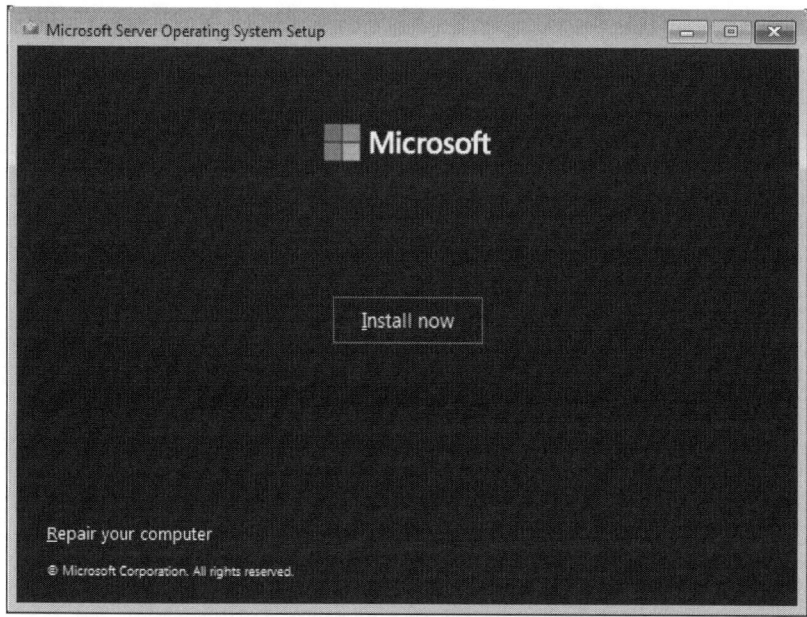

Figure 3.5 – Windows Server 2022 installation

5. Select **Windows Server 2022 Datacenter (Desktop Experience)** and click **Next**, as shown in *Figure 3.6*.

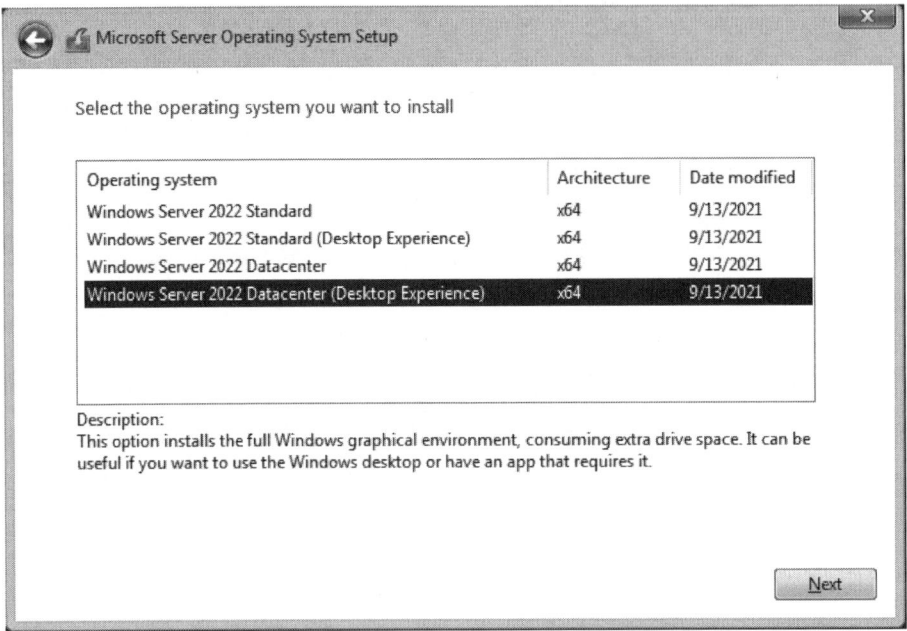

Figure 3.6 – The available Windows Server 2022 OSs for the installation

6. Take the time to read the license terms. When done, check the **I accept the license terms** checkbox and click **Next**.
7. Select **Custom: Install Microsoft Server Operating System only (advanced)** to run a clean installation, as shown in *Figure 3.7*.

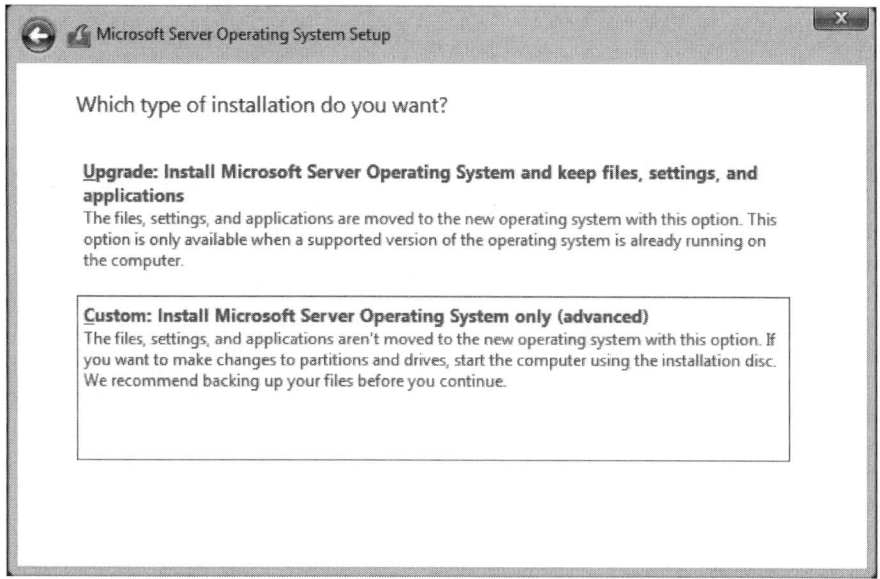

Figure 3.7 – The types of installation available

8. After preparing the drive, select the partition where you want to install Windows Server 2022. Then, click **Next**, as shown in *Figure 3.8*.

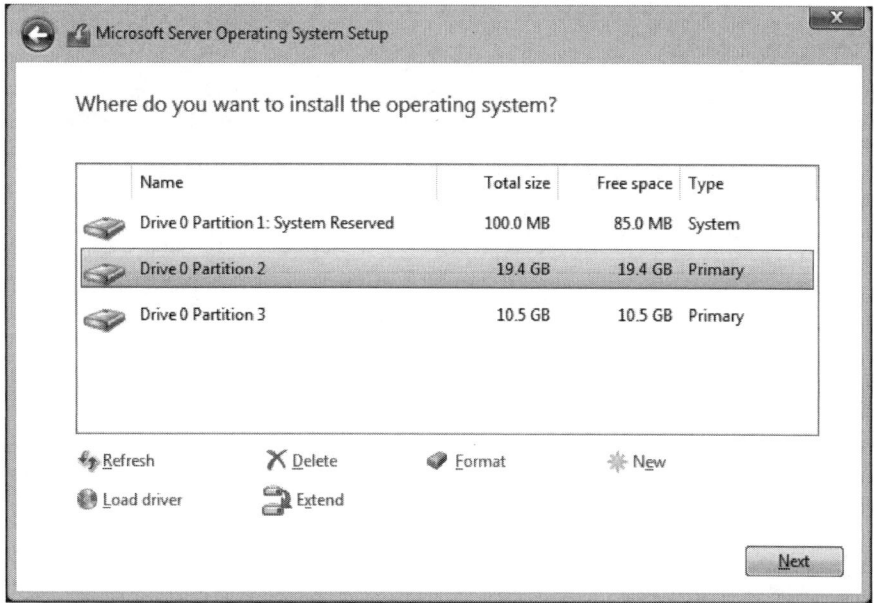

Figure 3.8 – Selecting the partition for installing the OS on a disk

9. Now, Windows Setup should start installing Windows Server 2022. Sit back and relax!
10. After getting the devices ready and performing a few restarts, set up the administrator password (see *Figure 3.9*) and click **Finish**.

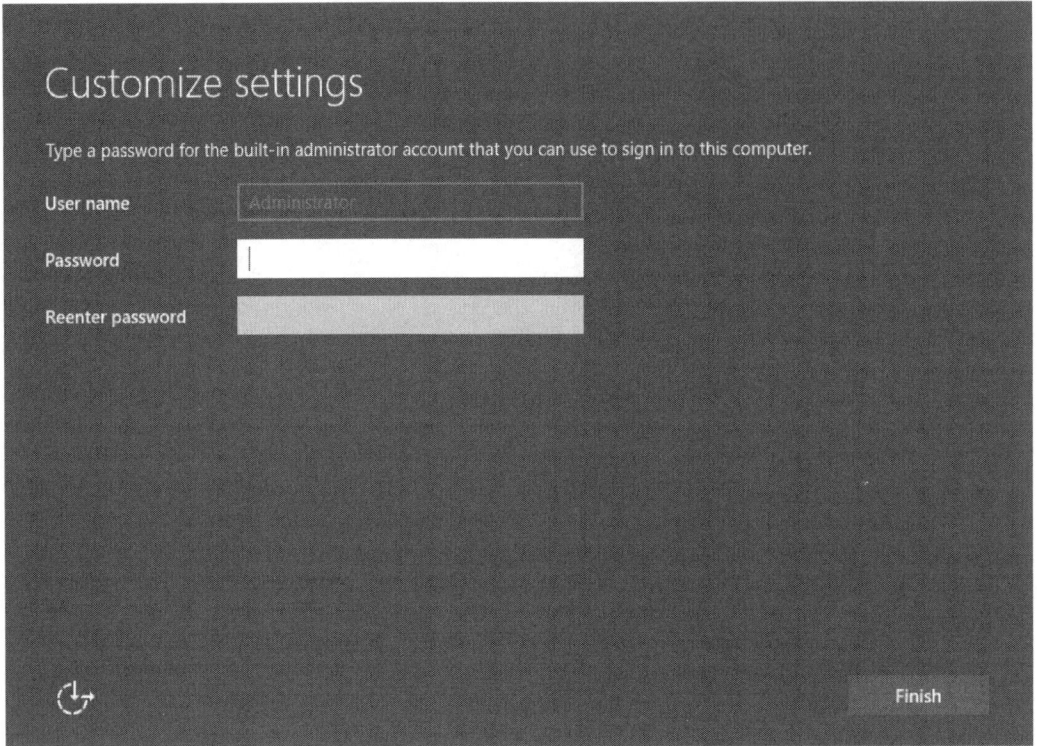

Figure 3.9 – Setting a password for the administrator account

11. Congratulations! You have successfully installed Windows Server 2022. First, press *Ctrl + Alt + Delete* to unlock the system. Then, provide the administrator's password to make the first login to Windows Server 2022.

> **Important note**
>
> If Microsoft's competitors had a single reason to praise it, then, without a doubt, it would be for Windows Installer. Windows Installer is an **application programming interface** (**API**) used by Windows for software installation, maintenance, and uninstalls.

Now that you have learned how to perform a clean installation, you are well equipped to apply that knowledge to various installation methods. Next, let's look at installation over a network using WDS.

Installing over a network

Often, organizations deploy hundreds of servers in their IT infrastructures and use WDS to enable installation over the network. Therefore, setting up WDS is relatively easy (as shown in the *Chapter exercise – setting up WDS* section later in this chapter): install and set up the WDS role on the server and then add the installation and boot images. However, as with a clean installation, installing over a network requires user interaction. Thus, the unattended option enables Windows Server 2022 installation with little or no user interaction. However, it requires an answer file to automate the deployment of an OS.

To perform the network installation of Windows Server 2022 using WDS, complete the following steps:

1. The **Preboot Execution Environment** (**PXE**) is created and establishes a communication with the WDS server, as shown in *Figure 3.10*.

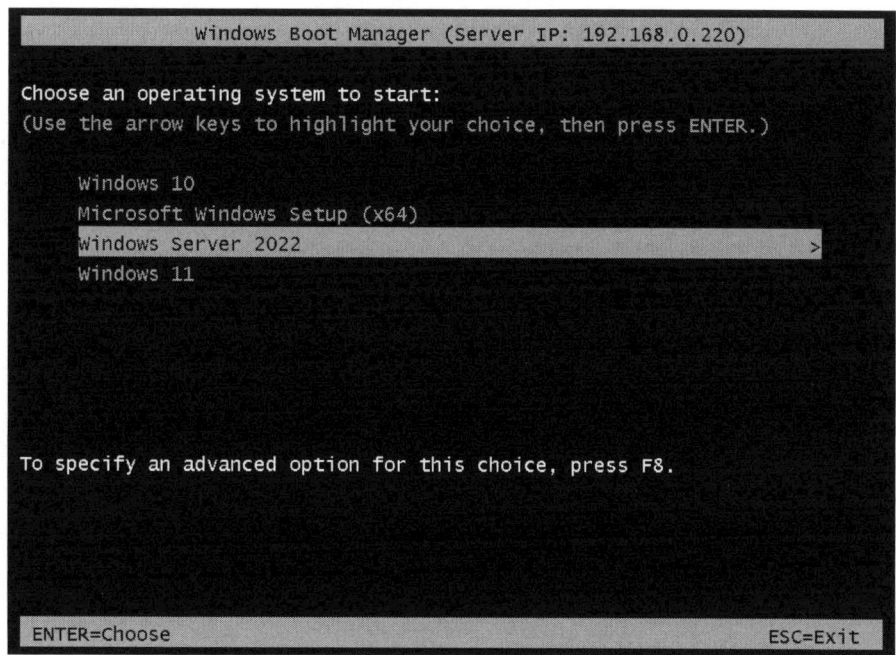

Figure 3.10 – The PXE enables a network installation of Windows Server 2022

2. Load the Windows Server 2022 installation files from the WDS server into the server's memory (i.e., the RAM).
3. In an attempt to support Microsoft Endpoint Configuration Manager or the MDT, Microsoft, as of Windows Server 2022, began partially depreciating the OS deployment functionality of WDS, as shown in *Figure 3.11*. Thus, Windows 11 cannot be deployed using WDS.

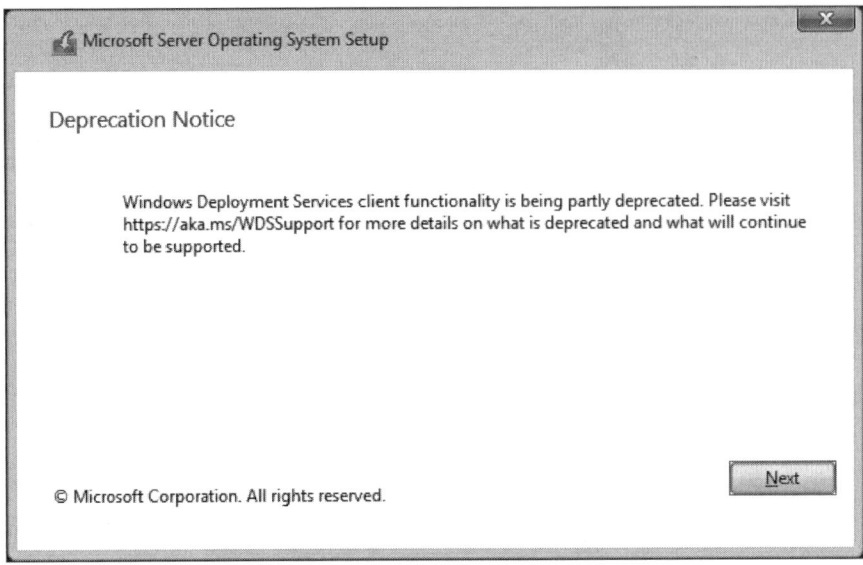

Figure 3.11 – A deprecation notice in Windows Server's 2002 WDS

4. Before choosing the appropriate settings for the **Locale**, **Keyboard,** or **input method** options, click **Next**, and enter the username and password of a WDS administrator, as shown in *Figure 3.12*.

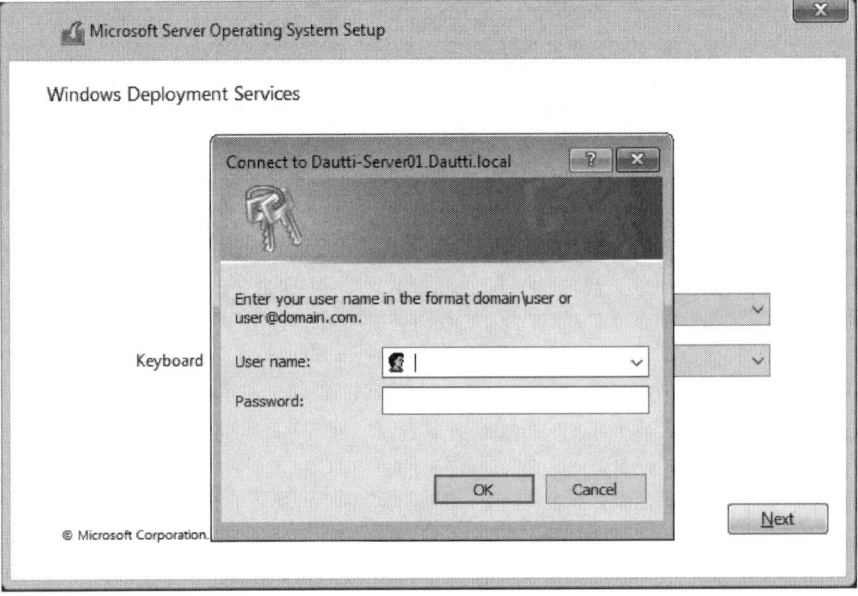

Figure 3.12 – Authentication with the WDS server

5. Select the OS you want to install and click **Next**, as shown in *Figure 3.13*.

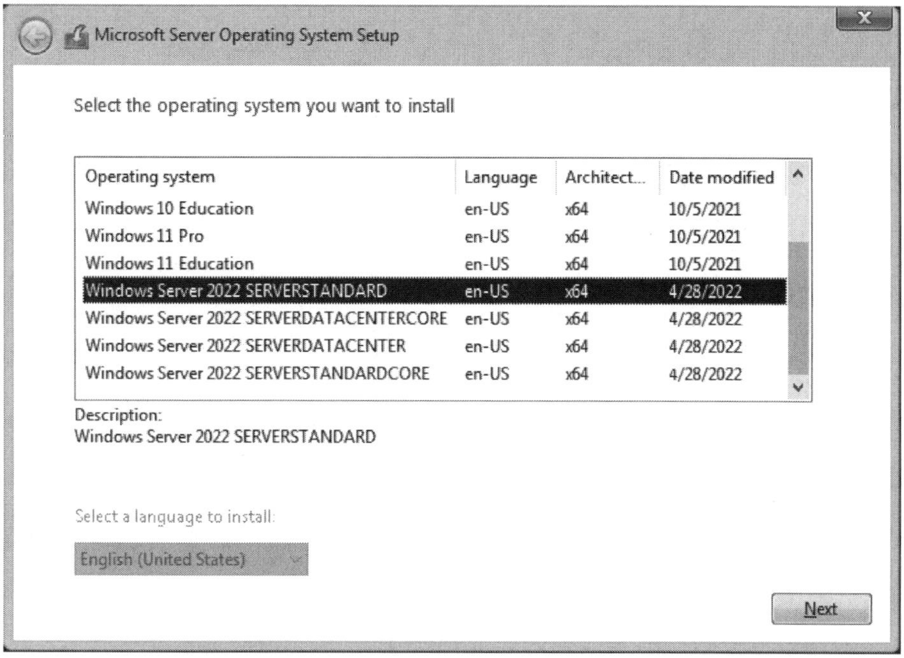

Figure 3.13 – Installation over a network (the WDS server)

6. Once the remaining installation steps are completed (similar to the clean installation steps shown in a previous section), the Windows Server 2022 Standard (Desktop Experience) is successfully installed over the network.

Next, let's look at unattended installation using ADK and MDT.

Unattended installation

In contrast to a clean installation, an unattended installation involves little or no interactivity during installation. It is an automated installation used to deploy many servers in a corporate network in conjunction with WDS. Part of an unattended installation is the *answer file*, an XML file that stores the answers for an installation prompt. You can use Notepad to create an answer file from scratch or download sample answer files from the internet. In addition, Microsoft provides several tools for automating the installation. Apart from WDS, as discussed in the *Installing over a network* section, ADK and MDT tools offer a unique platform for automating desktop and server deployments. Both tools are available for download.

To perform an unattended installation of Windows Server 2022, complete the following steps:

1. Download and install the Windows ADK on a Windows 11 computer, as shown in *Figure 3.14*.

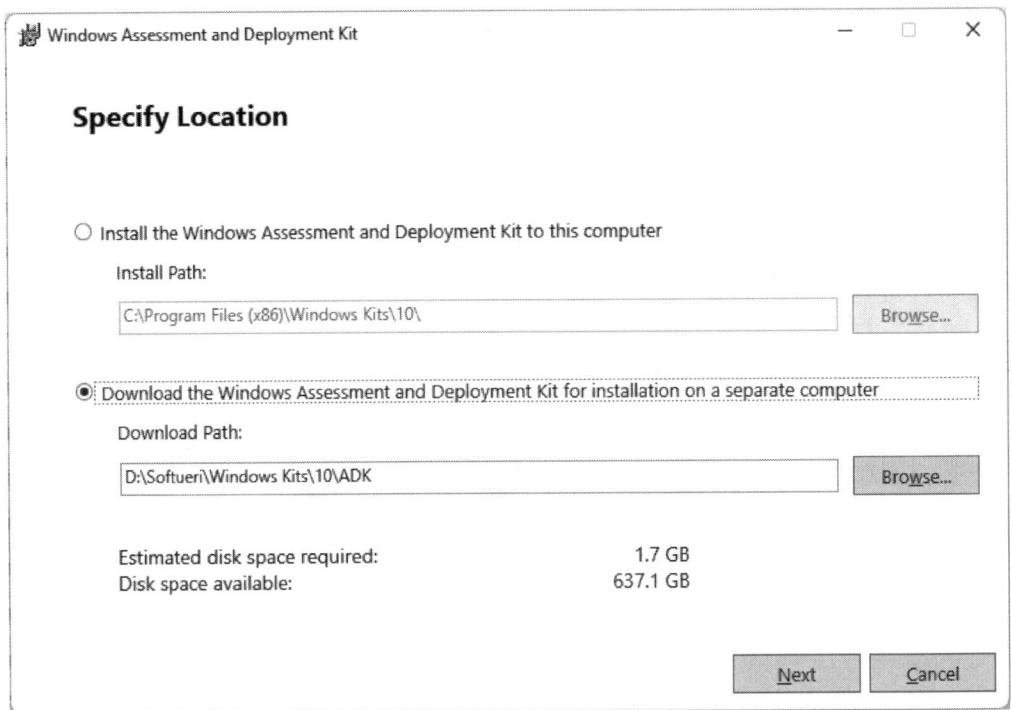

Figure 3.14 – Installing the Windows ADK

2. Next, download and install **Windows Preinstallation Environment** (**WinPE**), as shown in *Figure 3.15*, by running `adkwinpesetup.exe`.

Various Windows Server 2022 installation methods   55

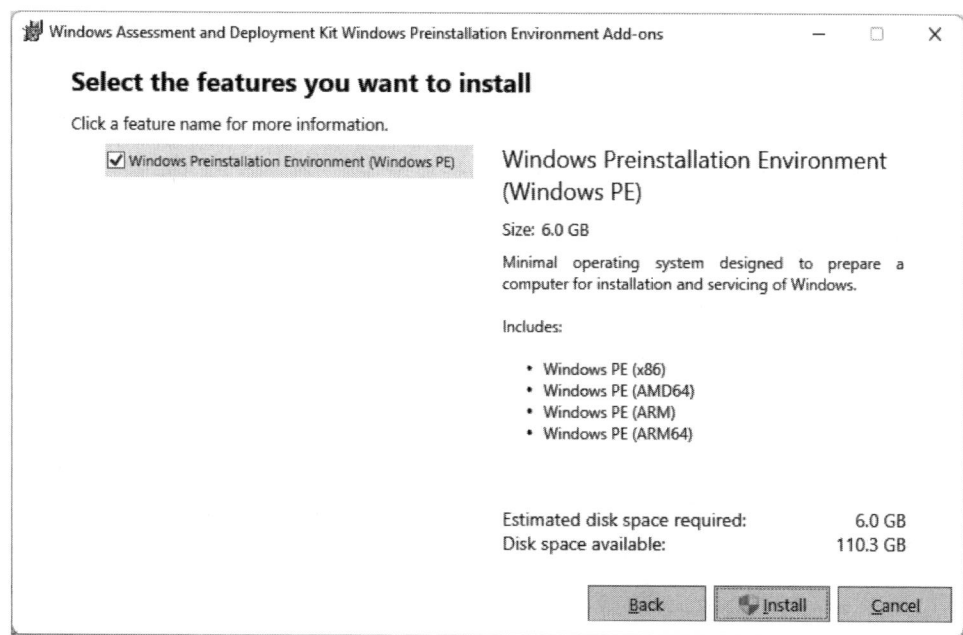

Figure 3.15 – Installing Windows PE

3. As with ADK, install MDT on a Windows 11 computer, as shown in *Figure 3.16*.

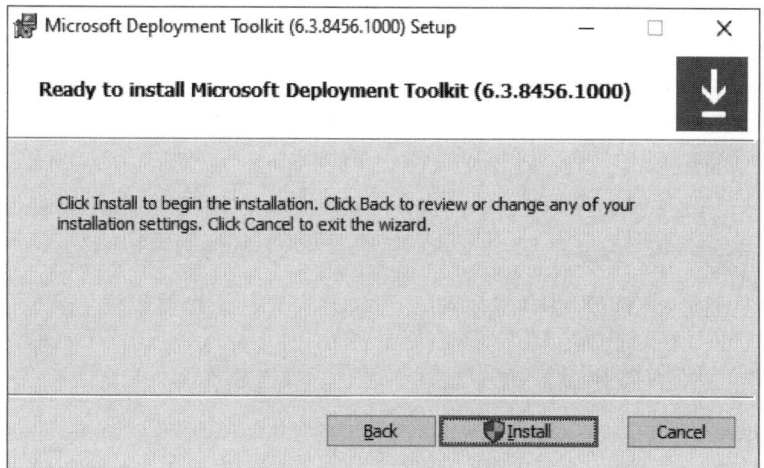

Figure 3.16 – Installing the MDT

4. After installing the Windows ADK, PE and MDT, run **Deployment Workbench** and select **New Deployment Share Wizard**, as shown in *Figure 3.17*.

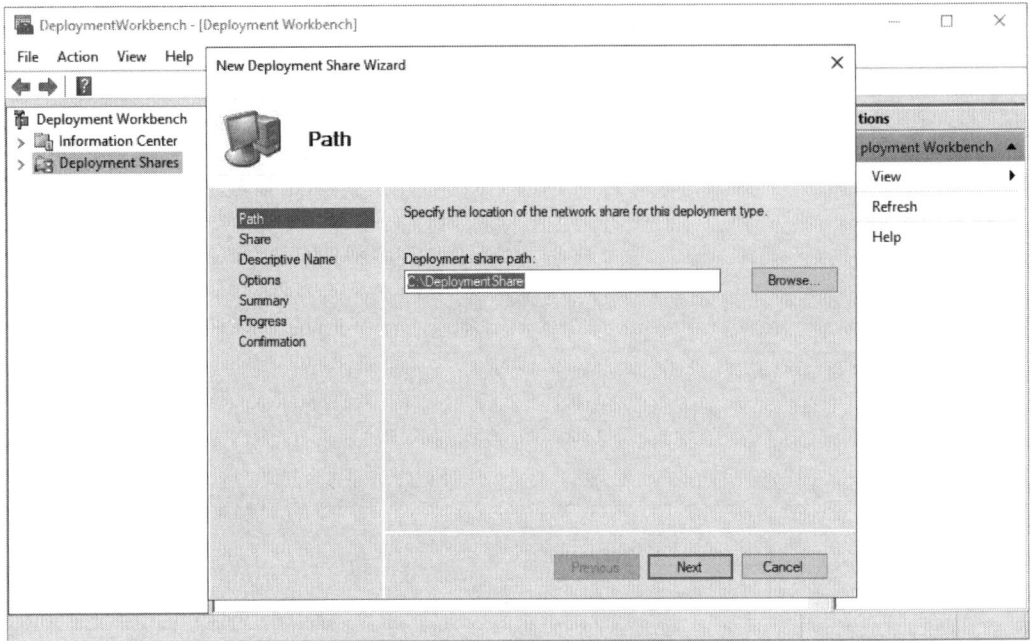

Figure 3.17 – New Deployment Share Wizard

5. After creating the deployment share, share the `DeploymentShare` folder through File Explorer by selecting **Everyone** and setting **Read permissions**.
6. Then, run **Import Operating System Wizard** to import the Windows Server 2022 files.
7. Afterward, run **New Task Sequence Wizard** to create the answer file for an unattended installation.
8. Then, update the deployment share to create a bootable PE image.
9. Boot the new server with the `LiteTouchPE_x64` image, located in the `Boot` subfolder of the `DeploymentShare` folder. After the successful boot, select **Run the Deployment Wizard to install a new Operating System**, as shown in *Figure 3.18*.

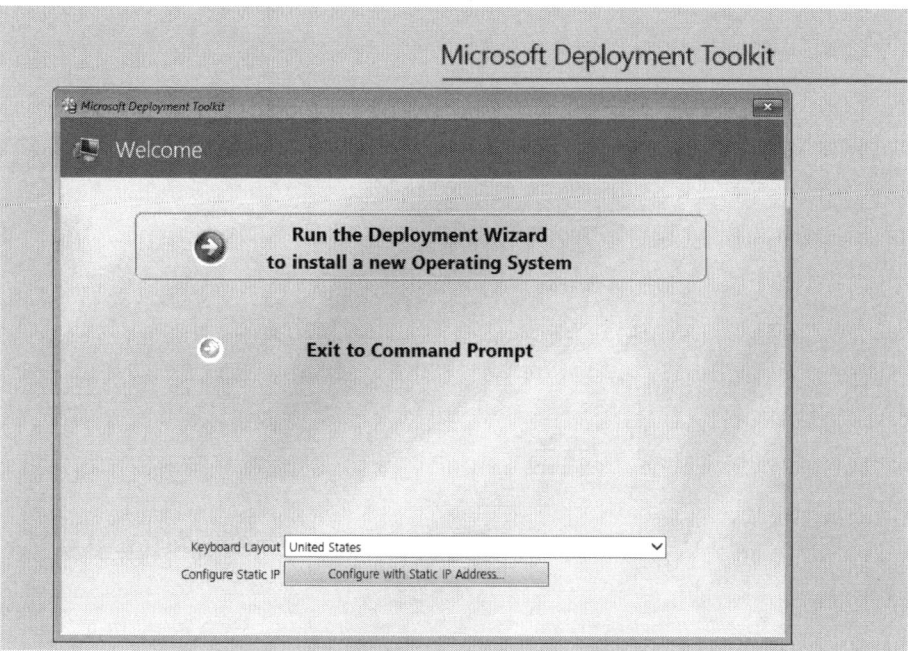

Figure 3.18 – Deploying Windows Server 2022 over the MDT

10. Provide credentials to access the `DeploymentShare` folder. Ensure that the provided user has full control of the `DeploymentShare` folder.
11. Select `Task Sequence`, the answer file created earlier with **Deployment Workbench**, and click **Next**.
12. After providing the **Computer Details**, **Locale**, and **Time** options, specifying whether to capture the image and the BitLocker configuration, you are ready to begin deploying Windows Server 2022.
13. Windows Server 2022 is then deployed through the MDT.
14. After the installation progress steps are complete, the installation prepares the devices, and Windows Server 2022 Standard (Desktop Experience) is deployed successfully.

> **Important note**
> You can download the Windows ADK at `https://developer.microsoft.com/en-us/windows/hardware/windows-assessment-deployment-kit` and the MDT at `https://www.microsoft.com/en-us/download/details.aspx?id=54259`.

Next, let's learn how to perform an in-place upgrade, which will help to upgrade the old OS to a new one while preserving user settings, applications, and data.

An in-place upgrade

An upgrade replaces your existing OS with a new one. This means that you retain your files and settings. That is often called an *in-place upgrade* because it happens in place on a machine with an OS already installed. It is recommended that you back up the Windows state, files, and folders before running an upgrade.

You can run an in-place upgrade to Windows Server 2022 if the actual server runs Windows Server 2019, Windows Server 2016, or Windows Server 2012 R2. Microsoft does not recommend the in-place migration from Windows Server 2012 R2. However, Windows Server 2012 R2 can be upgraded to Windows Server 2022 in two consecutive upgrade processes – first by upgrading to Windows Server 2016, and then finally upgrading Windows Server 2016 to Windows Server 2022. Windows Server 2016 and Windows Server 2019 can be upgraded to Windows Server 2022 in a single upgrade process.

To perform an in-place upgrade from Windows Server 2019 to Windows Server 2022, complete the following steps:

1. Insert the Windows Server 2022 installation disc, plug in the bootable USB flash drive, and run the `setup` file.
2. Shortly, the **Install Windows Server** window shows up. Click **Next** to continue, as shown in *Figure 3.19*.

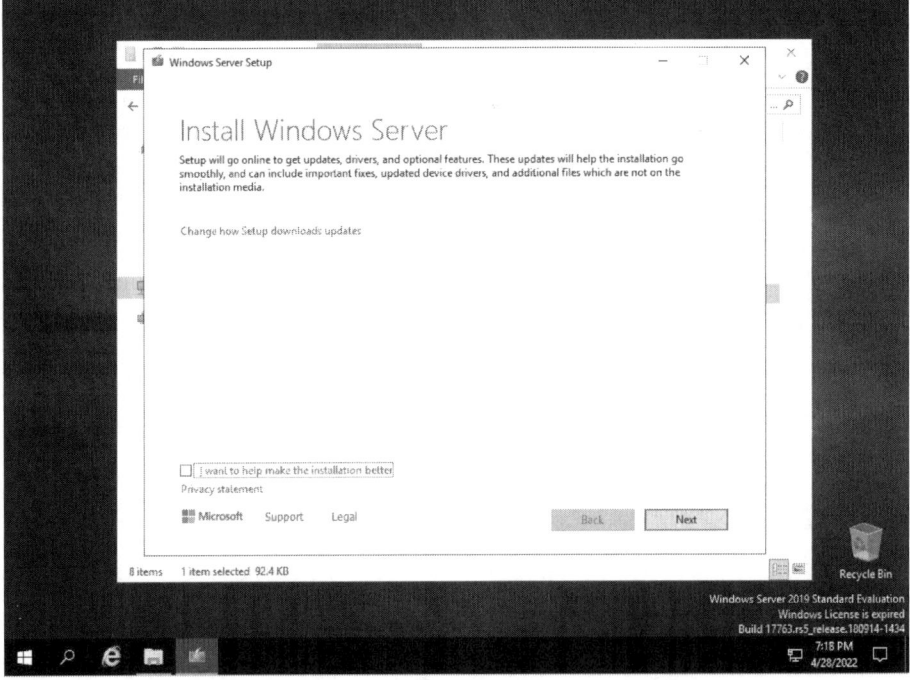

Figure 3.19 – Getting essential updates to help ease the upgrade

3. Type in the product key and click **Next**.
4. Select the Windows Server 2022 edition you want to install, as shown in *Figure 3.20*, and click **Next**.

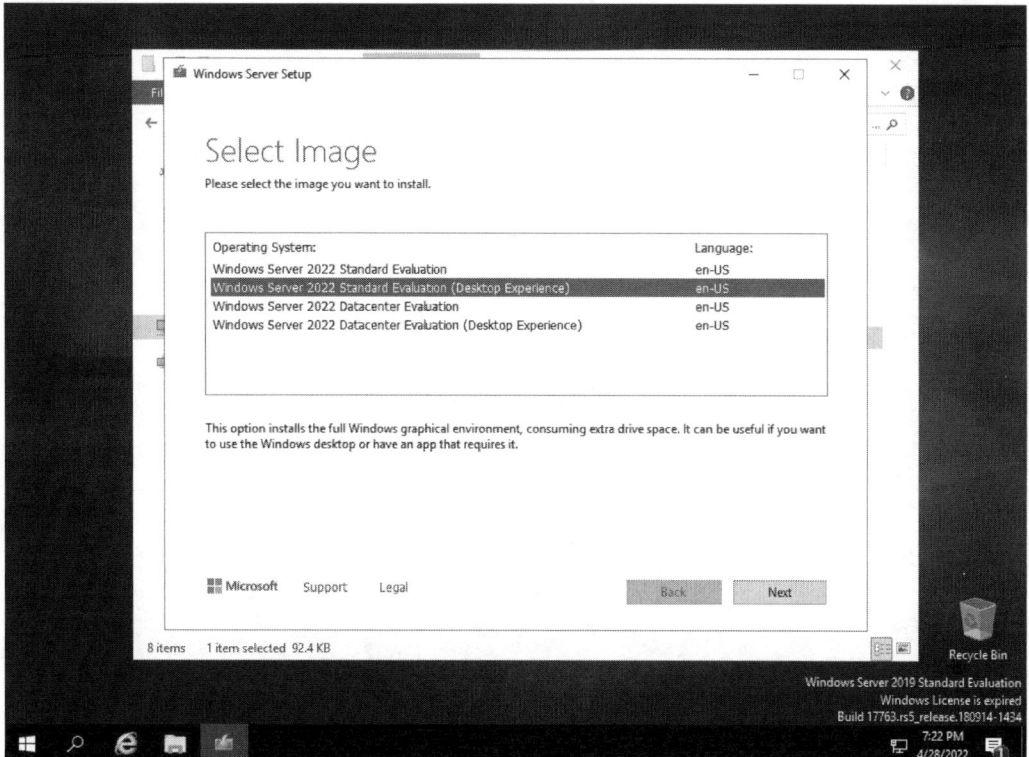

Figure 3.20 – Selecting the Windows Server 2022 edition to install

5. Click the **Accept** button in **Applicable notices and license terms** to accept the license terms.
6. Choose what to keep, and then click **Next** to continue.
7. Once the updates are downloaded and the Windows Server 2022 setup ensures enough disk space on the server, you are ready to install. Click the **Install** button to continue with the upgrade, as shown in *Figure 3.21*.

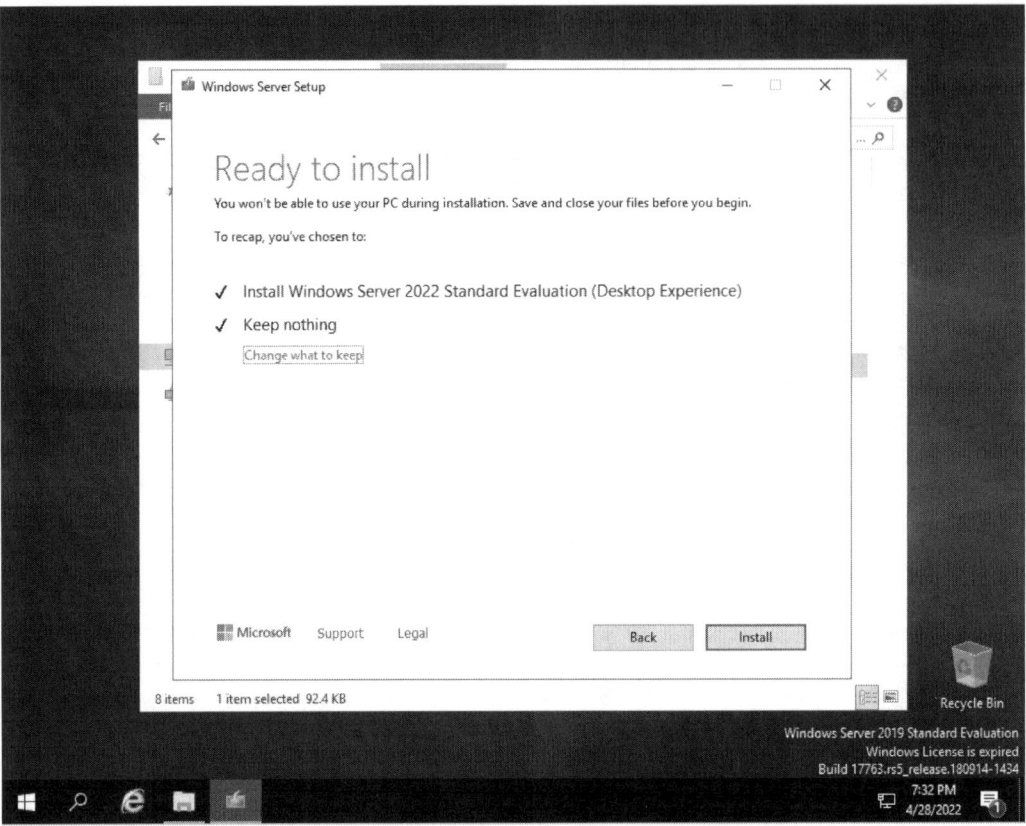

Figure 3.21 – Ready to run the in-place upgrade

8. The in-place upgrade begins. Either sit back and relax or do other work until the upgrade completes.

9. After several restarts, the upgrade from Windows Server 2019 to Windows Server 2022 is completed successfully, as shown in *Figure 3.22*.

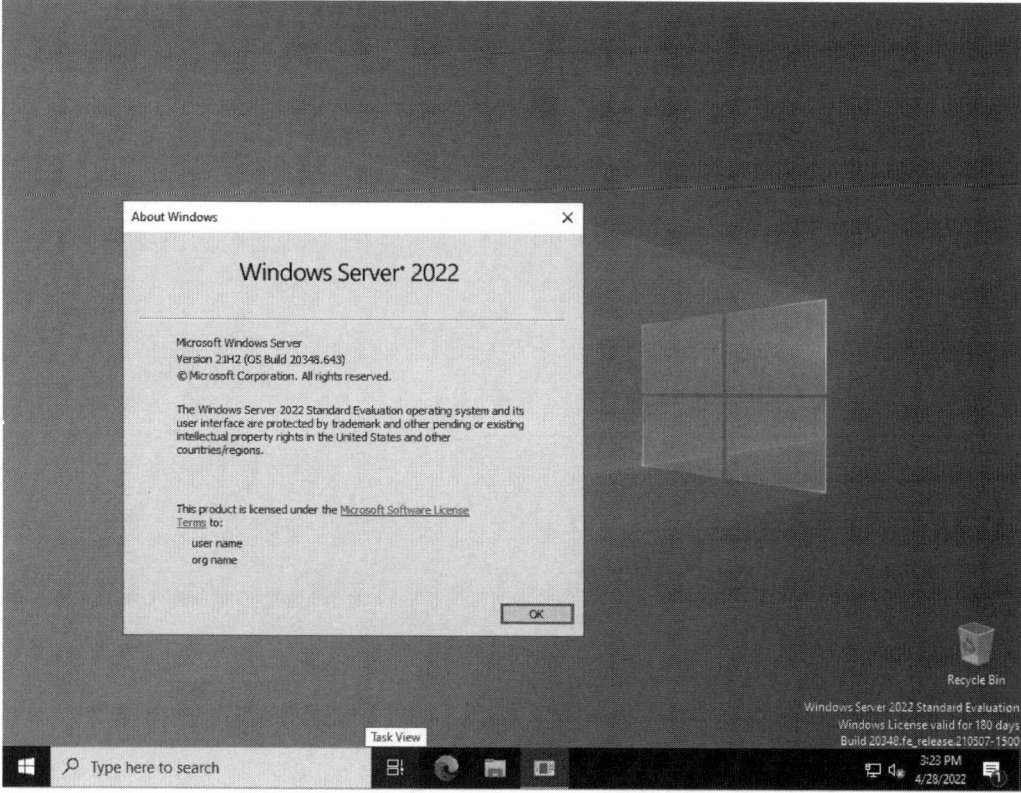

Figure 3.22 – The system properties confirm the in-place upgrade

Next, let's learn how to perform migration using **Windows Server Migration Tools** (**WSMT**). This will help migrate services from an old server to a new one.

Migrating network services

Migration occurs when you bring in a new machine (physical or virtual) and want to migrate the roles, features, apps, settings, and network services from an old server to a new one. First, you want to install the OS on the new server and then proceed with migration. Before migrating, check whether Windows Server 2022 supports your existing apps. Next, the WSMT feature is designed to facilitate the migration process whenever such a task is required. Alternatively, you may want to run the appropriate cmdlets to perform the migration of needed services, as presented in the following example.

To perform the migration of a DHCP server from an old server (Windows Server 2012 R2) to a new server (Windows Server 2022), complete the following steps:

1. In an old server (e.g., Windows Server 2012 R2), open Windows PowerShell with elevated admin rights and enter the following cmdlets (see *Figure 3.23*):

   ```
   Export-DhcpServer -File C:\DHCPdata.xml -Leases -Force
   -ComputerName <oldserver> -Verbose
   ```

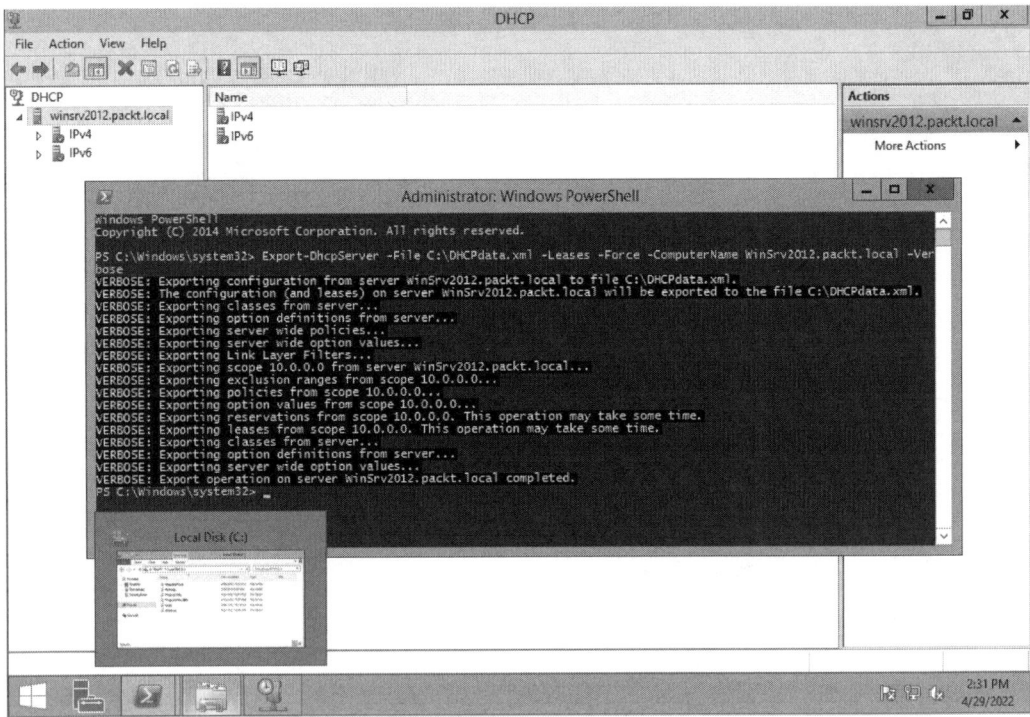

Figure 3.23 – Exporting the DHCP Server from an old server

2. After stopping the DHCP service, move the DHCPdata.xml file into a folder and share the folder with **Everyone**.

3. In a new server (e.g., Windows Server 2022), add the **DHCP Server** role using **Server Manager**, as shown in *Figure 3.24*.

Various Windows Server 2022 installation methods 63

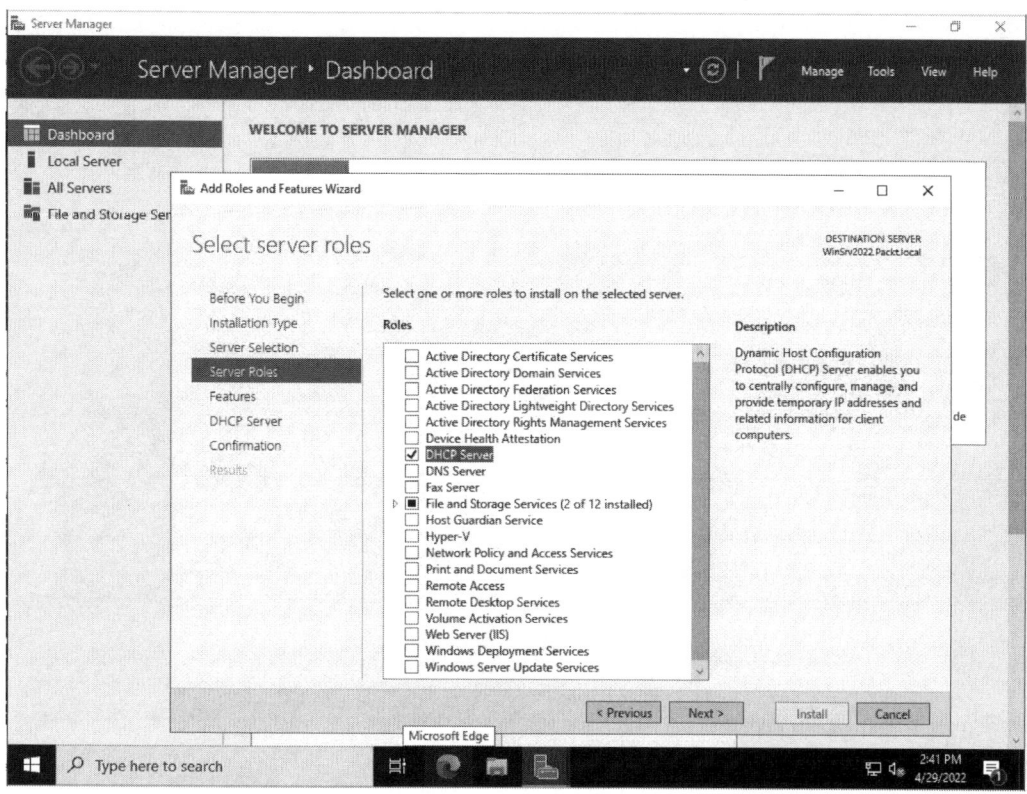

Figure 3.24 – Adding the DHCP Server to a new server

4. Try accessing the shared folder in an old server and copying `DHCPdata.xml` into a root disk in a new server.

5. Open Windows PowerShell with elevated admin rights and enter the following cmdlets (see *Figure 3.25*):

```
Import-DhcpServer -File C:\DHCPdata.xml -BackupPath C:\DHCP\
-Leases -ScopeOverwrite -Force -ComputerName <newserver> -
Verbose
```

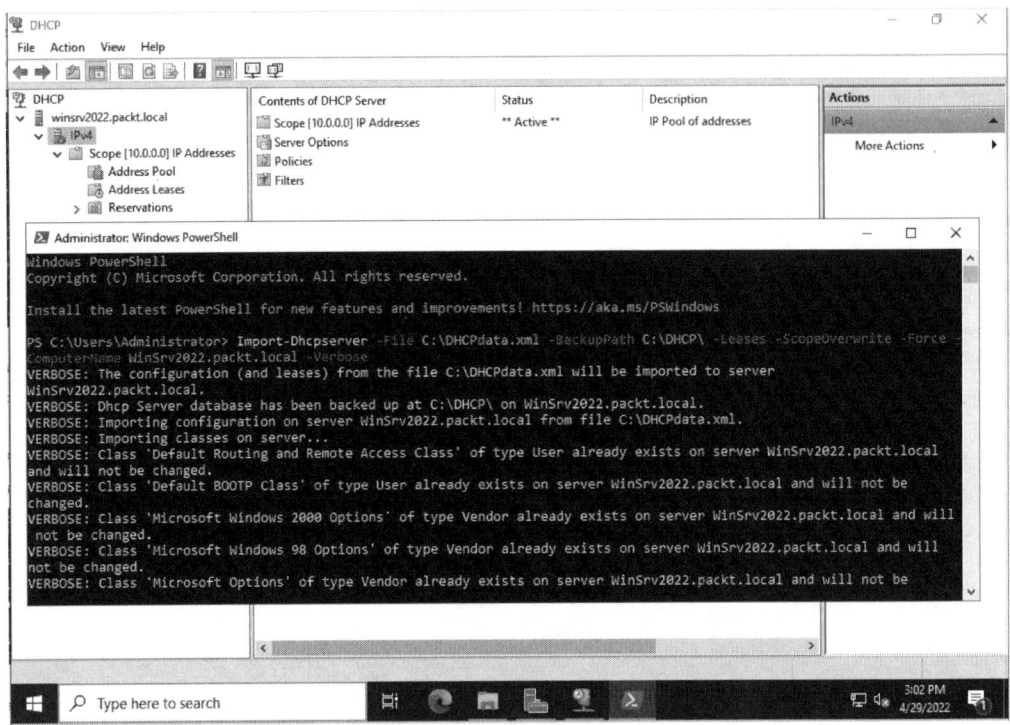

Figure 3.25 – Importing the DHCP Server into a new server

6. Restart the DHCP service, and once the restart completes, note that the DHCP Server is migrated successfully from an old server to a new server.

Next, let's learn how to install Windows Server 2022 in the cloud, such as **Microsoft Azure**. This will help to run services in the cloud.

Trying Windows Server 2022 in Azure

Today, when businesses provide multiple services from the cloud, they undoubtedly need an OS optimized for the cloud environment. For about 25 years, Microsoft has offered Windows Server for the midrange server. But now, in the cloud era, Microsoft is offering Windows Server 2022 in Azure for all businesses that want to trust their network services to this virtual machine's OS.

An Azure account and subscription are required to try out Windows Server 2022 in Azure. However, if you do not have any of them, you can get one from the following URL: https://azure.microsoft.com/en-us/free/. Once you have set up your Azure account and subscription, attempt to sign in to the Azure portal using this link: https://portal.azure.com/:

1. In Azure, search for **Virtual machines** and click **Create a Virtual Machine** to begin setting up a virtual machine.

2. Enter a subscription, and either use an existing resource group or create one.
3. Enter a name for the virtual machine, and choose the region where you want to deploy it.
4. For the OS image, choose **Windows Server 2022 Datacenter: Azure Edition - Gen2**, as shown in *Figure 3.26*.

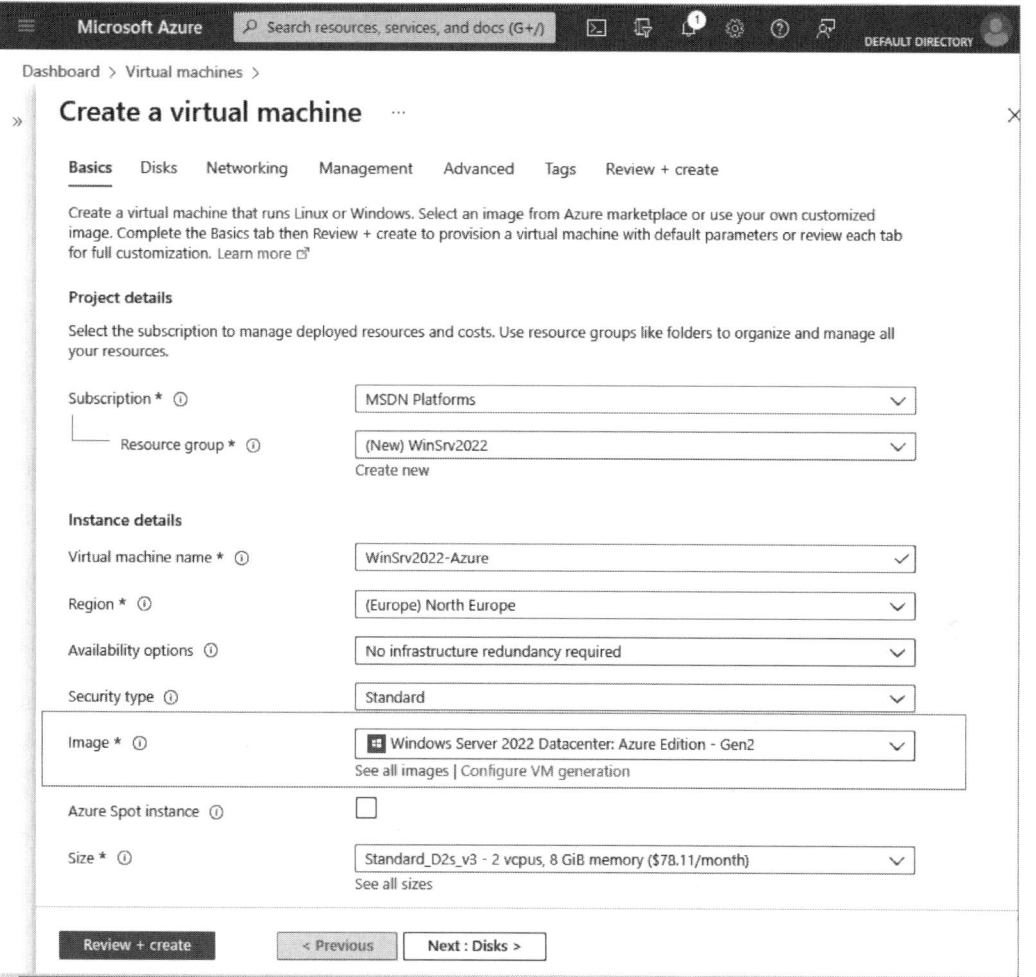

Figure 3.26 – Setting up the virtual machine with Windows Server 2002 in Azure

5. Once you have entered the values in the **Basic** tab, continue entering values in the other tabs, such as **Disks**, **Networking**, **Management**, **Advanced**, and **Tags**.

6. When you are done with entries, click **Review + create** to validate the entries, as shown in *Figure 3.27*.

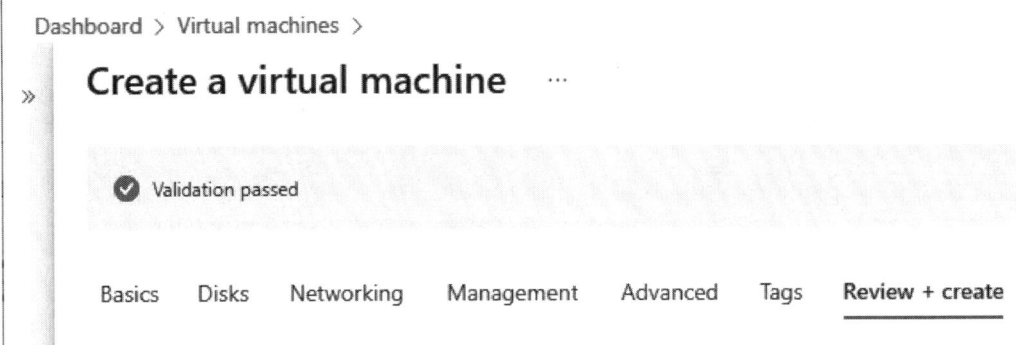

Figure 3.27 – Validating the virtual machine's entries

7. When validation is successful, click **Create** to deploy the virtual machine with Windows Server 2022 Azure edition, as shown in *Figure 3.28*.

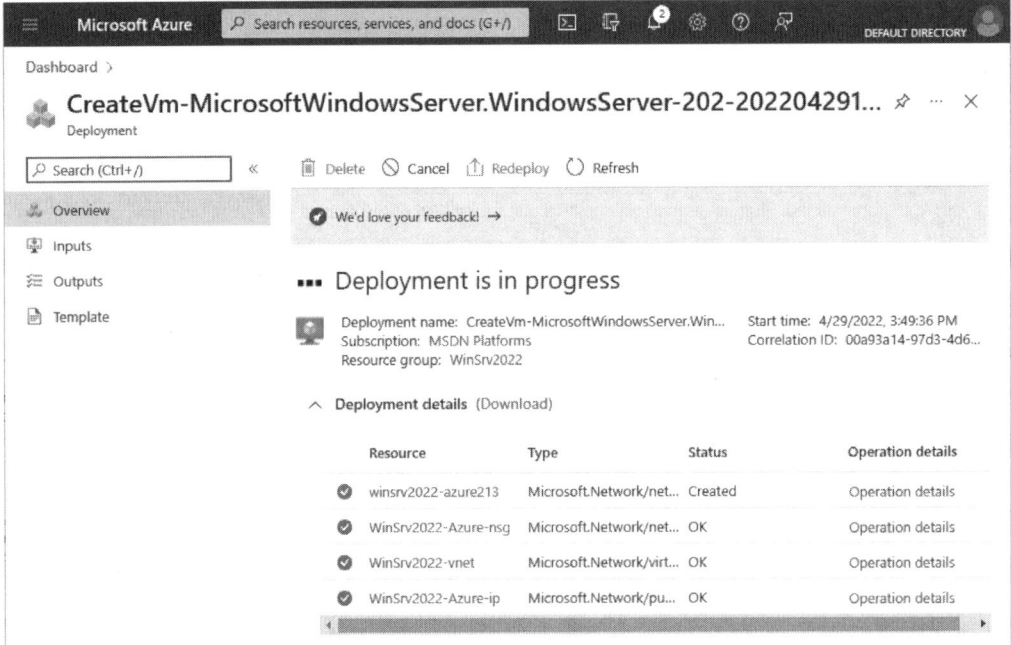

Figure 3.28 – Deploying the virtual machine in Azure

8. Once the deployment completes, go to **resource** and attempt to access the newly created virtual machine; use the **Connect>RDP** option from the Azure portal or **Remote Desktop Connections** by entering the public IP address.

In this section, you have learned about installation methods such as clean installation, installation over a network, unattended installation, in-place upgrade, migration, and trying Windows Server 2022 in Azure. All these are the various methods of installing and deploying Windows Server 2022 on-premises and in the cloud. In the following section, you will learn about setting up WDS.

Chapter exercise – setting up WDS

In this exercise, you will learn how to do the following:

- Installing WDS
- Setting up WDS

Installing WDS

To install WDS in Windows Server 2022, complete the following steps:

1. Click the **Start** button, and then, in the **Start** menu, click the **Server Manager** tile.
2. Click **Add roles and features** in the **WELCOME TO SERVER MANAGER** section in the **Server Manager** window.
3. When the **Add Roles and Features Wizard** opens, click **Next**.
4. Select the **Role-based or feature-based installation** option and click **Next**.
5. Select a server from the **Server Pool** option, which should be checked, and click **Next**.

6. Select the **Windows Deployment Services** role, as shown in *Figure 3.29*, and click **Next**.

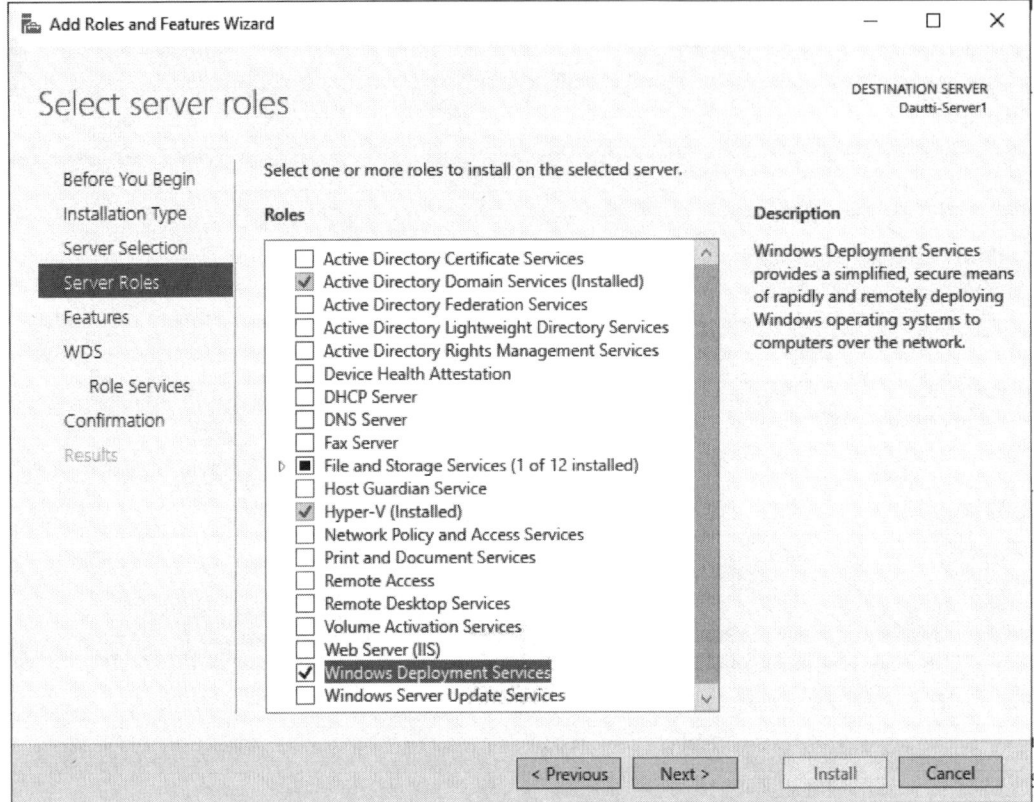

Figure 3.29 – Installing WDS in Windows Server 2022

7. Accept the default settings in the **Select features** step, and click **Next**.
8. Take your time to read the WDS description and the things to note regarding the WDS installation. Then, click **Next**.
9. Select **Role Services**, and then click **Next** as shown in *Figure 3.30*.

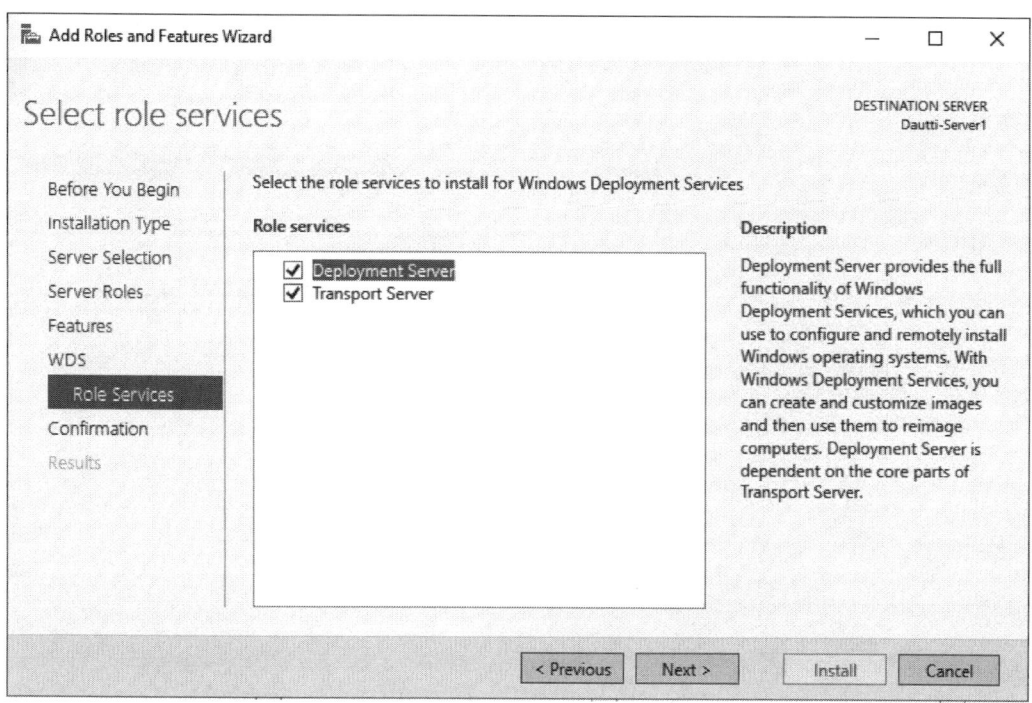

Figure 3.30 – Installing role services

10. Confirm the installation selections for the WDS role, and then click the **Install** button.
11. Either hit **Close** or wait until the installation progress reaches its end.
12. Click **Close** to close the **Add Roles and Features Wizard** window.

You have successfully installed WDS. Now, it is time to set it up.

Setting up WDS

To set up WDS on your Windows Server 2022 server, complete the following steps:

1. Click the **Start** button, and then on the Start menu, click **Windows Administrative Tools**.
2. From the list, double-click **Windows Deployment Services**. Shortly afterward, the **Windows Deployment Services** console will open, as shown in *Figure 3.31*:

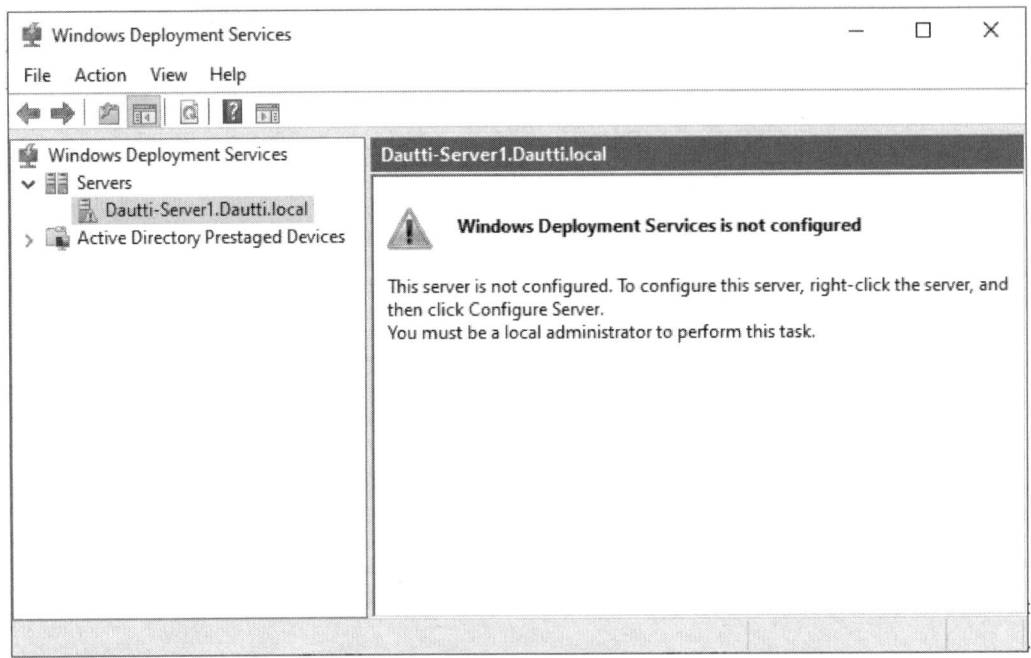

Figure 3.31 – Setting up WDS

3. In the WDS window, right-click the server and select **Configure Server** from the context menu.

4. In the **Windows Deployment Services Configuration Wizard** window, read the **Before you begin** message and click **Next**.

5. Select the **Integrated with Active Directory** option in the **Install Options** step and click **Next**.

6. In the **Remote Installation Folder Location** step, enter the path to the remote installation folder and then click **Next**.

7. In the **Proxy DHCP Server** step, select the options regarding **DHCP on your network infrastructure**, and then click **Next**.

8. In the *PXE* **Server Initial Settings** step, select the options to define which client computers this server will respond to, and then click **Next**.

9. In the **Task Progress** step, wait for the task to be completed and click **Finish**.

10. If the WDS service is not running, in the **Windows Deployment Services** window, right-click the server, and then, from the context menu, select **All Tasks | Start**, as shown in *Figure 3.32*.

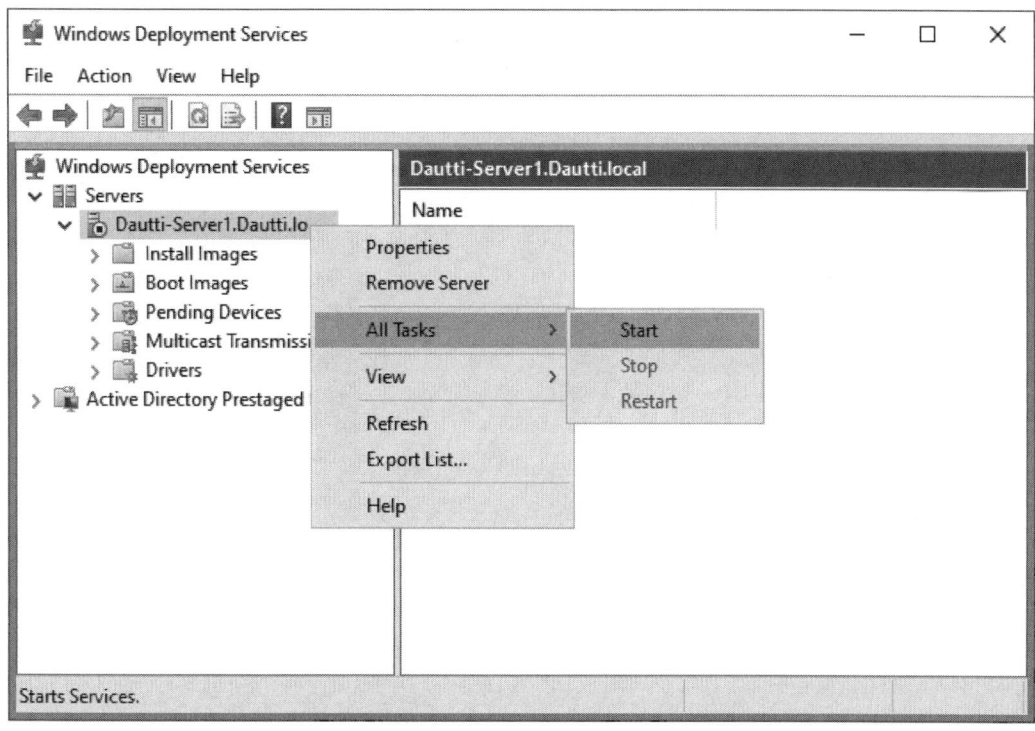

Figure 3.32 – Starting the WDS service

11. Before adding install and boot images, you need to extract the Windows Server 2022 ISO image.
12. Right-click the `Install Images` folder in the **Windows Deployment Services** window and select **Add Install Images...** from the context menu. The following window will open (see *Figure 3.33*).

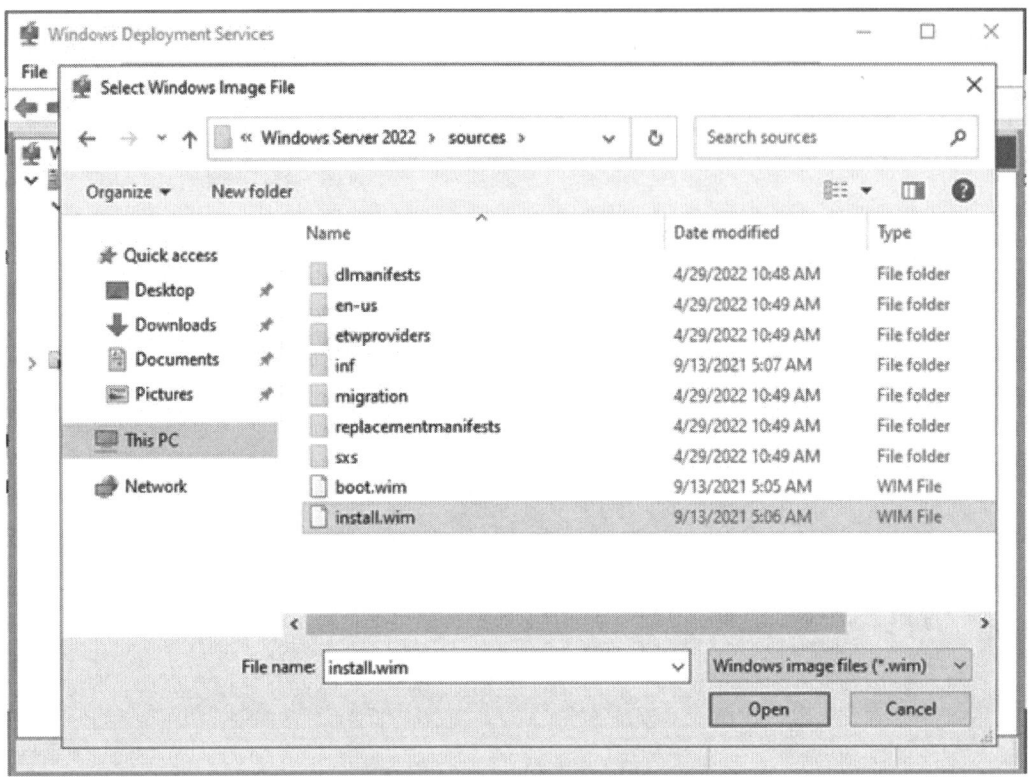

Figure 3.33 – Adding install images

13. Right-click the Boot Images folder in the **Windows Deployment Services** window and select **Add Boot Images...** from the context menu. The following window will open (see *Figure 3.34*).

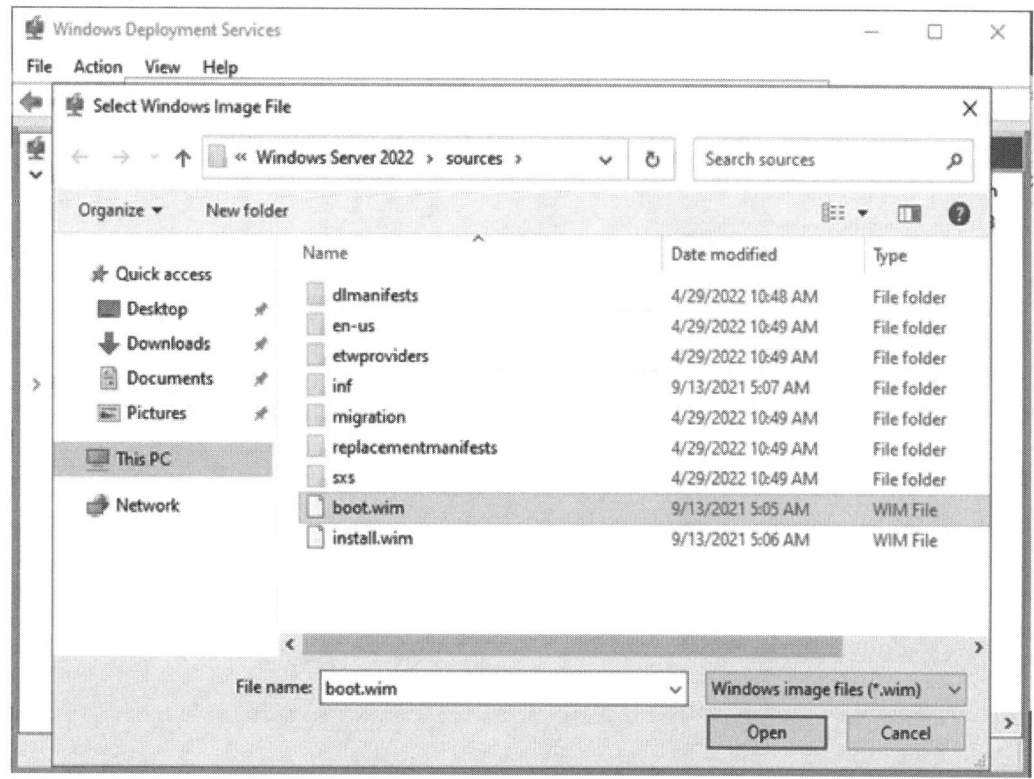

Figure 3.34 – Adding boot images

14. Once the install and boot images are added (see *Figure 3.35*), you can continue deploying Windows Server 2022 over the network using WDS.

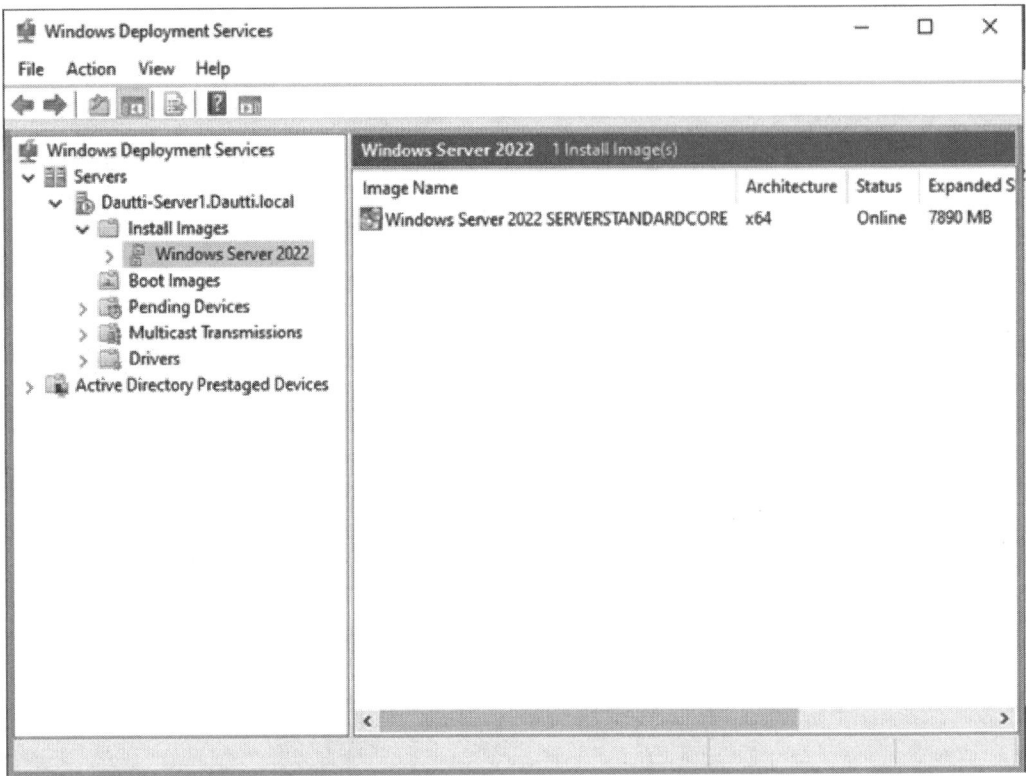

Figure 3.35 – Adding both install and boot images

The WDS configuration completes this chapter exercise, which helped you deploy Windows Server 2022 over the network.

Summary

In this chapter, you learned the various methods of installing and deploying Windows Server 2022. At the same time, we helped you determine which installation option you should have in your work environment depending on the scenario.

This module provided factual information about partition schemes, boot options, and advanced startup options, followed by detailed steps of the different installation methods for installing Windows Server 2022. These methods will help make a clean installation, run an upgrade, install over the network using WDS, automate the deployment, and migrate services from an old version of Windows Server to Windows Server 2022.

Finally, the chapter concluded with an exercise that provided instructions on setting up WDS. The next chapter will teach you about post-installation tasks in Windows Server 2022.

Questions

1. _____ is a new partition scheme that overcomes the limitations of the MBR partition scheme.
2. A clean installation enables automated installation over a network – true or false?
3. As a replacement for Server Core, _____ takes up far fewer hardware resources than the two other installation options, has more periodic updates, and supports only 64-bit applications.
4. Which of the following tools are provided by Microsoft to automate Windows Server 2022 installation? Choose two of the following:

    A. The Windows ADK

    B. MDT

    C. SharePoint Server 2022

    D. SQL Server 2022

5. An unattended installation requires interactivity during the installation of an OS – true or false?
6. _____ takes place when you bring in a new machine (physical or virtual) and you want to move the roles, features, apps, and settings into it.
7. Which of these are installation options in Windows Server 2022? Choose three of the following:

    A. Desktop Experience

    B. Server Core Nano Server

    C. KDE and GNOME

    D. Windows PowerShell

8. Discuss the pros and cons of the three boot options: installation media (DVD), USB flash drive, and a network boot.
9. Discuss the following installation types: a clean installation, a network installation, an unattended or automated installation, an in-place upgrade, and migration.

Further reading

To learn more about the topics that were covered in this chapter, take a look at the following resources:

- Boot to UEFI mode or legacy BIOS mode: `https://docs.microsoft.com/en-us/windows-hardware/manufacture/desktop/boot-to-uefi-mode-or-legacy-bios-mode?view=windows-11`
- Install, upgrade, or migrate to Windows Server: `https://docs.microsoft.com/en-us/windows-server/get-started/install-upgrade-migrate`

- Windows Deployment Services overview: https://docs.microsoft.com/en-us/previous-versions/windows/it-pro/windows-server-2012-r2-and-2012/hh831764(v%3Dws.11)

4
Post-Installation Tasks in Windows Server 2022

In this chapter, you will learn about Windows Server 2022 post-installation tasks. To make this more understandable, the content of this chapter is organized into three parts.

The first part explains the importance of device drivers after installing the Windows Server **operating system** (**OS**). Tasks such as installation, removing, disabling, updating/upgrading, rollback, and other related tasks concerning device drivers are included in this section. The second part covers the Windows Server OS registry, a hierarchical database, and programs that run in the background. Finally, the third part explains the importance of the initial configuration of Windows Server, which needs to be considered once Windows Server 2022 is installed.

Each topic is accompanied by step-by-step instructions illustrated with targeted, easy-to-understand graphics. Finally, this chapter also includes an exercise on performing an initial Windows Server configuration.

In a nutshell, the following topics will be covered in this chapter:

- Understanding devices and device drivers
- Understanding the registry and services
- Understanding Windows Server initial configuration
- Chapter exercise – performing an initial Windows Server configuration

Technical requirements

To complete the exercises in this chapter, you will need the following equipment:

- A PC with Windows 11 Pro with at least 16 GB of RAM, 1 TB of HDD, and access to the internet
- A virtual machine with Windows Server 2022 Standard (Desktop Experience) with at least 4 GB of RAM, 100 GB of HDD, and access to the internet

- A virtual machine with Windows Server 2022 Standard (Server Core) with at least 4 GB of RAM, 50 GB of HDD, and access to the internet

Understanding devices and device drivers

Computer hardware is more than a collection of physical components; the OS is just a collection of programmed instructions. Naturally, therefore, learning about the interaction between hardware and software is exciting. So, the question arises: how does the OS recognize the physical components?

Getting to know computer devices and device drivers

Today's PCs include a case, monitor, keyboard, and mouse in their default composition. However, among today's computers, we also find computers such as the all-in-one, in which the computer case is integrated with the monitor. All the visible differences between the physical parts of PCs and their general physical components are as follows:

- An *internal device* is any device that is located in the computer case. Examples of a computer's internal devices are the power supply, motherboard, accompanying components, hard drives, extension cards, and other internal hardware components that constitute the core computer architecture (any physical component inside a computer case).

- An *external device* is any device attached or connected to a computer case that becomes part of the computer system. Examples of a computer's external devices are the keyboard, monitor (see the dual monitor PC in *Figure 4.1*), mouse, speakers, earphones, webcam, microphone, and other external hardware components.

- A *peripheral device* is considered any device physically located near the computer and, as such, is not an essential part of the computer system. For example, peripheral devices are printers, scanners, projectors, plotters, and so on.

- A *network device* is a peripheral device connected to a computer over a network cable. Network devices include printers, network scanners, network backup libraries, **network-attached storage** (**NAS**), **storage area networks** (**SANs**), and more.

Figure 4.1 – A dual monitor computer

Another category of computer devices is *input* and *output* devices. These devices either create input or output for the computer core architecture. Lately, with technological advancements, some devices simultaneously act as input and output devices. Touch-enabled devices are an example of input/output devices.

> **Important note**
> In today's literature, often, external devices are referred to as peripheral devices too because these devices are connected to a computer case to add functionalities. Almost any device outside the core computer architecture is considered a peripheral device.

A *device driver* is a program that acts as a translator between computer hardware and an OS. Thus, an OS manages and operates the computer hardware through device drivers. Usually, device drivers come with installation media, such as a DVD that accompanies the device, or it can be downloaded from the manufacturer's website. However, do not be surprised if you recently purchased a hardware device that did not include any installation media containing a device driver. This is mainly because the current OSs, such as Windows 10 and 11, support **Plug and Play** (**PnP**). We will explain PnP later in the *Getting to know PnP, IRQ, DMA, and driver signing* section.

Next, let's look at working with devices and device drivers to add additional functionalities to the server.

Working with devices and device drivers

Windows Settings is a new administrative console used to manage devices, while **Device Manager** is a legacy applet that's used to work with device drivers. That being said, in Device Manager, depending on the status of the device driver, note that, other than the proper representation of device drivers, there are also the following representations (see *Figure 4.2*):

- *Generic* indicates that a generic (an alternative) device driver is installed. As such, the generic driver does not present the proper device driver.

- *A black exclamation point on a yellow triangle* indicates that either the device driver is missing or the installed device driver is unsuitable. As such, the correct device driver must be installed.

- *A downward black arrow* indicates a disabled device. The device has a device driver installed, but it is currently not enabled – right-click on a device driver and select **Enable** from the context menu to re-enable it.

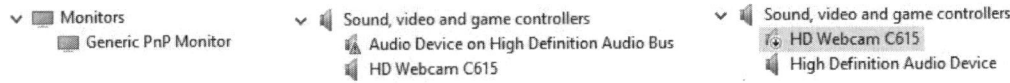

Figure 4.2 – Device drivers representation in Device Manager

Now that we have understood what devices and device drivers are, let's learn how to use them.

Accessing Devices and Device Manager

To access **Devices** from **Windows Settings**, follow these steps:

1. Click the **Start** button to open the **Start** menu.
2. In the **Start** menu, click the **Settings** icon.
3. In **Windows Settings**, click **Devices**.

To access **Device Manager** from the *secret* **Start** menu, follow these steps:

1. Right-click the **Start** button to open the *secret* **Start** menu.
2. In the *secret* **Start** menu, select **Device Manager**.
3. The **Device Manager** window will open shortly after, as shown here:

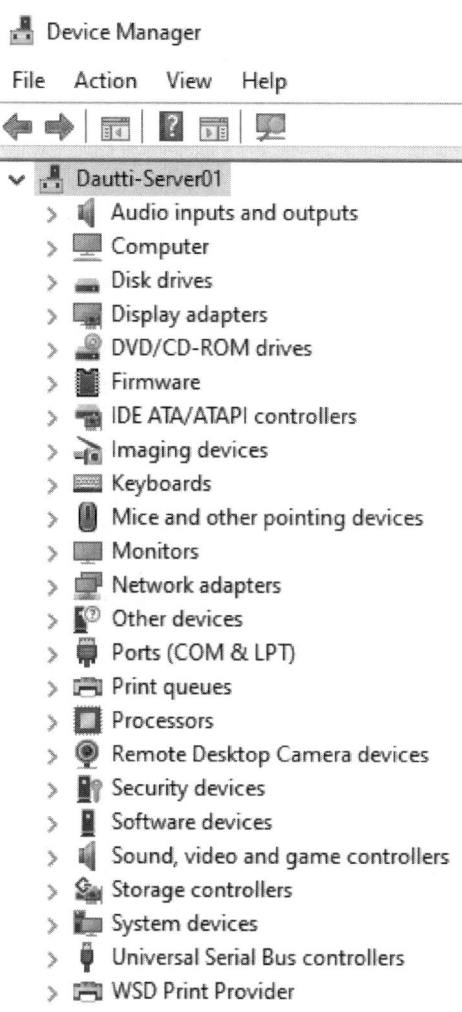

Figure 4.3 – Device Manager

> **Important note**
> Other than right-clicking the **Start** button to open the *secret* **Start** menu, you can use the *Windows key + X* combination. Similarly, you can use the *Windows key + I* combination to open **Windows Settings**. At the same time, to open **Device Manager**, enter `devmgmt.msc` in the **Run** dialog box.

Adding devices and installing device drivers

To add a device using **Windows Settings**, follow these steps:

1. Click the **Start** button to open the **Start** menu.
2. In the **Start** menu, click the **Settings** icon.
3. In **Windows Settings**, click **Devices**.
4. In the **Devices** navigation menu, click **Bluetooth & other devices**.
5. In the **Bluetooth & other devices** section, click **Add Bluetooth or other devices** to add a device.

To install a device driver using a file from installation media or downloaded from the internet, follow these steps:

1. Insert the DVD in a DVD drive or locate the downloaded *device driver* file on the server.
2. Through **File Explorer**, run the setup or installation file.
3. Follow the instructions in **Setup or Install Wizard**.

Now that the device drivers have been installed, let's learn how to update them.

Updating device drivers

To update the device driver using **Device Manager**, follow these steps:

1. Right-click the **Start** button to open the *secret* **Start** menu.
2. In the *secret* **Start** menu, select **Device Manager**.
3. In the **Device Manager** window, expand the device's category.
4. Right-click the device and choose **Update driver** from the context menu.
5. Select **Browse my computer for drivers**, as shown in the following screenshot. If you lack a device driver, let the **Update Drivers** wizard do the work for you by clicking on **Search automatically for drivers**:

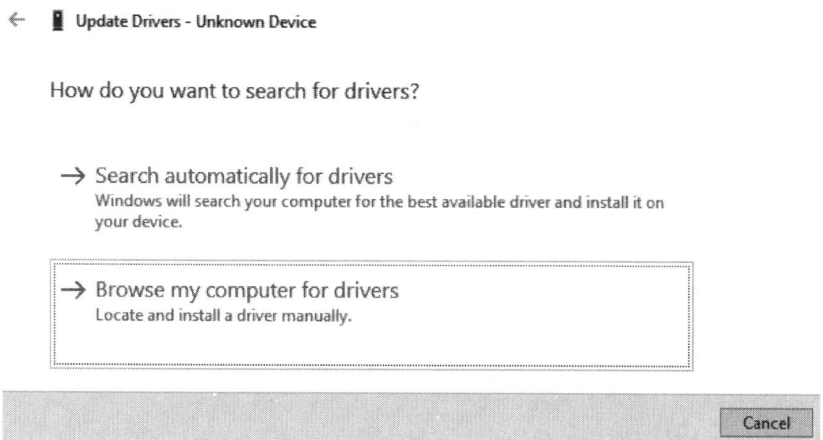

Figure 4.4 – Installing a device driver

> **Important note**
> You will always use the **Update driver** option from the context menu when installing or updating a driver through **Device Manager**.

Now, let's learn how to remove and uninstall device drivers.

Removing devices and uninstalling device drivers

To remove a device using **Windows Settings**, follow these steps:

1. Click the **Start** button to open the **Start** menu.
2. In the **Start** menu, click the **Settings** icon.
3. In **Windows Settings**, click **Devices**.
4. In the **Devices** navigation menu, click **Bluetooth & other devices** and select the device you want to remove.

5. Click the **Remove device** button, as shown in the following screenshot:

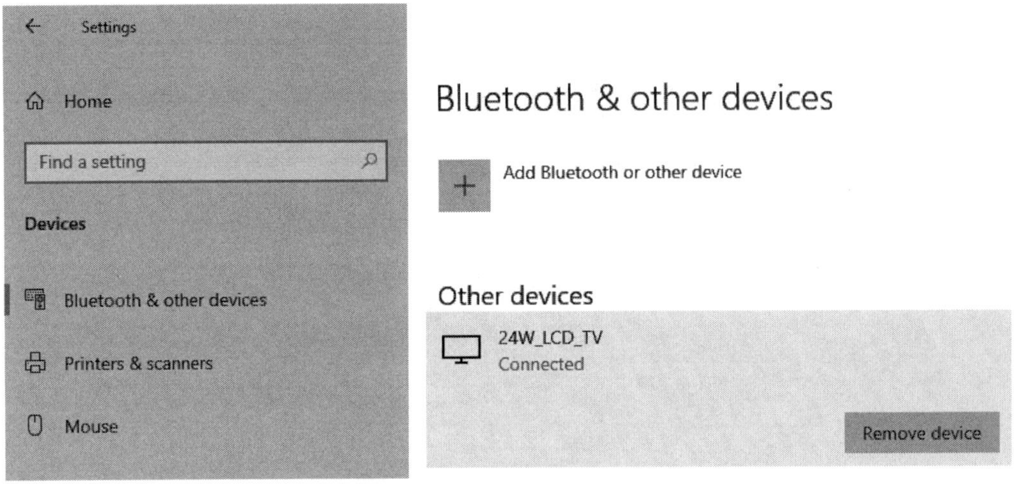

Figure 4.5 – Removing a device

To uninstall a device driver using **Device Manager**, follow these steps:

1. Right-click the **Start** button to open the *secret* **Start** menu.
2. In the *secret* **Start** menu, select **Device Manager**.
3. In the **Device Manager** window, expand the device's category.
4. Right-click the device and choose **Uninstall device** from the context menu.
5. Click the **Uninstall** button.

Now, let's learn how to manage and disable device drivers.

Managing devices and disabling device drivers

To manage a device using **Windows Settings**, follow these steps:

1. Click the **Start** button to open the **Start** menu.
2. In the **Start** menu, click the **Settings** icon.
3. In **Windows Settings**, click **Devices**.
4. In the **Devices** navigation menu, click **Printers & scanners** and select the device you want to manage.
5. Click the **Manage** button.

To disable a device driver using **Device Manager**, follow these steps:

1. Right-click the **Start** button to open the *secret* **Start** menu.
2. In the *secret* **Start** menu, select **Device Manager**.
3. In the **Device Manager** window, expand the device's category.
4. Right-click the device and select **Disable device** from the context menu.
5. In the confirmation pop-up box, click **Yes**, as shown in the following screenshot:

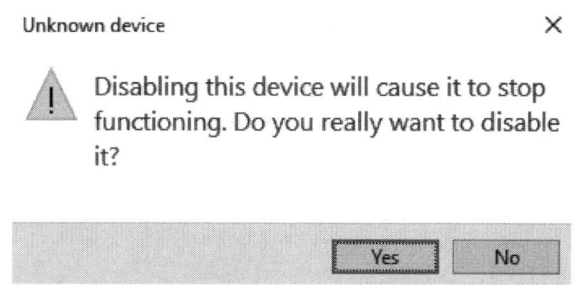

Figure 4.6 – Disabling the device driver

Now, let's learn how to roll back device drivers.

Rolling back device drivers

To roll back a device driver using **Device Manager**, follow these steps:

1. Right-click the **Start** button to open the *secret* **Start** menu.
2. In the *secret* **Start** menu, select **Device Manager**.
3. In the **Device Manager** window, expand the device's category.
4. Right-click the device and choose **Properties**.
5. Select the **Driver** tab and click the **Roll Back Driver** button, as shown in the following screenshot. Click **OK**:

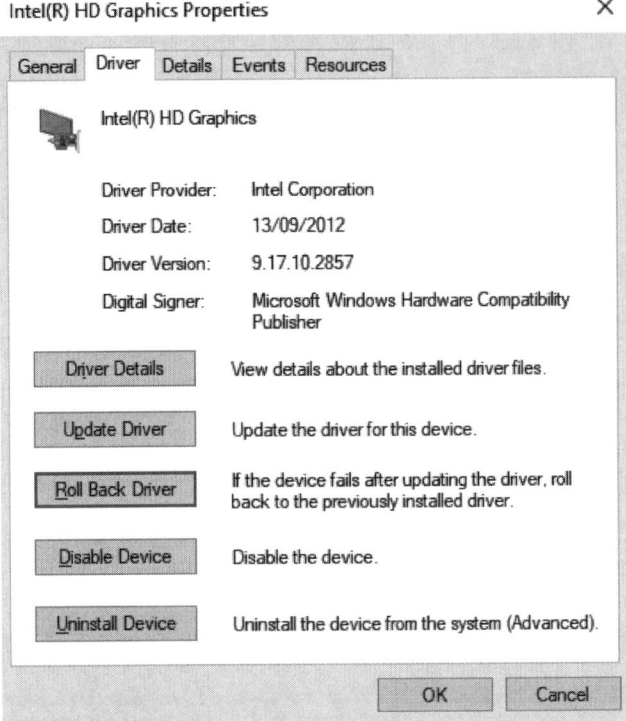

Figure 4.7 – Rolling back a device driver

> **Important note**
> You will want to roll back a driver if you have installed multiple device drivers or have updated the existing device driver with a newer one and the device is not performing as expected.

Now, let's learn about some solutions we can implement when troubleshooting device drivers.

Troubleshooting a device driver

If you encounter technical problems with device drivers, then there are several options you can choose from (see *Figure 4.8*) to overcome them:

- **Update Driver**: This allows you to update the driver automatically or browse the server for driver software.
- **Roll Back Driver**: This allows you to roll back the driver if your current driver is causing problems.
- **Disable Device**: This allows you to disable the driver if the current driver is causing significant issues, such as server instability.

- **Uninstall Device**: This allows you to uninstall the current driver if you have found the appropriate driver from the device manufacturer:

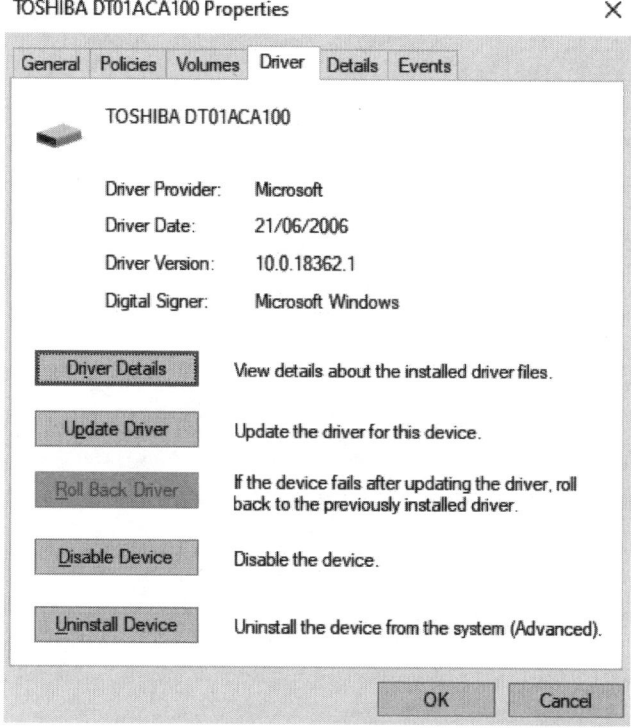

Figure 4.8 – Troubleshooting options for device drivers

In this section, you learned about the several ways in which you can handle devices. Now, let's learn about various system resources, such as PnP, IRQ, DMA, and driver signing.

Getting to know PnP, IRQ, DMA, and driver signing

Knowing that the two main components of a computer are hardware and software, the OS needs system resources to manage the hardware, such as CPU, memory, disk, input/output devices, and network connections. For example, with I/O ports and memory addresses, **Interrupt Request (IRQ)** and **Direct Memory Access (DMA)** are two well-known computer system resources that an OS must manage to use computer hardware.

PnP

It all started as a joint project between Intel and Microsoft long ago. Since then, PnP has massively simplified work with devices and device drivers. As the name suggests, all you need to do with a PnP-enabled computer is plug a device into a computer, which the Windows OS immediately recognizes. Once the

Windows OS identifies the device, it uses its **Driver Store** to install the device driver. For example, in Windows Server 2022, the Driver Store is located at `C:\Windows\System32\DriverStore`.

IRQ and DMA

In modern computers, an IRQ is identified by a decimal number from 0 to 31. From a technical point of view, it is a signal sent by a device through communication channels to get the attention of a processor when that device requires processing. In contrast, DMA, identified by a number from 0 to 8, represents a system resource used by a device to bypass the processor whenever such a device needs direct access to the RAM.

To view the IRQ and DMA resource settings using **Device Manager**, follow these steps:

1. Right-click the **Start** button to open the *secret* **Start** menu.
2. In the *secret* **Start** menu, select **Device Manager**.
3. In the **Device Manager** window, expand the device's category.
4. Right-click the device and choose **Properties** from the context menu.
5. Click the **Resources** tab and check out the **Resource settings:** section, as shown in the following screenshot:

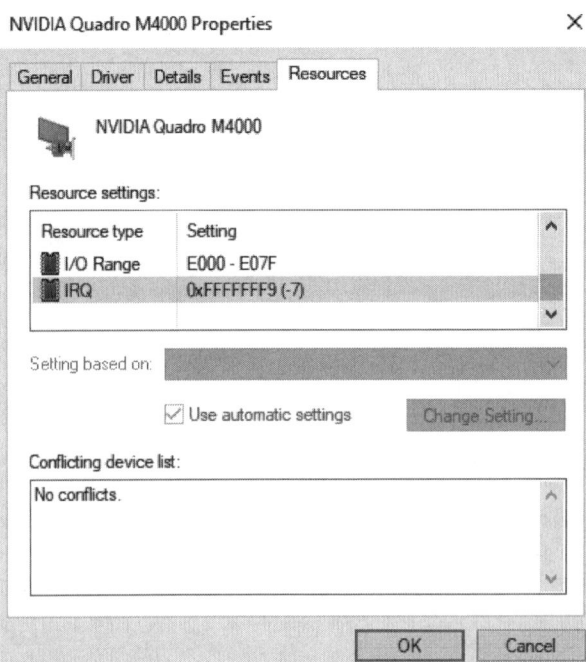

Figure 4.9 – The driver's resource information

Next, look at the driver's signature, which helps verify both the driver's integrity and identity.

Driver signing

Driver signing is a driver's digital signature to identify the publisher of the driver package. In addition, a driver's digital signature proves that Microsoft has tested and approved the driver package, ensuring its installation will not cause reliability or security issues. To view a driver's digital signing information in Windows Server 2022, follow these steps:

1. Right-click the **Start** button to open the *secret* **Start** menu.
2. In the *secret* **Start** menu, select **Device Manager**.
3. In the **Device Manager** window, expand the device's category.
4. Right-click the device and select **Properties** from the context menu.
5. Click the **Driver** tab and the **Driver Details** button to view the **Driver File Details** window, as shown in the following screenshot:

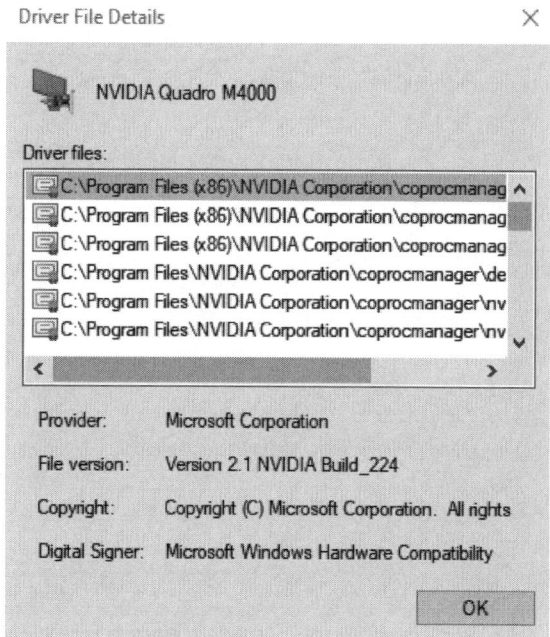

Figure 4.10 – The driver's digital signing information

In this section, you learned about computer devices and drivers and the various ways of handling them. You also learned about computer system resources that devices use to communicate. The next section will teach you how to work with and manage the Windows Registry and its services.

Understanding the Windows Registry and its services

In many IT books, the Windows Registry is often portrayed as the heart of the Windows OS, and services are referred to as *the background programs* of the Windows OS. However, the Windows Registry and Windows Services comprise the OS's core architecture, regardless of the language used to describe them.

Windows Server Registry

Whatever hardware or software change is made to the server is stored on the registry. The **Windows Registry** is a hierarchical database that stores the hardware/software configuration and system security information. Once you access the Windows Registry, you will notice that its console tree (left-hand side) consists of five registry keys known as **hives** (**HKEYs**). Note that the syntax of the registry keys and sub-keys follows the standard of the Windows file path separated by a backslash. For example, in Windows Server 2022, there are five HKEYs:

- **HKEY_CLASSES_ROOT**: This stores information about installed applications and their extensions.
- **HKEY_CURRENT_USER**: This stores information about the user that is currently logged in.
- **HKEY_LOCAL_MACHINE**: This stores information specific to the local computer.
- **HKEY_USERS**: This contains information on logged user profiles.
- **HKEY_CURRENT_CONFIG**: This contains information that was gathered during the boot process.

Windows Server services

Whether you are running an application or a network service, working behind the scenes are services that support their execution. These background programs can be started, stopped, restarted, and paused through the **Services Control Manager**.

Getting to know service startup types

When accessing services through Control Manager, you will notice that each service has the following startup types:

- **Automatic**: The service starts automatically when the OS starts.
- **Automatic (Delayed Start)**: The service begins approximately 2 minutes after all marked automated services.
- **Manual**: The service must be initiated by a user or dependent services.
- **Disabled**: The service cannot be started by the OS, user, or dependent services.

These can be seen in the following screenshot:

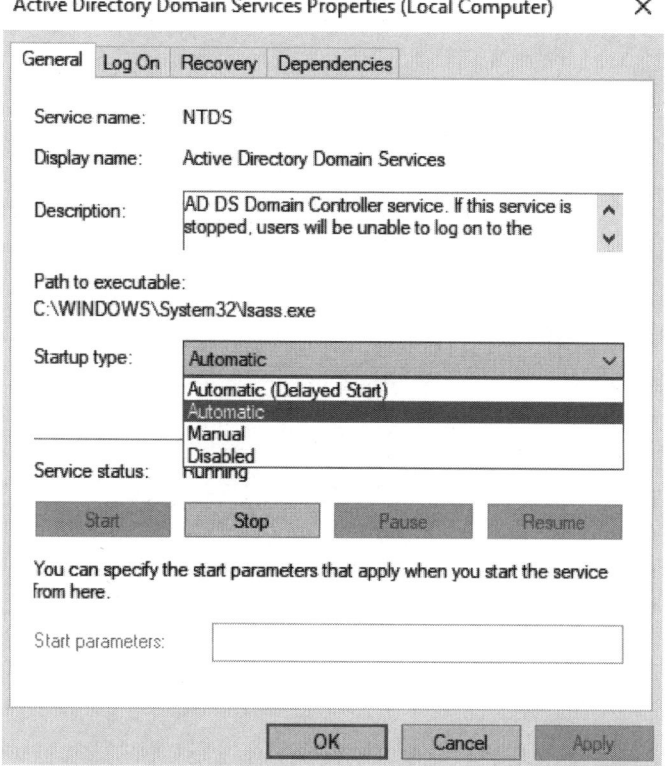

Figure 4.11 – Windows Server service startup types

Now that we have looked at services and registries, let's learn how to access and manage them.

Working with Windows Registry and its services

While Windows Registry is accessed and managed by the **Registry Editor**, Windows services can be accessed and managed via **Control Manager**. So, first, let's learn how to access the Windows Registry.

Accessing and managing Windows Registry keys and values

To access the Windows Registry using **Registry Editor**, follow these steps:

1. Click the search box in the taskbar, enter `regedit`, and press *Enter*.
2. After a short time, **Registry Editor** will open, as shown here:

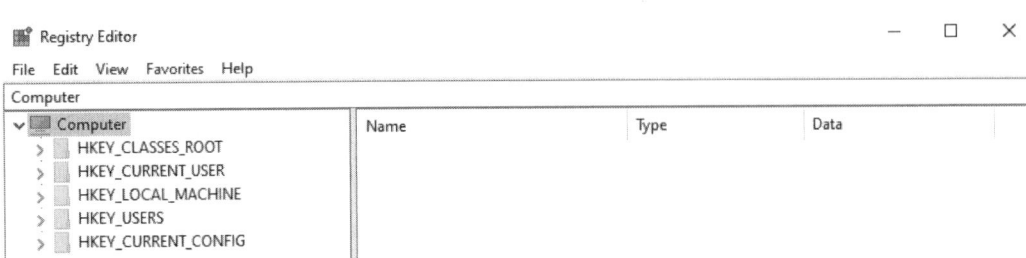

Figure 4.12 – Windows Server Registry Editor

Next, let's learn how to modify registry values.

Modifying a registry value

To modify a registry value using **Registry Editor**, follow these steps:

1. Click the search box in the taskbar, enter regedit, and press *Enter*.
2. On the left-hand side of **Registry Editor**, locate the registry key and sub-key(s).
3. On the right-hand side of **Registry Editor**, right-click the registry value you want to change and select **Modify....** Then, modify the **Value data** field, as shown here:

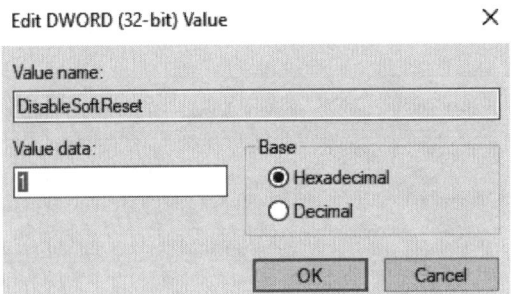

Figure 4.13 – Modifying a registry value

Next, let's learn how to rename a registry value.

Renaming a registry value

To rename a registry value using **Registry Editor**, follow these steps:

1. Click the search box in the taskbar, enter *regedit*, and press *Enter*.
2. On the left-hand side of **Registry Editor**, locate the registry key and sub-key(s).

3. On the right-hand side of **Registry Editor**, right-click the registry value you want to rename and select **Rename**, as shown here:

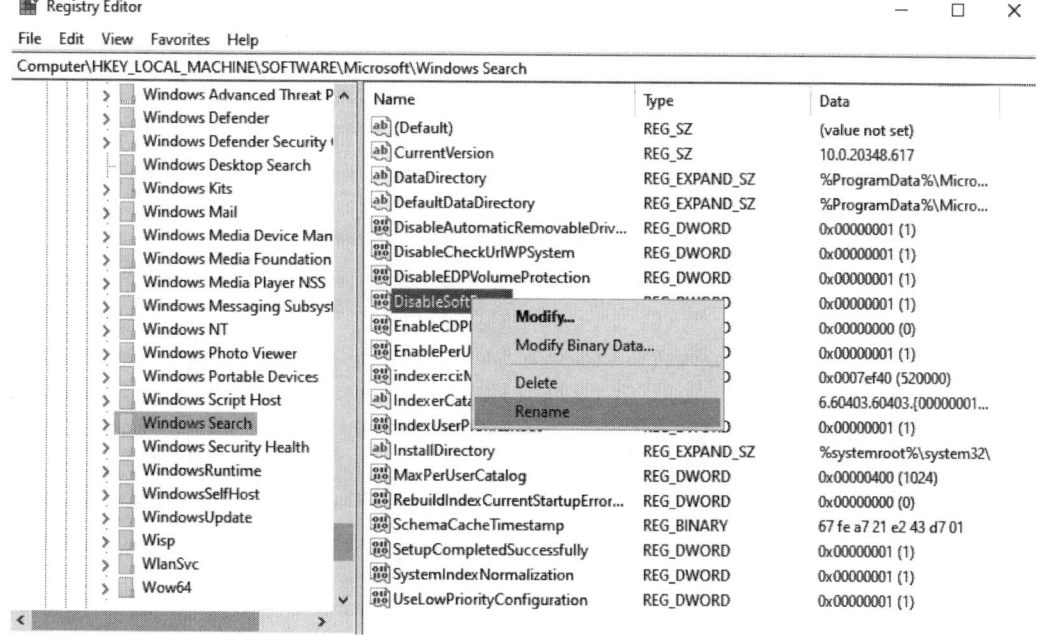

Figure 4.14 – Renaming a registry value

Now, let's look at how to delete a registry value.

Deleting a registry value

To delete a registry value using **Registry Editor**, follow these steps:

1. Click the search box in the taskbar, enter *regedit*, and press *Enter*.
2. On the left-hand side of **Registry Editor**, locate the registry key and sub-key(s).
3. On the right-hand side of **Registry Editor**, right-click the registry value you want to delete and select **Delete**, as shown here:

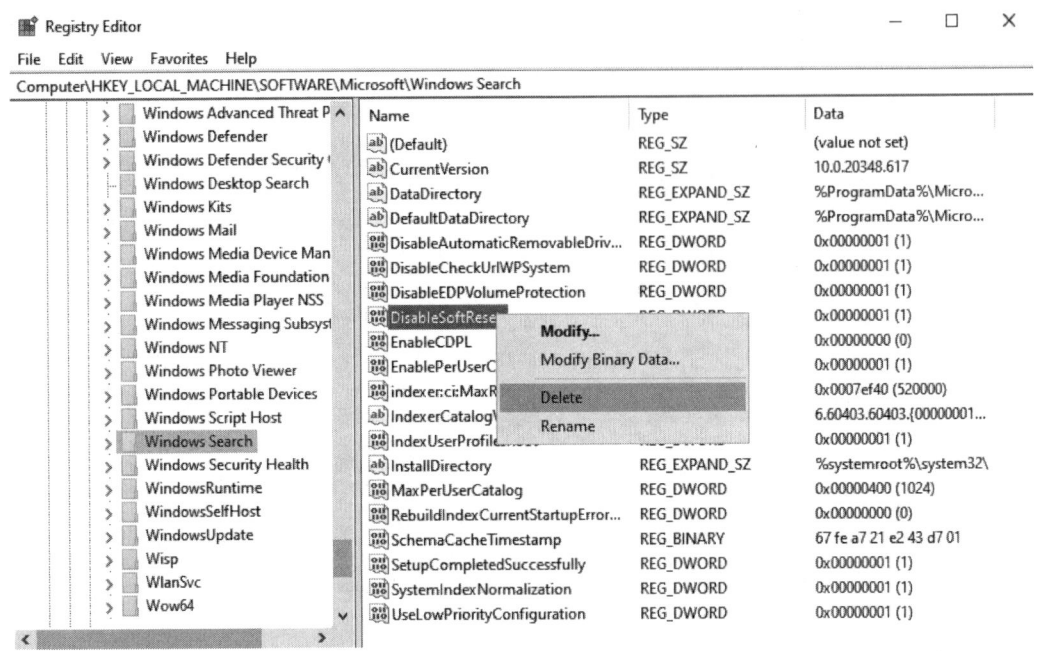

Figure 4.15 – Deleting a registry value

> **Important note**
> Working with registry keys happens to be the same as working with registry values. Deleting, renaming, and exporting are some operations you can accomplish with registry keys.

Accessing and managing Windows services

To access Windows services, follow these steps:

1. Click the **Start** button.
2. In the **Start** menu, select the **Windows Administrative Tools** option.
3. Scroll down and select **Services**.
4. **Windows Services Control Manager** will be displayed shortly after, as shown here:

Understanding the Windows Registry and its services

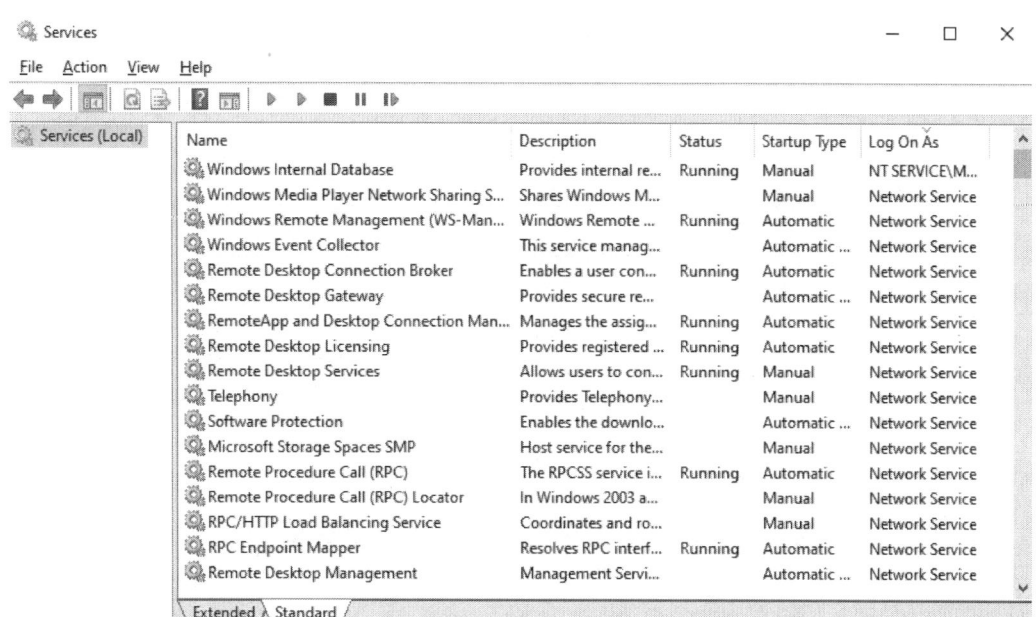

Figure 4.16 – Windows Services Control Manager

Now, let's learn how to set up service recovery options using Control Manager.

Setting up service recovery options

To set up service recovery options using Control Manager, follow these steps:

1. Click the **Start** button.
2. In the **Start** menu, select the **Windows Administrative Tools** option.
3. Scroll down and select **Services**.
4. On the right-hand side of the **Services** window, right-click the service you want to set up recovery options for.
5. In the context menu, select **Properties**.
6. From the window that opens, click the **Recovery** tab.
7. Select the computer's response if the service fails by specifying the **First failure**, **Second failure**, and **Subsequent failures** actions, as shown here:

Figure 4.17 – Setting up service recovery options in Windows Server 2022

8. Click **OK** to close the dialog box.

Now, let's learn how to delay the start of a service.

Delaying the start of a service

To delay the start of a service using Control Manager, follow these steps:

1. Click the **Start** button.
2. In the **Start** menu, select **Windows Administrative Tools**.
3. Scroll down and select **Services**.
4. Right-click the service you want to delay on the right-hand side of the **Services** window.
5. In the context menu, select **Properties**.
6. From the **General** tab, click the **Startup type** drop-down list.

7. Select **Automatic (Delayed Start)**, as shown here:

Figure 4.18 – Setting up the delayed startup of a service in Windows Server 2022

8. Click **OK** to close the dialog box.

Now, let's learn how to set up the logon settings for a service.

Run as settings for a service

To set up the logon settings for a service using Control Manager, follow these steps:

1. Click the **Start** button.
2. In the **Start** menu, select **Windows Administrative Tools**.
3. Scroll down and select **Services**.
4. Right-click the service you want to delay on the right-hand side of the **Services** window.
5. In the context menu, select **Properties**.
6. In the window that opens, click the **Log On** tab.

7. Click the **This account:** option in the **Log on as:** section.
8. Enter a user account, including the domain with a backslash, and fill in the **Password** and **Confirm password** areas, as shown here:

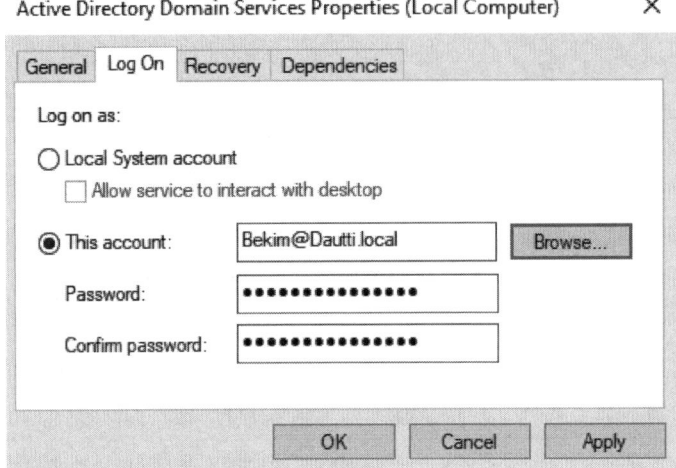

Figure 4.19 – Setting up the log settings for a service in Windows Server 2022

9. Click **OK** to close the dialog box.

Now, let's learn how to start the service.

Starting the service

To start the service using Control Manager, follow these steps:

1. Click the **Start** button.
2. In the **Start** menu, select **Windows Administrative Tools**.
3. Scroll down and select **Services**.
4. Right-click the service you want to start on the right-hand side of the **Services** window.
5. In the context menu, select **Start**, as shown here:

Understanding the Windows Registry and its services

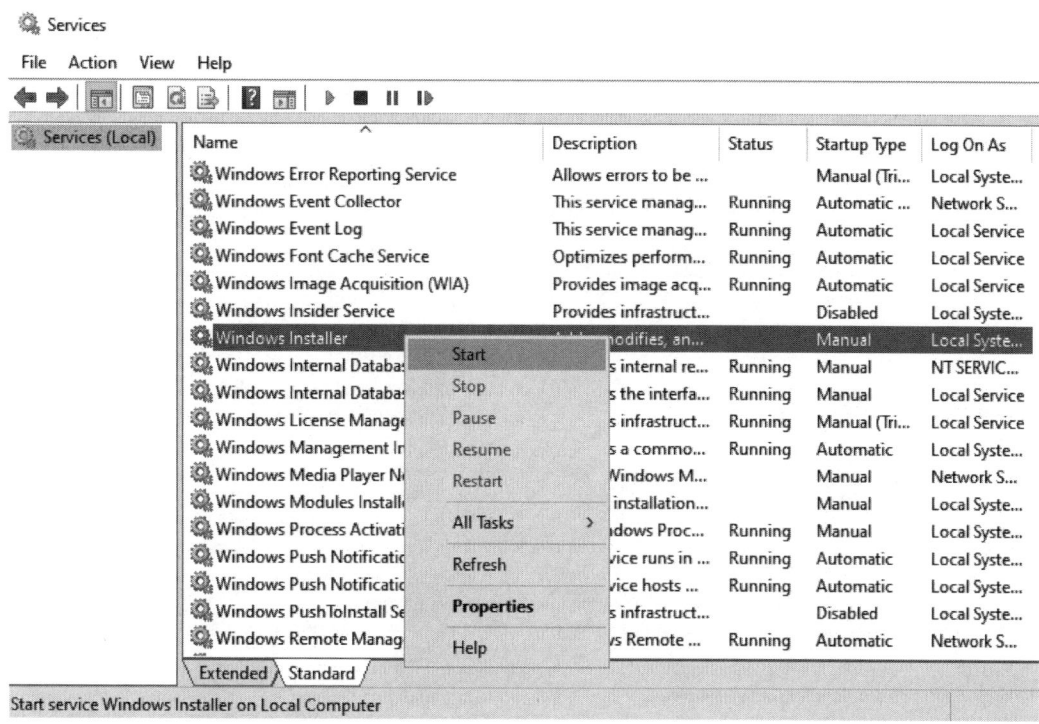

Figure 4.20 – Starting the service

Now that you have learned how to start a service, the next section will teach you how to stop service.

Stopping a service

To stop a service using Control Manager, follow these steps:

1. Click the **Start** button.
2. In the **Start** menu, select **Windows Administrative Tools**.
3. Scroll down and select **Services**.
4. Right-click the service you want to stop on the right-hand side of the **Services** window.
5. In the context menu, select **Stop**, as shown here:

Post-Installation Tasks in Windows Server 2022

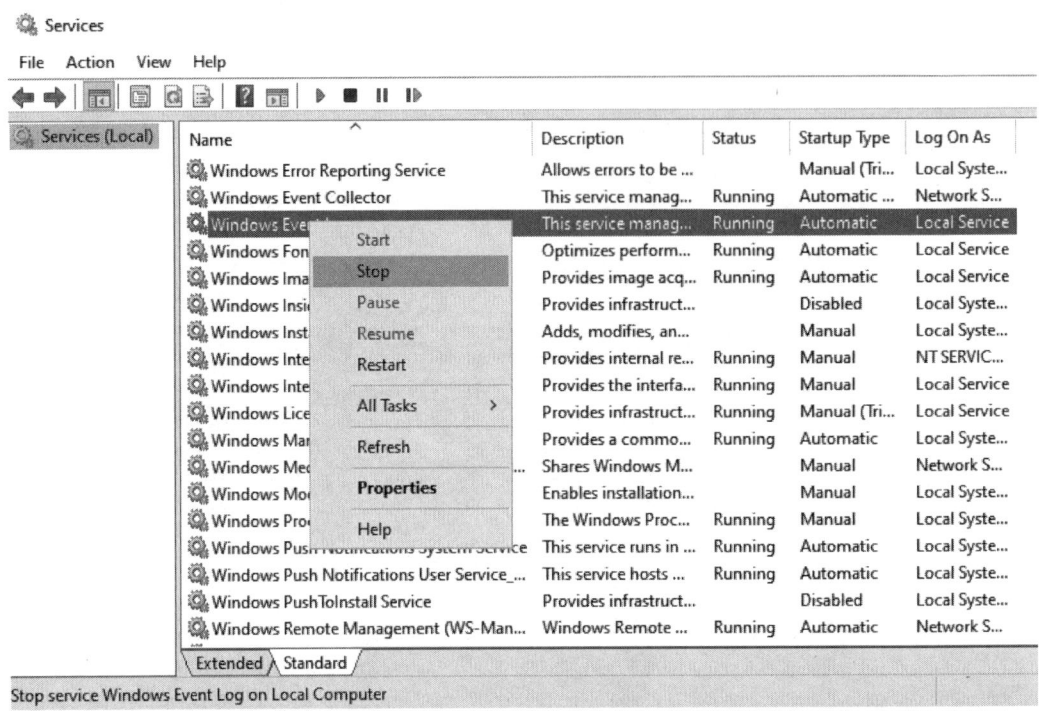

Figure 4.21 – Stopping the service

Now, let's learn how to restart the service.

Restarting the service

To restart the service using Control Manager, follow these steps:

1. Click the **Start** button.
2. In the **Start** menu, select **Windows Administrative Tools**.
3. Scroll down and select **Services**.
4. Right-click the service you want to restart on the right-hand side of the **Services** window.
5. In the context menu, select **Restart**, as shown here:

Understanding the Windows Registry and its services   101

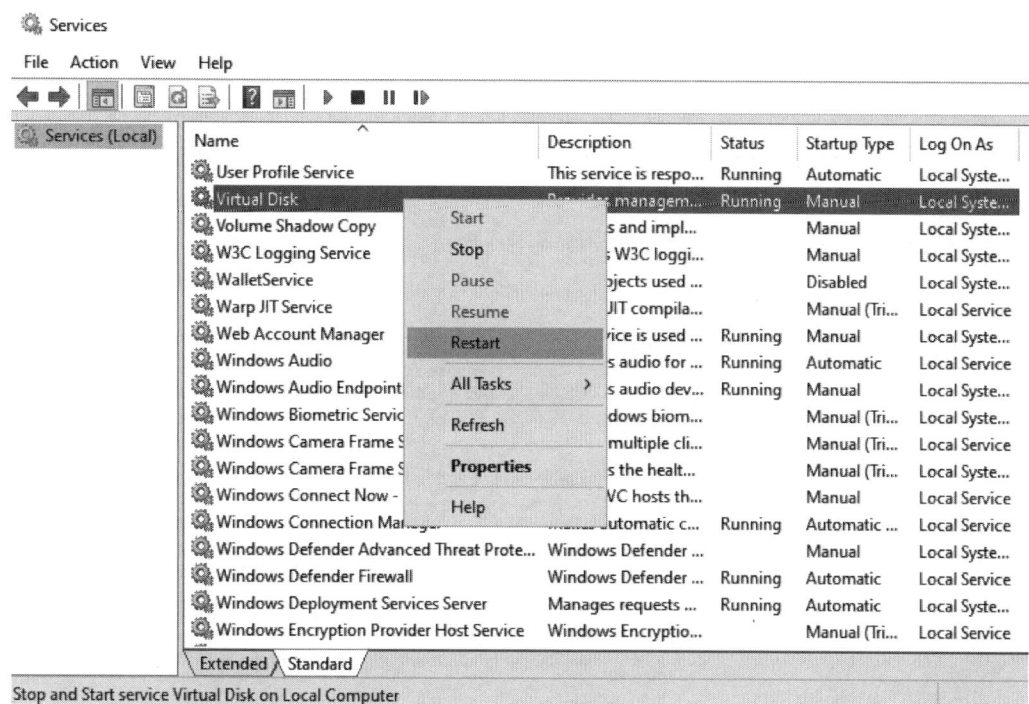

Figure 4.22 – Restarting the service

Next, you will learn about registry entries, service accounts, and dependencies.

Explaining registry entries, service accounts, and dependencies

In most cases, you will add a new registry key or a registry value while fixing an issue or adding a new feature to Windows Server. Therefore, it would help to be careful when working with Windows Registry, no matter how you do it. Concerning services, the service account is the Windows Server native account or an account created by you to manage running services. This service account enables services to access local and network resources from a security standpoint. Regarding native statements that services are running, as shown in the following screenshot, the following service accounts are available in Windows Server 2022:

- **Local system** is a built-in account with the most privileges in a Windows OS. It is also known as a **superuser**, and this account is more powerful than an admin account.
- **NT Authority\LocalService** is a built-in account with the same privileges as user group members.
- **NT Authority\NetworkService** is a built-in account with more privileges than user group members.

As far as service dependency is concerned, applications use more than one service. So, if you try to stop a dependent service, you must stop a few others. Conversely, others need to be activated if you try to start a dependent service:

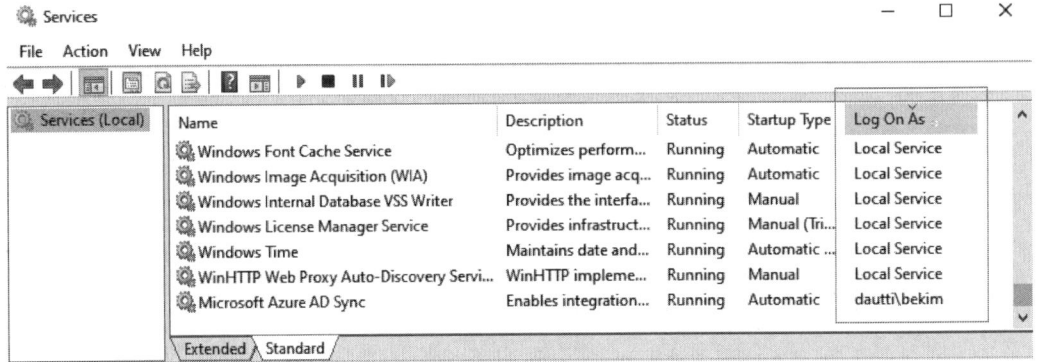

Figure 4.23 – Native service accounts in Windows Server 2022

Let's move on and learn how to add a new registry key.

Adding a new registry key

To add a new registry key using **Registry Editor**, follow these steps:

1. Click the search box in the taskbar, enter `regedit`, and press *Enter*.
2. On the left-hand side of **Registry Editor**, right-click the registry key or its sub-key(s).
3. In the context menu, select **New | Key**, as shown here:

Understanding the Windows Registry and its services

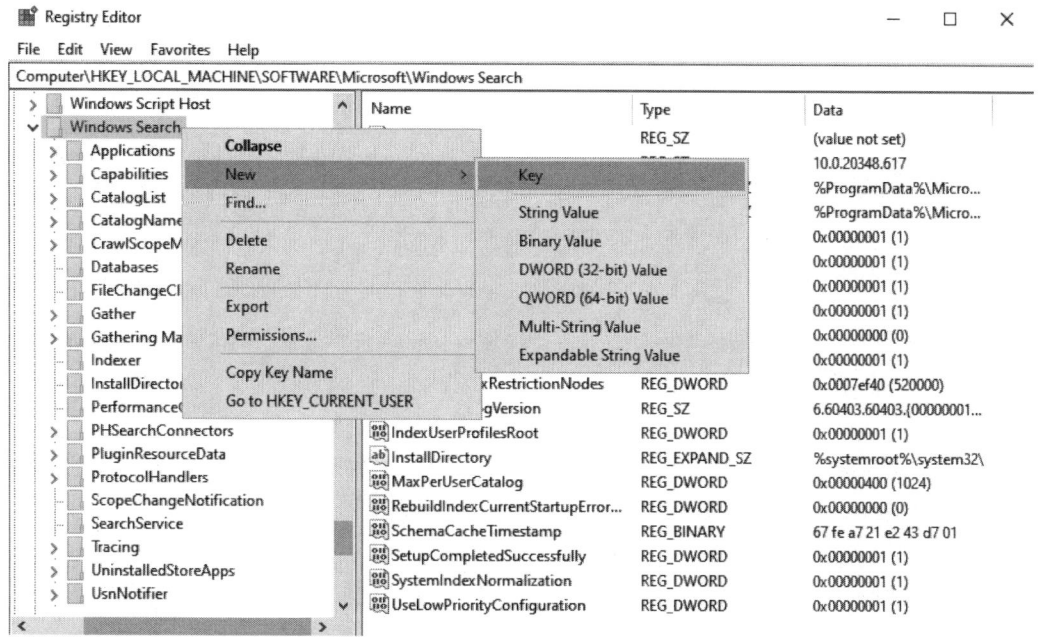

Figure 4.24 – Adding a registry entry

> **Important note**
> To add a new registry value, right-click the space on the right-hand side of **Registry Editor** once you have located the registry key or its sub-key(s). Then, in the context menu, select **New** and enter the value you want to add.

Next, let's learn how to add a service account.

Adding service accounts

To add a service account using **Control Manager**, follow these steps:

1. Click the **Start** button.
2. In the **Start** menu, select **Windows Administrative Tools**.
3. Scroll down and select **Services**.
4. On the right-hand side of the **Services** window, right-click the service you want to add a service account for.
5. In the context menu, select **Properties**.

6. In the **Properties** window, click the **Log On** tab, as shown in *Figure 4.25*.
7. Select the **This account** option in the **Log on** section and click the **Browse** button.
8. Specify the service account in your organization's Active Directory (note that only administrators can be set as **Log on account**).
9. Enter the service account password and confirm the password.
10. Click **OK** to close the **Properties** window:

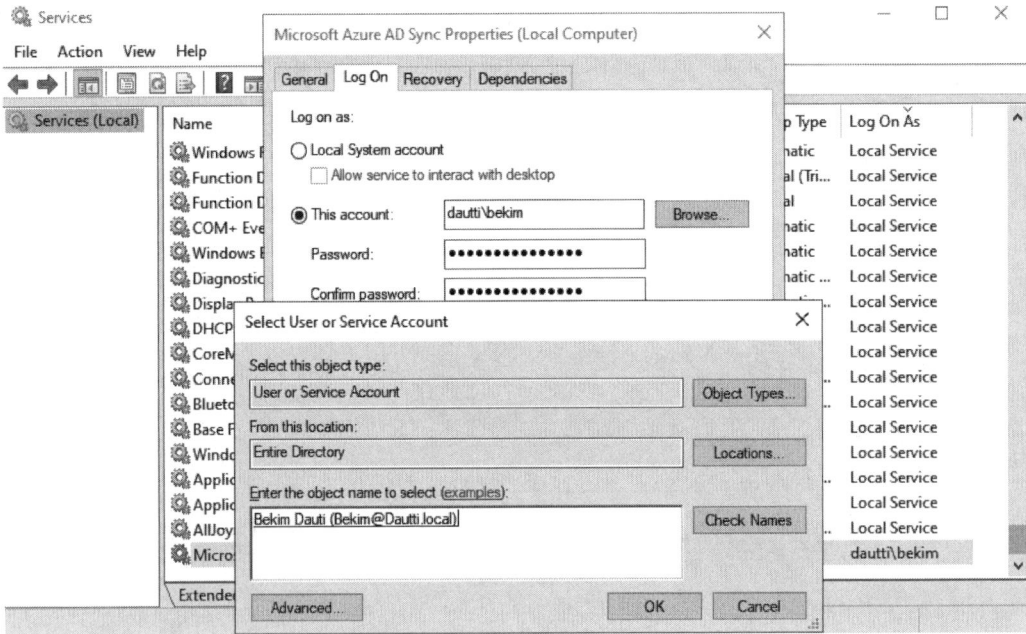

Figure 4.25 – Adding a service account

Now, let's learn how to add a service dependency.

Adding a service dependency

To add a service dependency using **Registry Editor**, follow these steps:

1. Click the search box in the taskbar, enter `regedit`, and press *Enter*.
2. On the left-hand side of **Registry Editor**, locate the service (`HKEY_LOCAL_MACHINE\SYSTEM\CurrentControlSet\Services\`) that you want to add a dependency to.
3. On the right-hand side, modify the value if there is a `DependOnService` value.
4. If not, right-click in an empty area in side the Registry Editor and select **Multi-string value** to create a `DependOnService` value.

5. Rename the `DependOnService` value with the exact name of the service you want to create a dependency for.
6. Restart the server.
7. With the **Services** window open, locate the service you have created a dependency for and right-click on it to select **Properties**.
8. In the **Properties** window, click the **Dependencies** tab to see the added dependency.

In this section, you learned about Windows Server Registry and its services. You also now know how to manage and work with them. In the next section, you will learn how to perform the initial server configuration for the Desktop Experience and Server Core installation options.

Understanding Windows Server initial configuration

The initial server configuration is necessary after setting up the device drivers and ensuring that the OS services are up and running. It is an activity that involves changing the server name and joining a domain that depends on the server's role. This includes enabling Remote Desktop, setting up a static IP address, changing the time zone, activating Windows Server 2022, turning off **Internet Explorer** (**IE**) enhanced security, and checking for updates. This ensures that the server is ready to take on a new role in its IT infrastructure.

The server's initial configuration is a crucial task as it determines the functional status of the server just before it takes on the task of adding roles. Thus, from my experience, first, you will want to set up the IP address, change the time zone, and activate Windows Server 2022. Then, you can check for updates, change the default server name, join the domain, enable Remote Desktop, and finally turn off IE enhanced security. With that in mind, let's perform the server's initial configuration for the Desktop Experience and Server Core installation options.

Using Server Manager in Desktop Experience

In Desktop Experience, the initial server configuration can be accomplished using Server Manager, as shown in the following screenshot. After logging into Windows Server 2022, Server Manager starts automatically. It will always open up automatically unless you change its configuration.

To run the server's initial configuration using Server Manager in Desktop Experience, click **Configure this local server** in the **WELCOME TO SERVER MANAGER** section:

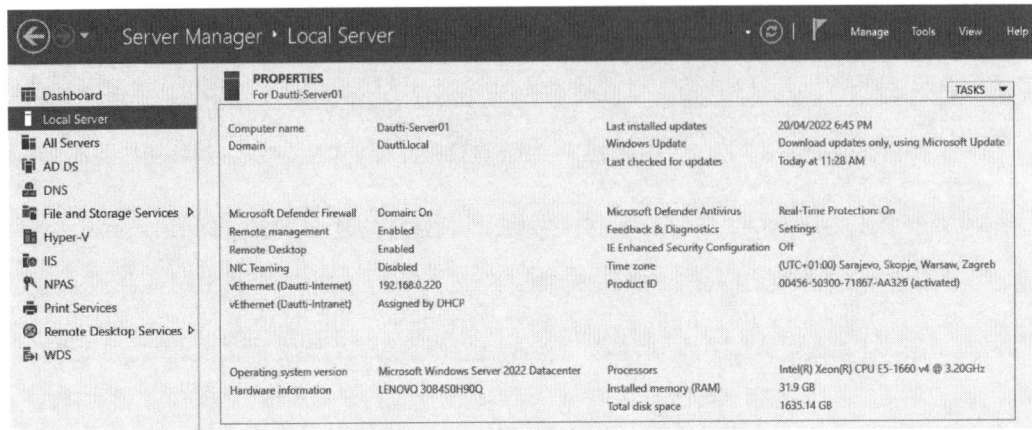

Figure 4.26 – Server Manager in Windows Server 2022

Next, look at the Server Configuration tool, which is another way to run an initial server's configuration in the CLI.

Using Server Configuration in Server Core

In Server Core, the initial server configuration can be accomplished through the Server Configuration tool, as shown in the following screenshot. In contrast to Server Manager, the Server Configuration tool can be accessed by entering `SConfig.cmd` in Command Prompt:

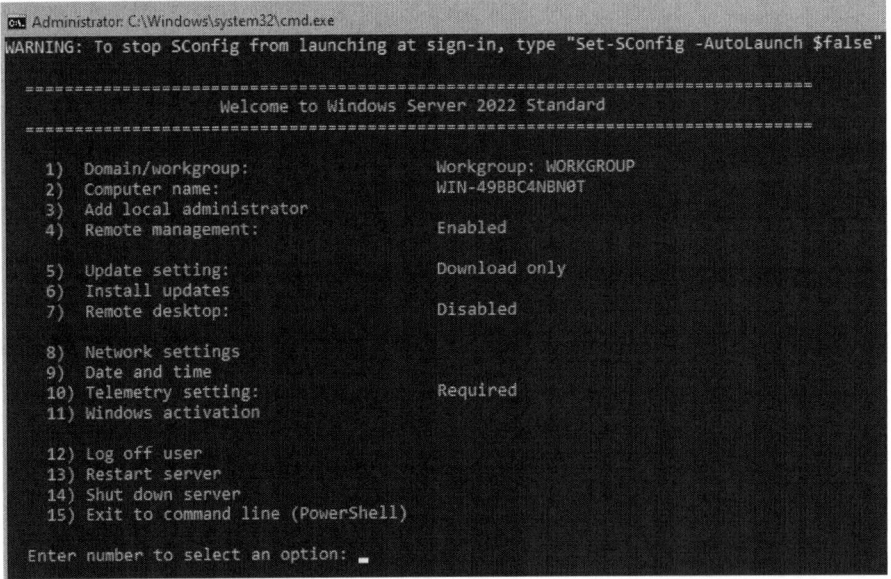

Figure 4.27 – Server Configuration in Windows Server 2022

At this point, you are well equipped for this chapter's exercise, where we will be performing an initial Windows Server configuration. So, let's get straight into it.

Chapter exercise – performing an initial Windows Server configuration

This chapter's exercise will teach you how to perform Windows Server initial configuration using Server Manager, as well as Windows Server initial configuration using Server Configuration.

Performing Windows Server initial configuration using Server Manager

This section explains the initial configuration of Windows Server 2022 Standard (Desktop Experience) using Server Manager.

Changing the server name

To change the server name, as shown in the following screenshot, follow these steps:

1. In the **Properties** section, click the highlighted default computer name.
2. In the **System Properties** window, click the **Change** button.
3. In the **Computer Name/Domain Changes** window, delete the existing **Computer name**, provide the server's name, and click **OK**:

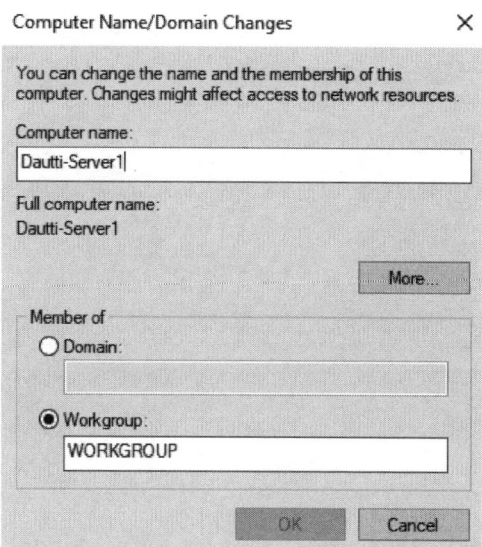

Figure 4.28 - Changing the server's name

4. Click **OK** to confirm that you will restart the server to apply these changes.
5. In the **System Properties** window, click the **Close** button.
6. In the **Microsoft Windows** dialog box, click **Restart Now**.

Now, let's learn how to join the server to a domain.

Joining the server to a domain

Before joining the server to a domain, evaluate the role of the server. For example, if the server will be a **domain controller** (**DC**), there is no need to join a domain, as adding the AD DS role will automatically make the server a DC. But if the server will have a role other than AD DS, then, as a domain member, it must join the domain, as shown in the following screenshot. To do so, follow these steps:

1. In the **Properties** section, click the highlighted workgroup.
2. In the **System Properties** window, click the **Change** button.
3. In the **Computer Name/Domain Changes** window, select the **domain:** option, click the text box to provide your organization's domain, and click **OK**:

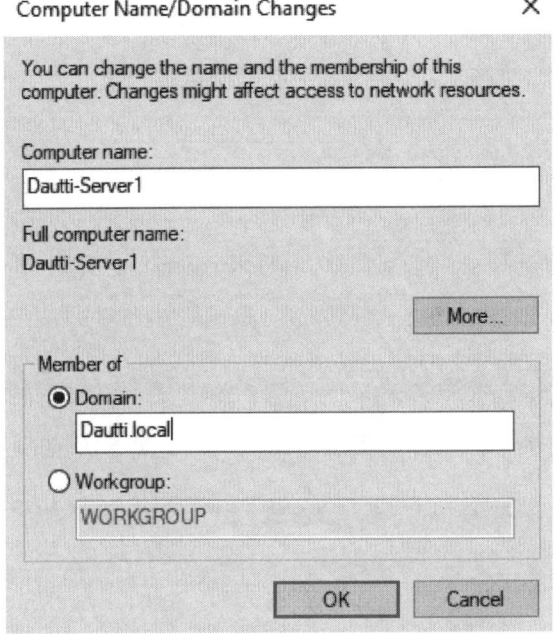

Figure 4.29 – Joining a server to a domain

4. In the **Windows Security** window, enter the name and password of an account with permission to join the domain, and then click **OK**.

5. The **Computer Name/Domain Changes** dialog box will welcome the server to your organization's domain. Click **OK** to close it.
6. Click **OK** to confirm that you will restart the server to apply these changes.
7. In the **System Properties** window, click the **Close** button.
8. In the **Microsoft Windows** dialog box, click **Restart Now**.

Next, let's learn how to enable **Remote Desktop**, which allows remote access to the server.

Enabling Remote Desktop

To enable Remote Desktop, as shown in the following screenshot, follow these steps:

1. In the **System Properties** dialog box, click the **Remote** tab.
2. Select the **Allow remote connections to this computer** option in the **System Properties** window.
3. The **Remote Desktop Connection** dialog box informs you that **Remote Desktop Firewall exception will be enabled**. Click **OK** to close it:

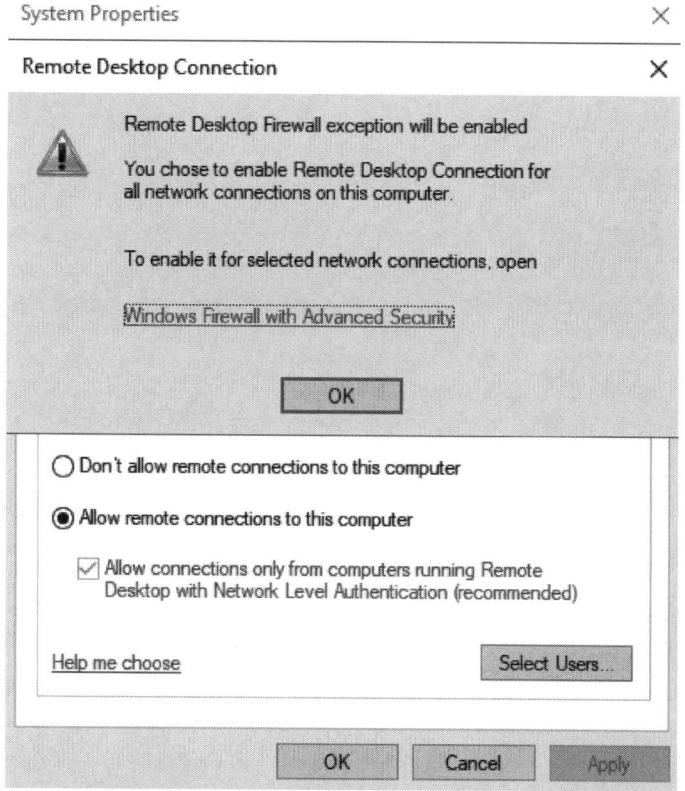

Figure 4.30 – Enabling Remote Desktop

4. To add Remote Desktop users, click the **Select Users...** button.

5. In the **Remote Desktop Users** window, click the **Add** button to add users. Select users or groups from your AD DS. When you have finished adding Remote Desktop users, click **OK** to close the **Remote Desktop Users** window.

6. Again, click **OK** to close the **System Properties** window.

Now, let's learn how to set up the IP address since servers are recommended to have static IP addresses assigned.

Setting up the IP address

To set up the IP address, as shown in the following screenshot, follow these steps:

1. In the **Properties** section, click the highlighted Ethernet setting:

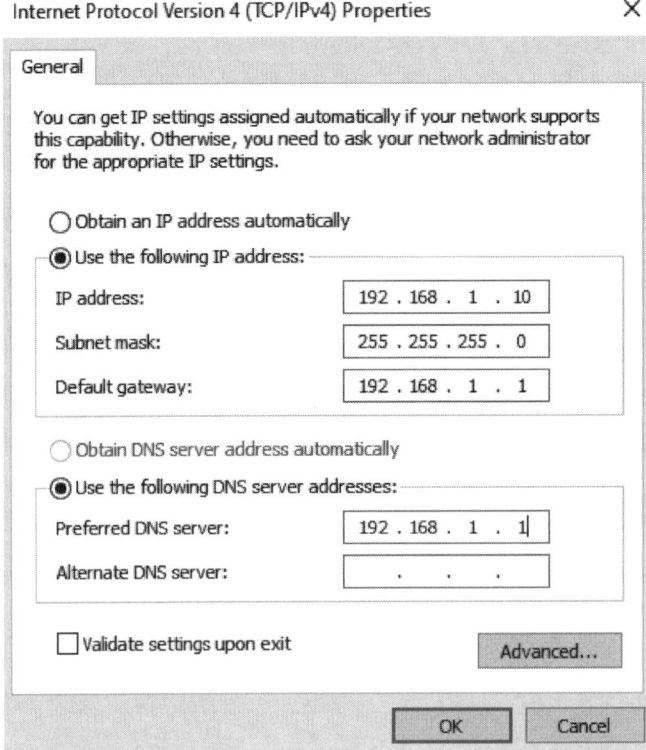

Figure 4.31 – Setting up the IP address

2. Right-click the server's Ethernet in the **Network Connections** window and select **Properties**.

Chapter exercise – performing an initial Windows Server configuration   111

3. Select **Internet Protocol Version 4 (TCP/IPv4)** in the **Ethernet Properties** window and click the **Properties** button.

4. In the **Internet Protocol Version 4 (TCP/IPv4) Properties** window, select **the following IP address:** option and enter the **IP address**, **Subnet Mask**, and **Default Gateway** fields. Additionally, select the **use the following DNS server addresses:** option and fill in the **Preferred DNS server** and **Alternate DNS server** fields. Click **OK** to close.

5. Click the **Close** button to close the **Ethernet Properties** window.

6. Click the *Close* button (the red X) in the top-right corner to close the **Network Connections** window.

Now, let's learn how to check for updates, which is considered among the first steps once Windows Server 2022 has been installed.

Checking for updates

To check for updates, as shown in the following screenshot, follow these steps:

1. In the **Properties** section, click the highlighted **Last checked for updates** setting.

2. In the **Settings** window on the right-hand side of the **Windows Update** section, the available updates are listed (if any). If any updates are ready to be installed, click the **Install now** button:

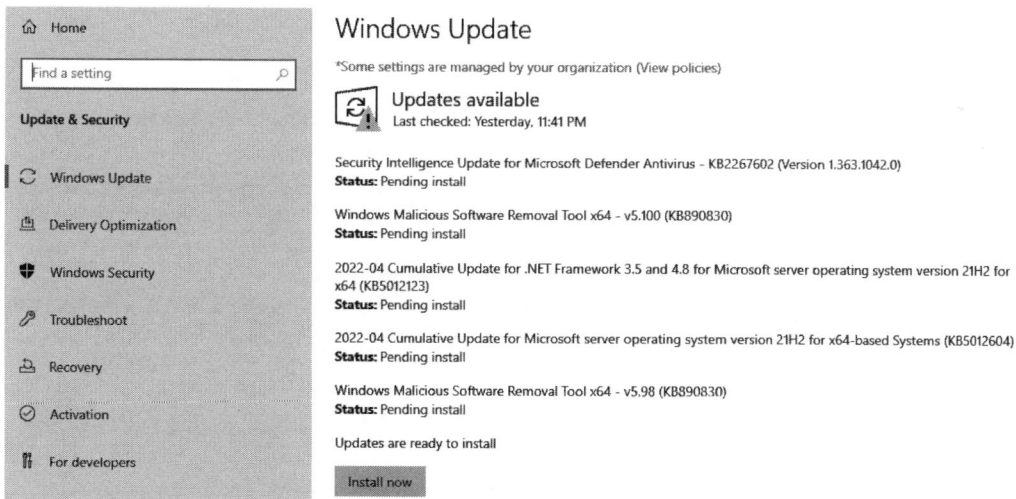

Figure 4.32 – Checking for updates

3. Installing updates may take some time! In addition, when the installation is done, you will often be asked to restart the server for the updates to occur.

Now, let's learn how to turn off the IE enhanced security settings in Windows Server 2022, even though Microsoft has officially retired Internet Explorer.

Turning off IE enhanced security

To turn off the IE enhanced security settings, as shown in the following screenshot, follow these steps:

1. In the **Properties** section, click the highlighted IE enhanced security configuration setting.
2. Select the **Off** option in the **Internet Explorer Enhanced Security Configuration** window within the **Administrators**: section.
3. Click **OK** to close the **Internet Explorer Enhanced Security Configuration** window:

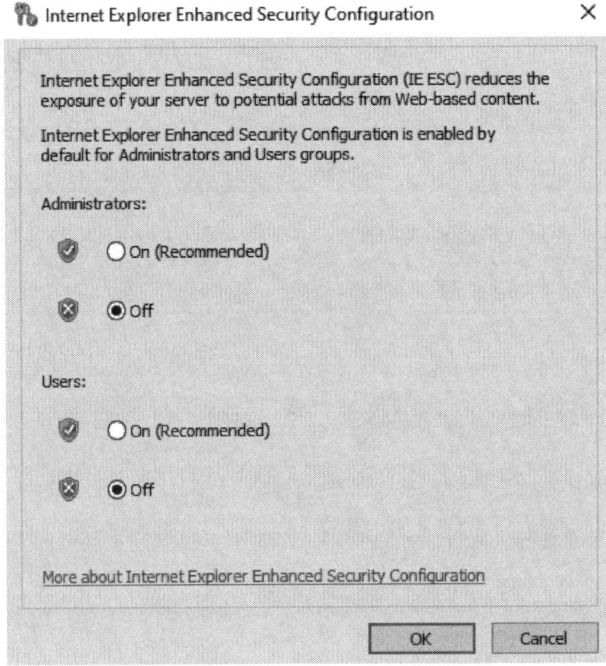

Figure 4.33 – Turning off IE enhanced security

Now, let's learn how to change the time zone, a critical setting for network services.

Changing the time zone

To change the time zone, as shown in the following screenshot, follow these steps:

1. In the **Properties** section, click the highlighted time zone setting.
2. Click the **Change time zone...** button in the **Date and Time** window.

3. Click the drop-down list in the **Time Zone Settings** window to select your time zone.
4. Click **OK** to close the **Time Zone Settings** window.
5. Again, click **OK** to close the **Date and Time** window:

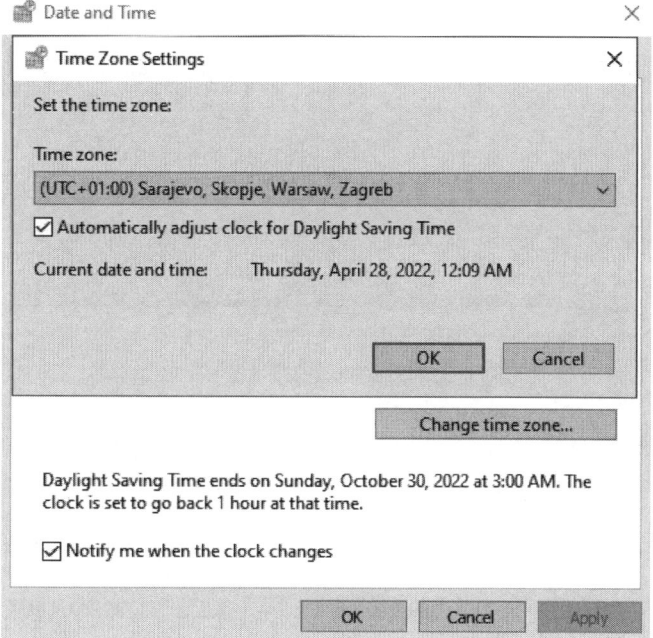

Figure 4.34 – Changing the time zone

Now, let's learn how to activate Windows Server, a mandatory requirement in Windows OSs, including Windows Server 2022.

Activating Windows Server

To activate Windows Server 2022 (Desktop Experience), as shown in the following screenshot, follow these steps:

1. In the **Properties** section, click the highlighted **Not activated** setting of the product ID:

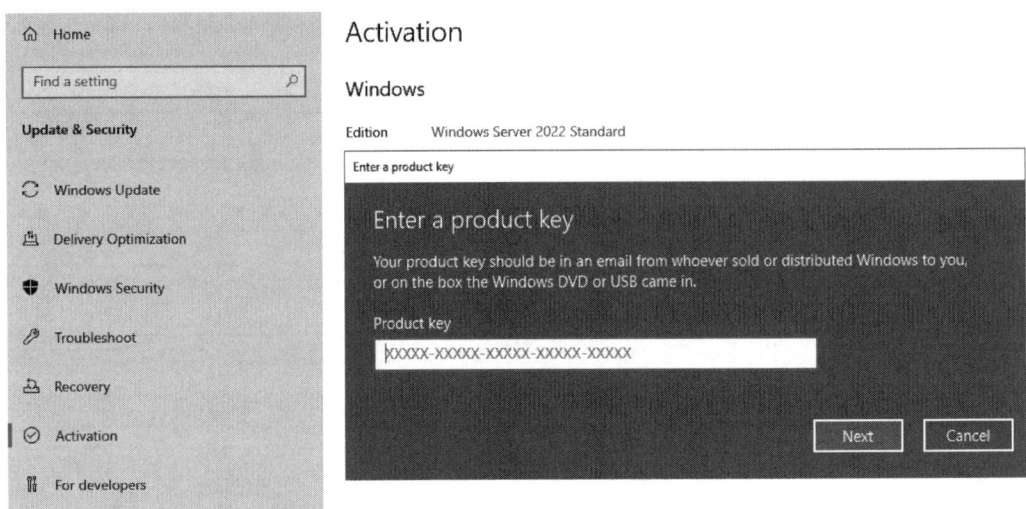

Figure 4.35 – Activating Windows Server 2022

2. In the **Enter a product key** window, enter your Windows Server 2022 product key, and press *Enter*.

3. Microsoft's Activation Server will check the product key you entered. If it is valid, click **Next** in the **Activate Windows** window.

4. When activation finishes, click **Close** to close the **Thank you for activating** window.

In this section, you learned how to perform Windows Server initial configuration using Server Manager. Now, let's learn how to accomplish this configuration using Server Configuration.

Performing Windows Server initial configuration using Server Configuration

This section explains the initial configuration of Windows Server 2022 Standard (Server Core) using Server Configuration.

Changing the server name

To change the server name, as shown in the following screenshot, follow these steps:

1. In the **Server Configuration** menu prompt, enter 2 as a selected option and press *Enter*.

2. Enter the new server name and press *Enter*.

3. In the **Restart** dialog box, click **Yes** to restart the server:

```
=================================================================
                          Computer name
=================================================================

Current computer name: WIN-49BBC4NBN0T

Enter new computer name (Blank=Cancel): Dautti-Server2
Changing computer name...
WARNING: The changes will take effect after you restart the computer
WIN-49BBC4NBN0T.
    Restart now? (Y)es or (N)o: Y_
```

Figure 4.36 – Changing the server name

4. The server will restart so that it can apply the server name change.

Now that we know how to change the server name, let's learn how to join the server to a domain.

Joining the server to a domain

Before joining the domain, consider the notes mentioned earlier when performing the Server Manager configuration. Then, to join the domain, as shown in the following screenshot, follow these steps:

1. In the **Server Configuration** menu prompt, enter 1 as a selected option and press *Enter*.
2. To join the server to your organization domain, enter D and press *Enter*.
3. Enter your organization domain and press *Enter*.
4. Enter the authorized domain user and press *Enter*.
5. Enter the password and press *Enter*.
6. Click **No** in the **Change Computer Name** dialog box when you're asked to change the server's name:

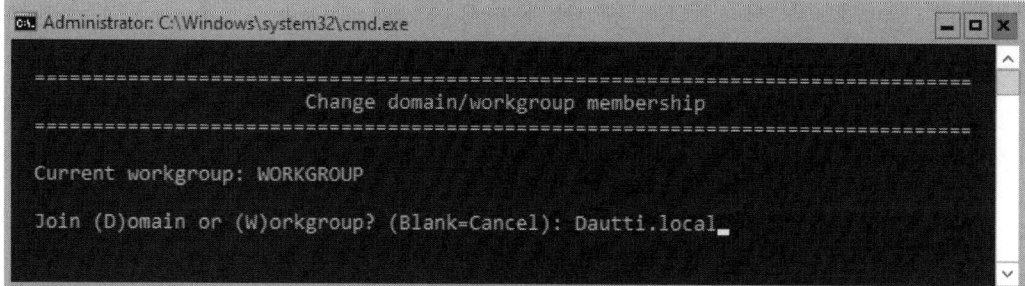

Figure 4.37 – Joining the server to a domain

Now, let's learn how to enable Remote Desktop, which allows remote access to Windows Server 2022 Server Core edition.

Enabling Remote Desktop

To enable Remote Desktop, as shown in the following screenshot, follow these steps:

1. In the **Server Configuration** menu prompt, enter 7 as a selected option and press *Enter*.
2. To enable Remote Desktop, enter E and press *Enter*.
3. Enter 1 and press *Enter* for more secure access.
4. Click **OK** to confirm whether Remote Desktop is enabled in the **Remote Desktop** dialog box:

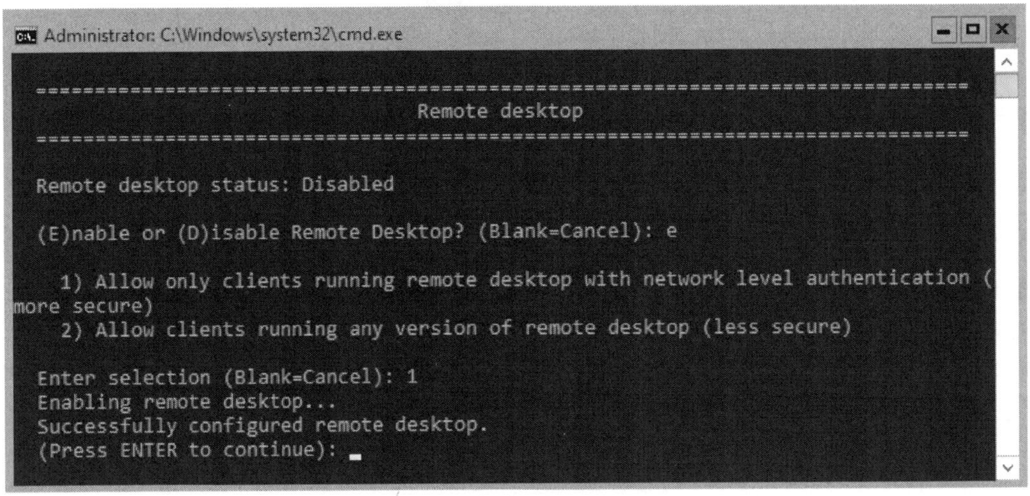

Figure 4.38 – Enabling Remote Desktop

Now, let's learn how to set up the IP address since even Windows Server 2022 Server Core edition requires a static IP address.

Setting up the IP address

To set up the IP address, as shown in the following screenshot, follow these steps:

1. In the **Server Configuration** menu prompt, enter 8 as a selected option and press *Enter*.
2. Enter the number of the network adapter you want to set up the IP address for and press *Enter*.
3. Enter 1 in the sub-menu to set the network adapter address and press *Enter*.
4. Enter S for the static IP address and press *Enter*.
5. Enter the static IP address and press *Enter*.

6. Enter the subnet mask and press *Enter*.
7. Enter the default gateway and press *Enter*.
8. Enter 2 in the sub-menu to set the DNS servers and press *Enter*.
9. Enter the new preferred DNS server and press *Enter*.
10. In the **Network Settings** dialog box, click **OK** to close it.
11. Enter the alternate DNS server and press *Enter*.
12. Enter 4 in the sub-menu to exit and **Return to Main Menu**:

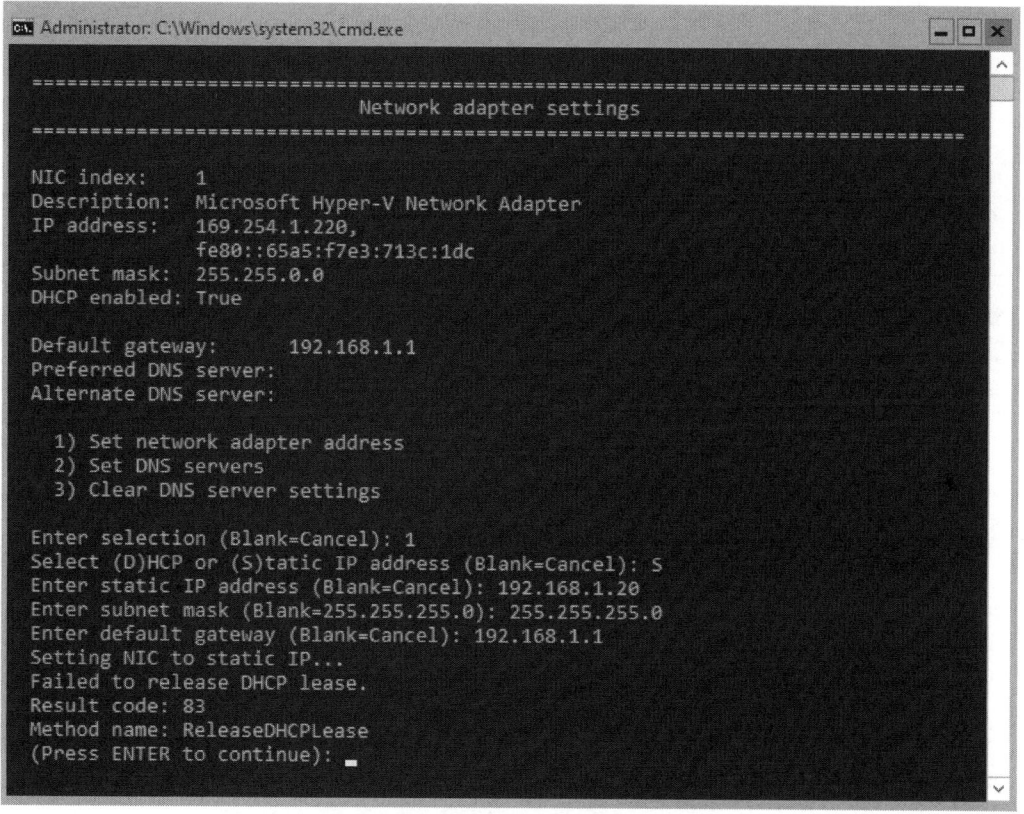

Figure 4.39 – Setting up the IP address

Now, let's learn how to check for updates to update Windows Server 2022 Server Core edition.

Checking for updates

To check for updates, as shown in the following screenshot, follow these steps:

1. In the **Server Configuration** menu prompt, enter 5 as a selected option and press *Enter*.
2. Enter A for all or R for recommended updates in a new window and press *Enter*.
3. After a while, Windows Update will start searching for updates.
4. If applicable updates are found, enter A for all, N for no updates, or S for a single update, and press *Enter*.
5. Once the updates have been downloaded, the installation takes place. Click **Yes** to restart the server:

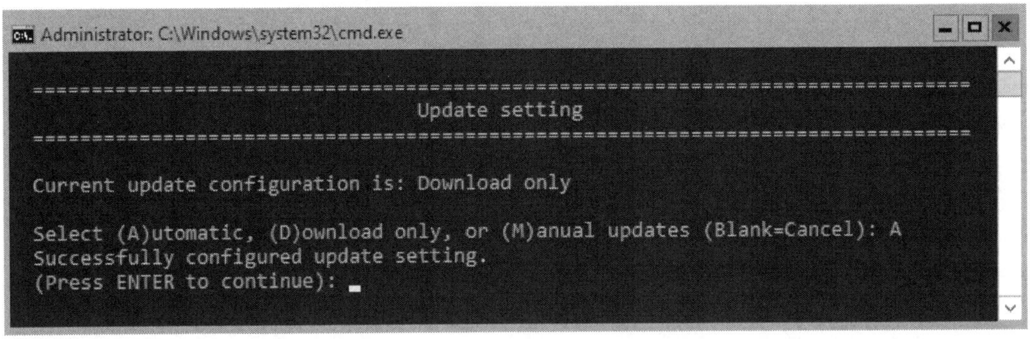

Figure 4.40 – Checking for updates

Now, let's learn how to change the time zone in Windows Server 2022 Server Core edition.

Changing the time zone

To change the time zone, as shown in the following screenshot, follow these steps:

1. In the **Server Configuration** menu prompt, enter 9 as a selected option and press *Enter*.
2. Click the **Change date and time...** button in the **Date and Time** window.
3. Click on the **Date** and **Time** sections to set the date and time, respectively.
4. Click **OK** to close the **Date and Time** window.
5. Again, click **OK** to close the **Date and Time** window:

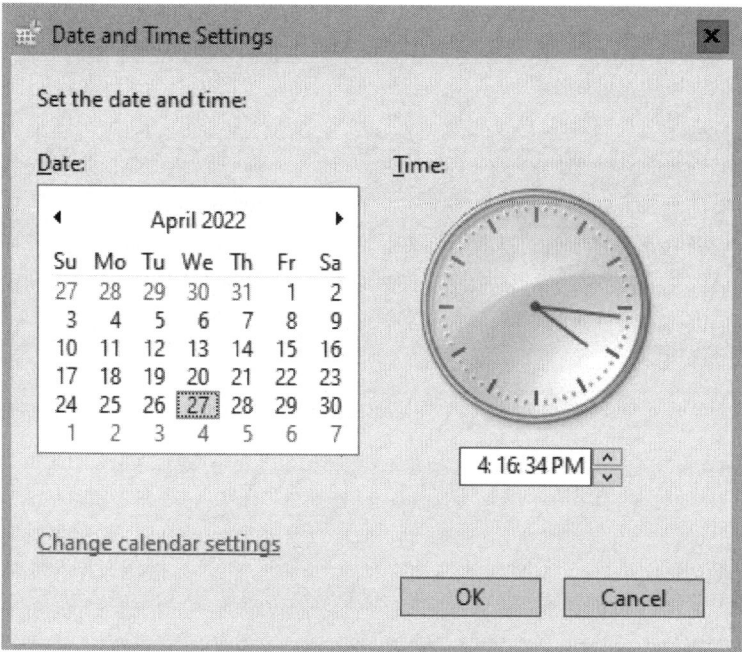

Figure 4.41 – Setting the date and time

Now, let's learn how to activate Windows Server 2022 Server Core edition.

Activating Windows Server

To activate Windows Server 2022 Server Core edition, as shown in the following screenshot, follow these steps:

1. In the **Server Configuration** menu prompt, enter 11 as a selected option and press *Enter*.
2. Enter 3 in the sub-menu to install the product key and press *Enter*.
3. Enter the Windows Server 2022 product key and click **OK** in the **Enter Product Key** window.
4. Enter 2 from the sub-menu to activate Windows and press *Enter*.
5. After a while, Windows Server 2022 will start. Enter Exit to close the activation window.
6. Enter 4 from the sub-menu to exit and return to the main menu:

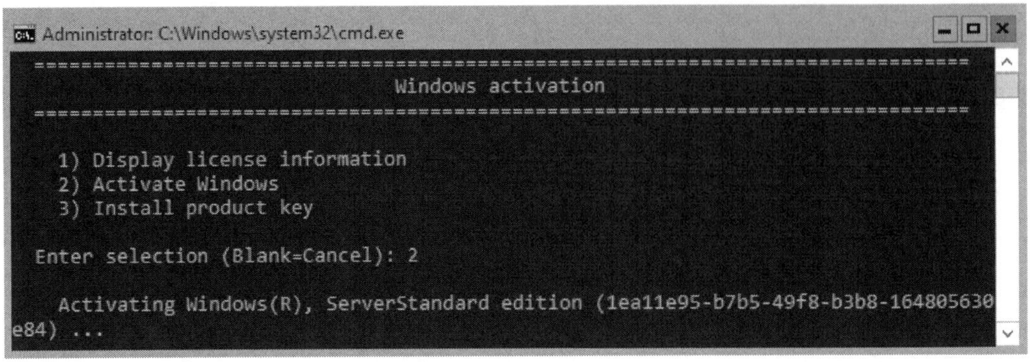

```
============================================================
                    Windows activation
============================================================

    1) Display license information
    2) Activate Windows
    3) Install product key

Enter selection (Blank=Cancel): 2

    Activating Windows(R), ServerStandard edition (1ea11e95-b7b5-49f8-b3b8-164805630
e84) ...
```

Figure 4.42 – Activating Windows Server 2022

With that, we can conclude this chapter's exercise, which has helped you learn how to complete the server's initial configuration using Server Manager and Server Configuration.

Summary

In this chapter, you learned about various post-installation tasks in Windows Server 2022, including device drivers, registries, and services.

You became familiar with how devices are organized in a computer system and what a device driver is. This will help you work with drivers, such as installing, uninstalling, and updating them. Furthermore, you learned about system resources and the Windows Server Registry and its services. You also learned about registry keys and service startup types. Hopefully, you are now familiar with stopping, starting, and restarting Windows Services and, as with the Windows Registry, acquainted with regedit and some basic activities such as adding a new key and modifying and deleting it.

Finally, you learned about the concept of Windows Server initial configuration; thus, you were able to run Windows Server 2022 post-installation tasks in this chapter's exercise.

In the next chapter, you will learn about Directory Services in Windows Server 2022.

Questions

Answer the following questions to test your knowledge of this chapter:

1. A device driver is a program that acts as the translator between computer hardware and an OS. (True/False)
2. _____ works on the principle that when a device is plugged into a computer, the device is immediately recognized by the OS.

3. Which two of the following are known as a computer's system resources?

   A. IRQ

   B. DMA

   C. SAN

   D. NAS

4. A driver's digital signature identifies its publisher. (True/False)

5. _____ is a hierarchical database that stores hardware and software configurations and system security information.

6. Which two Windows Server tools are used to operate devices and device drivers?

   A. Devices

   B. Device Manager

   C. Registry Editor

   D. Control Manager

7. Which two Windows Server tools are used to operate the Windows Registry and its services?

   A. Services Control Manager

   B. Registry Editor

   C. Device Manager

   D. Devices

8. The _____ is the Windows Server native account or an account created by you to manage running services.

9. Discuss Windows Registry keys.

10. Discuss Windows service startup types.

Further reading

To learn more about the topics that were covered in this chapter, take a look at the following resources:

- *How to Use the Windows Device Manager for Troubleshooting*: `https://www.howtogeek.com/167094/how-to-use-the-windows-device-manager-for-troubleshooting/` *Structure of the Registry*: `https://docs.microsoft.com/en-us/windows/desktop/ sysinfo/structure-of-the-registry`

- *Understanding and Managing Windows Services*: `https://www.howtogeek.com/school/using-windows-admin-tools-like-a-pro/lesson8/`

Part 2: Setting Up Windows Server 2022

Part 2 covers the roles in Windows Server 2022. Upon completing this part, you will set up a domain and build up the network services on Windows Server 2022, such as **Domain Name System (DNS)**, **Dynamic Host Configuration Protocol (DHCP)**, **Print and Document Services (PDS)**, **Internet Information Services (IIS)**, **Windows Deployment Service (WDS)**, and **Windows Server Update Services (WSUS)**.

This part of the book comprises the following chapters:

- *Chapter 5, Directory Services in Windows Server 2022*
- *Chapter 6, Adding Roles to Windows Server 2022*

5
Directory Services in Windows Server 2022

Now that you have learned how to install Windows Server 2022 and run the initial server configuration, it is time to set up the first services in your organization's IT infrastructure. With that in mind, this chapter is designed to explain the domain services as the most critical services in a Windows-based domain network. Specifically, this chapter covers Microsoft's **Active Directory Domain Services (AD DS)** and **Domain Name System (DNS)** roles in Windows Server 2022. In addition, you will become acquainted with concepts such as domain, forest, tree domain, child domain, **domain controller (DC)**, functional level, trust relationship, forward and reverse lookup zones, DNS record, and many others.

Furthermore, this chapter provides instructions on installing AD DS and DNS roles. You will also learn about **organizational units (OUs)**, default containers, user accounts, and group scopes and types, which you can use to organize the user and computer accounts in a Windows Server domain-based network.

Finally, this chapter concludes with an exercise on installing the AD DS and DNS roles and promoting the server to a **DC**.

The following topics will be covered in this chapter:

- Understanding the Active Directory infrastructure
- Understanding DNS
- Understanding OUs and containers
- Understanding accounts and groups
- Chapter exercise – installing the AD DS and DNS roles and promoting the server to a DC

Technical requirements

To complete the exercises in this chapter, you will need the following equipment:

- A PC with Windows 11 Pro with at least 16 GB of RAM, 1 TB of HDD space, and access to the internet
- Virtual machine 1 (tree domain: `Dautti.local`) with Windows Server 2022 Standard (Desktop Experience) with at least 4 GB of RAM, 100 GB of HDD space, and access to the internet
- Virtual machine 2 (tree domain: `ITTrainings.local`) with Windows Server 2022 Standard (Desktop Experience) with at least 4 GB of RAM, 100 GB of HDD space, and access to the internet
- Virtual machine 3 (child domain: `Programming.Dautti.local`) with Windows Server 2022 Standard (Desktop Experience) with at least 4 GB of RAM, 100 GB of HDD space, and access to the internet

Understanding the Active Directory infrastructure

Active Directory (**AD**) is Microsoft's technology for representing a distributed database that stores objects in a hierarchical, structured, and secure format. AD objects represent users, computers, peripheral devices, network services, and security settings. Each object is uniquely identified by its name and attributes. The domain, the forest, and the tree represent the main tiers of an AD infrastructure. AD uses the following protocols and services:

- **Lightweight Directory Access Protocol** (**LDAP**) is used to access the directory services data.
- **Kerberos** securely authenticates and proves the identity between users and servers on the network.
- **DNS** translates domain names into IP addresses.

AD generally offers centralized management for administrators for various services it provides. However, there are several administrative consoles (snap-ins) in **Microsoft Management Console** (**MMC**) (`mmc.exe`) for managing AD services:

- **Active Directory Administrative Center** (`dsac.exe`), as shown in *Figure 5.1*, is the administrative console used to manage Windows Server's directory services. Active Directory Users and Computers (`dsa.msc`) is the administrative console that's used to manage users, computers, OUs, and relevant information.
- **Active Directory Domains and Trusts** (`domain.msc`) is the administrative console used to manage domains, trusts, and relevant information.
- **Active Directory Sites and Services** (`dssite.msc`) is the administrative console used to manage replication and the services between sites.

- **Active Directory Module for Windows PowerShell** is the administrative console that manages Windows Server's directory services via cmdlets.

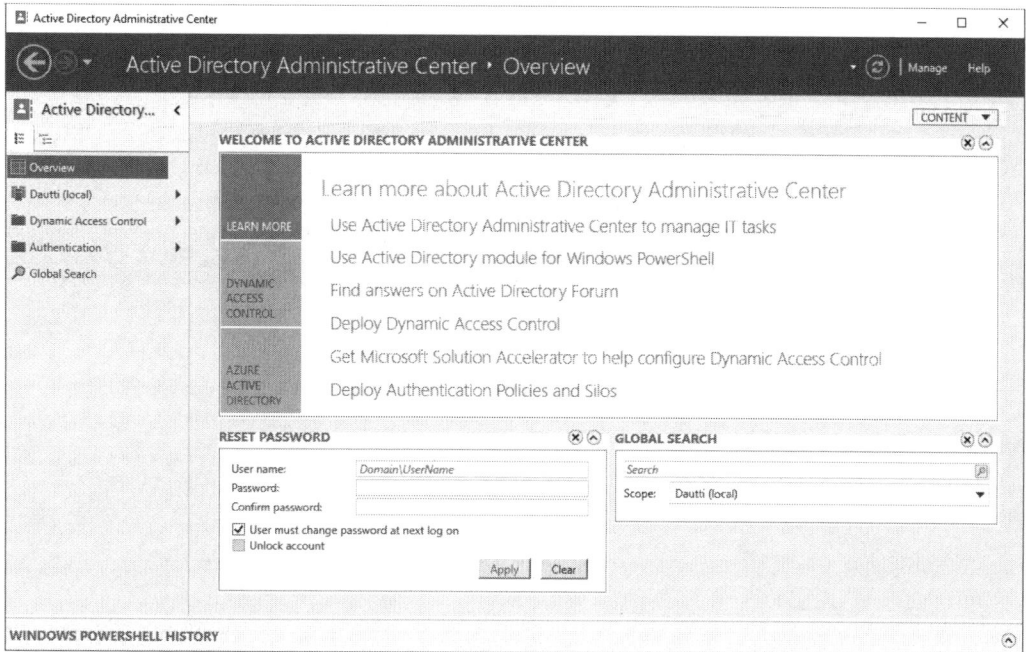

Figure 5.1 – Active Directory Administrative Center in Windows Server 2022

> **Important note**
> You can access Microsoft's Script Center at https://technet.microsoft.com/en-us/scriptcenter/bb410849.aspx and PowerShell Gallery at https://www.powershellgallery.com/. Both are well-known repositories that contain free and public domain PowerShell scripts. Additionally, substantial collections of AD- and DNS-related entries are included.

To set up directory services in your organization's IT infrastructure, you must add and configure the AD DS role in a server. AD DS is considered the leading service of AD. While you can find more information about server roles in *Chapter 6, Adding Roles to Windows Server 2022*, AD DS is a role in Windows Server 2022 that lets system administrators manage and store a network's information resources. See the *Chapter exercise – installing the AD DS and DNS roles and promoting the server to the DC* section later in this chapter to learn how to install AD DS on a server.

Now, let's learn about the various elements of an AD infrastructure. Let's begin with DCs.

DC

A DC, as shown in *Figure 5.2*, is a server that's responsible for securely authenticating users so that they can access an organization's network resources. In a former Windows NT, one DC per domain was configured and referred to as a **primary domain controller** (**PDC**). All other DCs acted as **backup domain controllers** (**BDCs**). In contrast, in Windows Server 2022, no primary and backup references are used; instead, numbers are used next to DCs to identify the order – for example, DC1 and DC2.

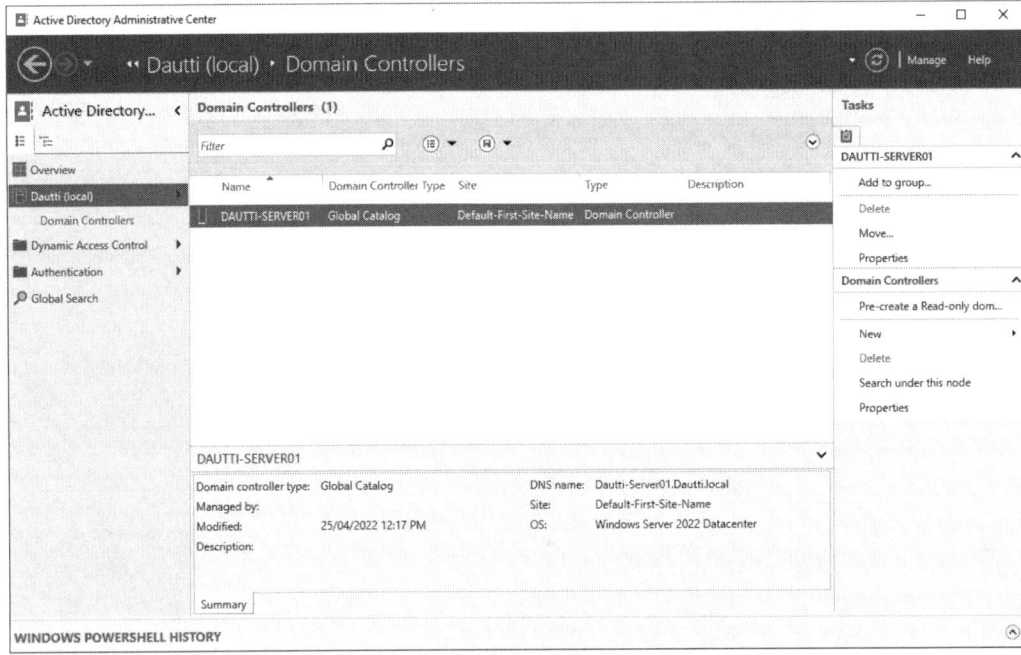

Figure 5.2 – Accessing DCs through Active Directory Administrative Center

> **Important note**
> A server that has joined a domain in an organization's network is considered a member server.

The DC is responsible for granting access to domains. With that, let's learn about domains.

Domains

A domain is a logical grouping of users, computers, peripheral devices, and network services. From the perspective of network architecture, usually, domains are centralized network environments where a DC governs authentication. For example, in Windows Server-based networks, the domain is powered by the AD DS role.

The following screenshot shows the step in the **Active Directory Domain Services Configuration Wizard** in which the domain is set:

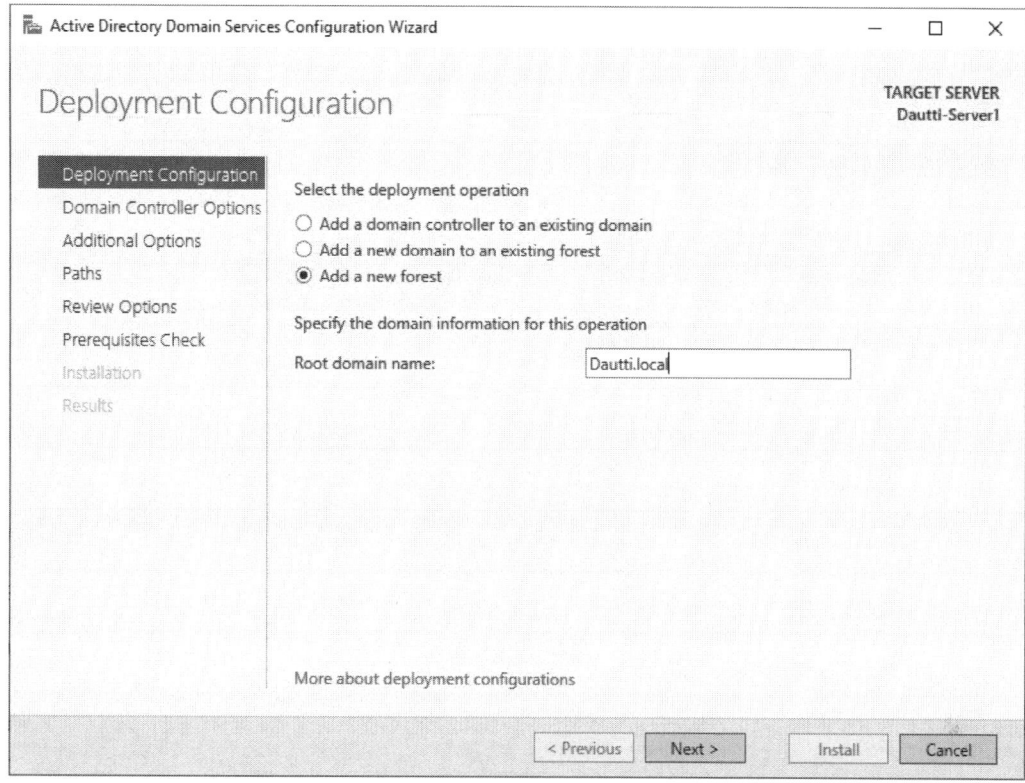

Figure 5.3 – Setting up a root domain in Windows Server 2022

> **Important note**
>
> A profound difference exists between the meaning of a *domain* and a *domain name*. The former means a logical organization of users, servers, devices, and resources within a specific collection of such things. The latter represents the logical naming system that governs the internet, including web servers and websites.

Several domains come together to make up a tree domain. Thus, next, let's learn about the tree domain.

Tree domain

In an AD structure, a tree domain comprises one or more domains. Domains in a tree are linked through transitive trust. For example, with transitive trust, if A trusts B, and B trusts C, then A trusts C. In a tree domain, when a new domain joins an existing tree, the new domain automatically trusts all existing domains in the tree. Similar to adding a new domain, the tree domain is configured during the *Promote this server to a domain controller* process:

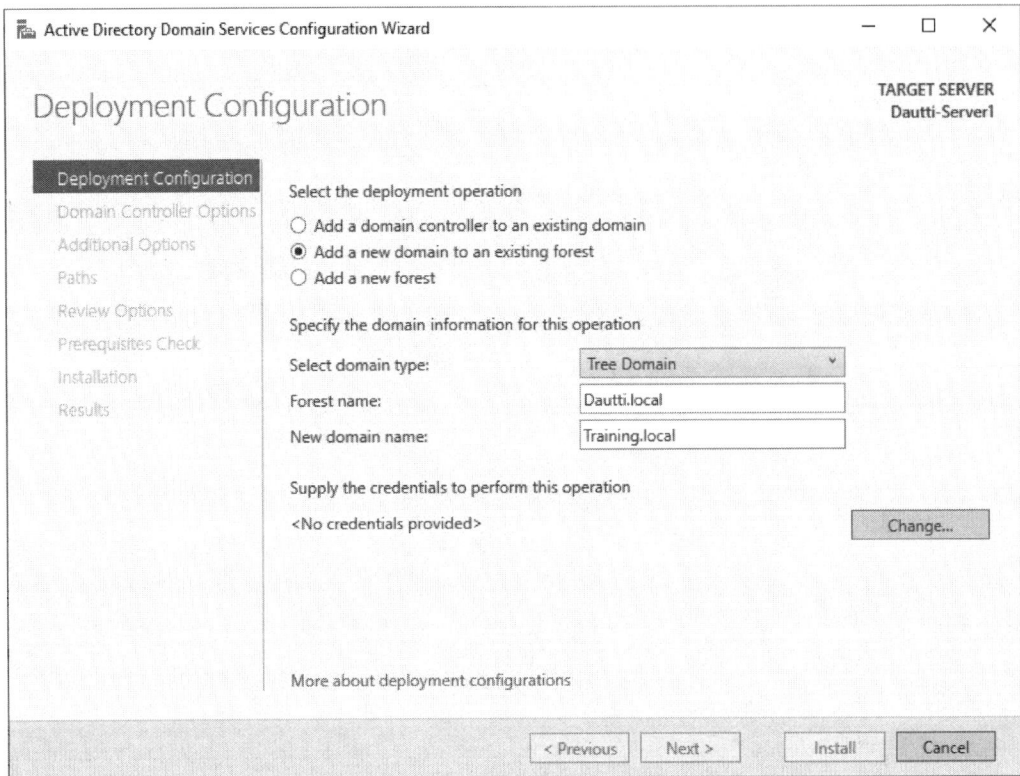

Figure 5.4 – Setting up a tree domain in Windows Server 2022

When several tree domains come together, the result is a forest. But what exactly is a forest? Let's find out in the next section.

Forest

As you know, in real life, a forest consists of trees. Likewise, in AD, a **forest** consists either of a single tree domain or a collection of tree domains. From that, it can be concluded that the tree domain represents a domain that collects domains in an AD DS hierarchy. Common schema and configuration are shared by several domains, thus forming the contiguous namespace.

Because of that, a forest is considered a domain too. Isn't that a closed circle?

To set up a forest in Windows Server 2022, as with tree domains, you must use the **Active Directory Domain Services Configuration Wizard**, as shown in *Figure 5.3*.

Child domain

To understand the child domain, let's illustrate this with an example, as shown in *Figure 5.6*. `Dautti.local` and `Training.local` are both tree domains. However, since `Dautti.local` is the root domain, it represents a forest. From that, `Administration.Dautti.local` represents a child domain of the `Dautti.local` tree domain.

To set up a child domain in Windows Server 2022, as in the case of a domain and tree domain, you can use the **Active Directory Domain Services Configuration Wizard** area, as shown here:

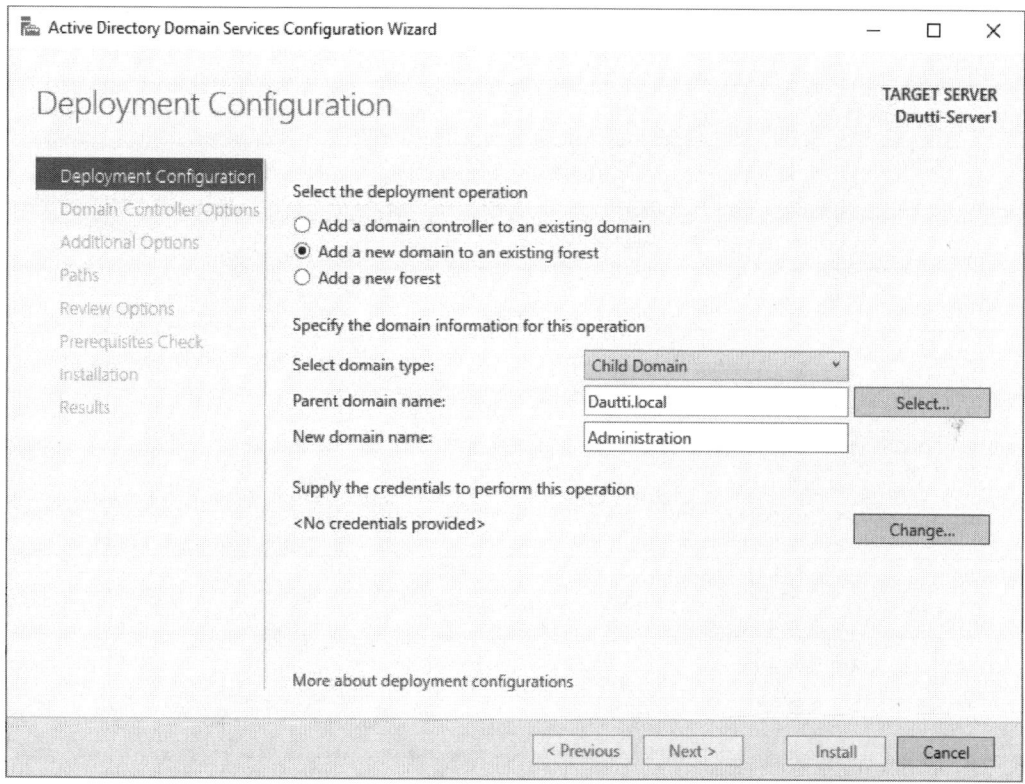

Figure 5.5 – Setting up a child domain in Windows Server 2022

As you can tell, the tree and child domain concept is similar to the tree data structure since it contains the parent-child relationship. Next, let's learn about the operations master role.

Operations master roles

By its very nature, AD DS is complex! However, once you begin deploying it, everything becomes more apparent. Let's look at the operations master role this way. In the *Domains* section, in *Figure 5.3*, we created the root domain, `Dautti.local`, which represents a forest. In fact, in that example, the machine that hosts the `Dautti.local` forest is a DC for that network.

As mentioned previously, when the AD DS role is being installed, and the server gets promoted to a DC, AD DS automatically assigns five master operations roles. The first two, *master schema* and *domain naming master*, are forest-wide operations master roles. The remaining three, the **relative identifier** (**RID**), PDC emulator, and infrastructure master, are tree domain-wide operations master roles. While the schema master and domain naming master manage AD's schema, the read-write copy ensures that only one unique domain is in the forest. On the other hand, RID assigns **security identifiers** (**SIDs**) to DCs, the PDC emulator deals with password updates, and the infrastructure master keeps track of the changes made to other domain objects.

Let's relate this to our example, as shown in the following diagram. The root domain, such as `Dautti.local`, is the master schema and domain naming master in the whole forest (`Dautti`). `Dautti.local` and `Training.local` are both tree domains in a forest, and as such, they have RID masters, PDC emulators, and infrastructure masters:

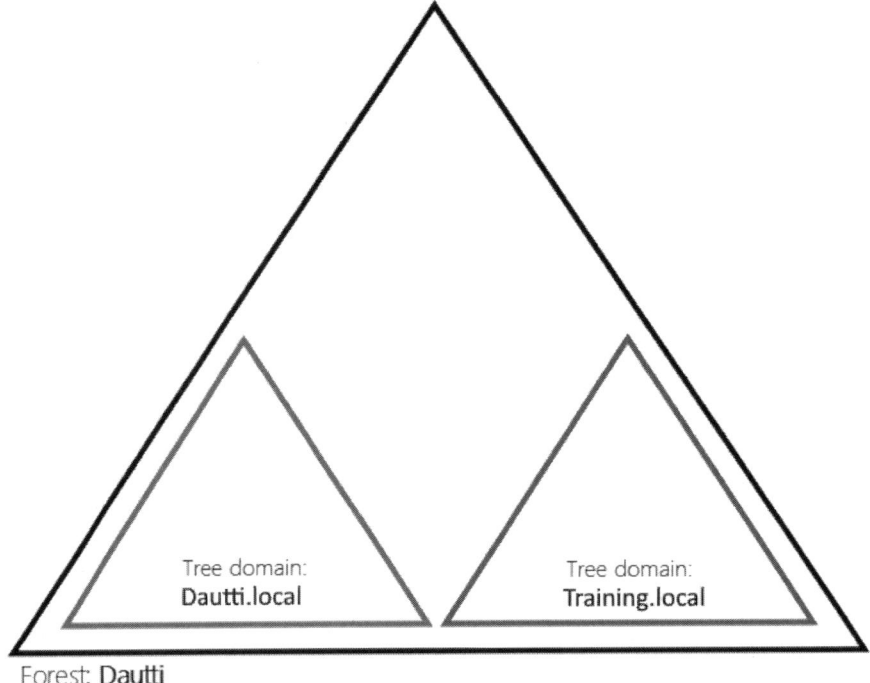

Figure 5.6 – AD DS structure

According to Microsoft, the previously explained roles represent the five roles of AD's operations: **flexible single master operation (FSMO)**. Next, let's learn the difference between a domain and a workgroup.

Comparing a domain with a workgroup

Let's consider network architectures such as **peer-to-peer (P2P)** networking and the client/server to illustrate the difference between a domain and a workgroup better. The following table provides a comparison between a domain and a workgroup:

Domain	Workgroup
A dedicated server is used to provide services	Computers equally share resources among themselves without needing a dedicated server
An example is the client/server network	An example is a P2P network

Table 5.1 – Domains versus workgroups

Now that you have become familiar with the difference between workgroups and domains, let's examine the trust relationship between a computer and a DC.

Trust relationship

In AD, a **trust relationship** exists between a computer, a DC, and domains. Once a computer joins a domain, the **Security Account Manager (SAM)** in the local computer trusts AD's authentication mechanism, Kerberos, on a DC. Hence, the user is authenticated by a DC in a network and not the local SAM.

Similarly, the authentication mechanism of each tree domain is trusting every other authentication mechanism of other trusted tree domains within a forest. So, for example, from *Figure 5.6*, the `Dautti.local` domain authenticates `user.local`. Then, its authentication is accepted by `Training.local` too since these tree domains are part of the same forest; the root domain, such as `Dautti.local`.

To understand this better, try to link the trust relationship with the administration and communication links between domains. Next, let's learn about the functional levels of the domain and forest and how to check them.

Functional levels

The **forest functional level (FFL)** and the **domain functional level (DFL)** are two functional levels. The FFL controls which versions of Windows Server can run in the DCs of the forest, and simultaneously enables the available capabilities across all domains in a forest. In contrast, DFL controls which versions of Windows Server can run in the DCs of that domain, and simultaneously enables the available capabilities in that domain only. In Windows Server 2022, the DFL and FFL should be lowered to at least Windows Server 2008 since Windows Server 2003 is no longer supported. Furthermore, at most, the DFL and FFL can be raised to Windows Server 2016.

To check the DFL and FFL in Windows Server 2022, follow these steps:

1. Click the **Start** button. Then, from the **Start** menu, click on the **Server Manager** tile.
2. Click **Tools** from the menu bar and select **Active Directory Domains and Trusts** in the **Server Manager** window.
3. Right-click the root domain in the **Active Directory Domains and Trusts** window and select **Properties** from the context menu.
4. In the **Properties** dialog box, you will find **Domain functional level** and **Forest functional level** under the **General** tab, as shown in the following screenshot:

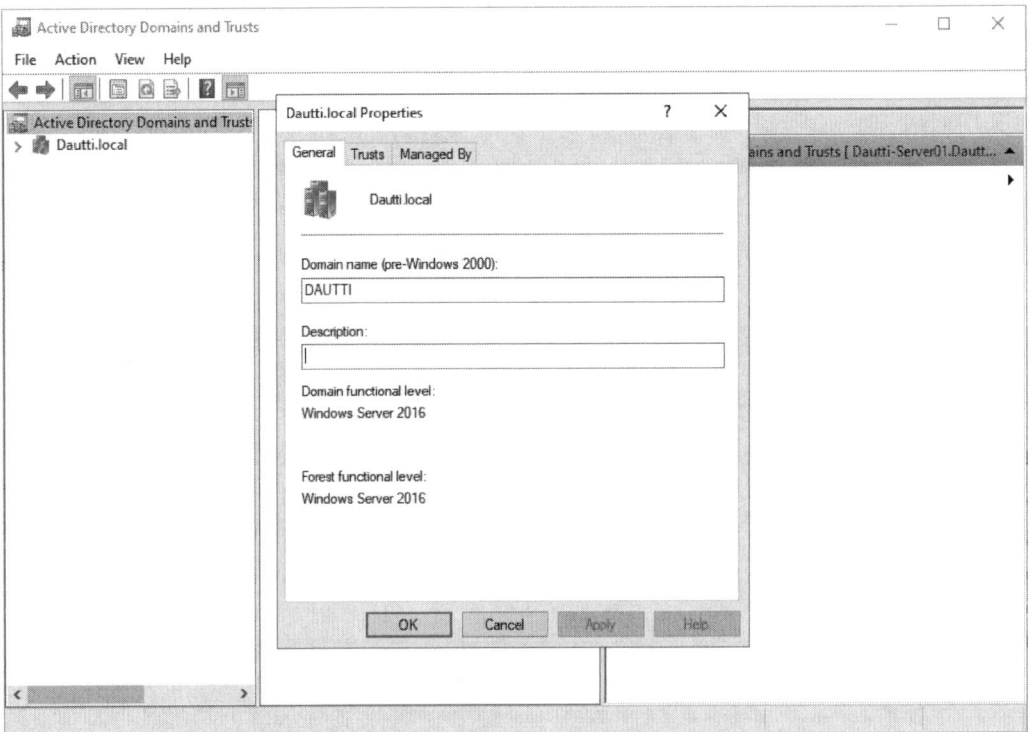

Figure 5.7 – Checking the DFL and FFL in Windows Server 2022

> **Important note**
> Be aware that there is no functional level for Windows Server 2022. Instead, the highest functional level offered is **Windows Server 2016**.

Now that you have understood the roles of FFL and DFL, let's learn about the contiguous namespace, which links the child domain to its parent domain (the tree domain).

Namespaces

In the following diagram, `Dautti.local` represents a forest and a root domain simultaneously, whereas `Dautti.local` and `ITTrainings.local` are tree domains in the forest, as mentioned earlier. Furthermore, the `Dautti.local` tree domain contains a child domain such as `Programming.Dautti.local`. The child and tree domains share a common namespace, `Administration.Dautti.local`, within `Dautti.local`. This is known as a **contiguous namespace**.

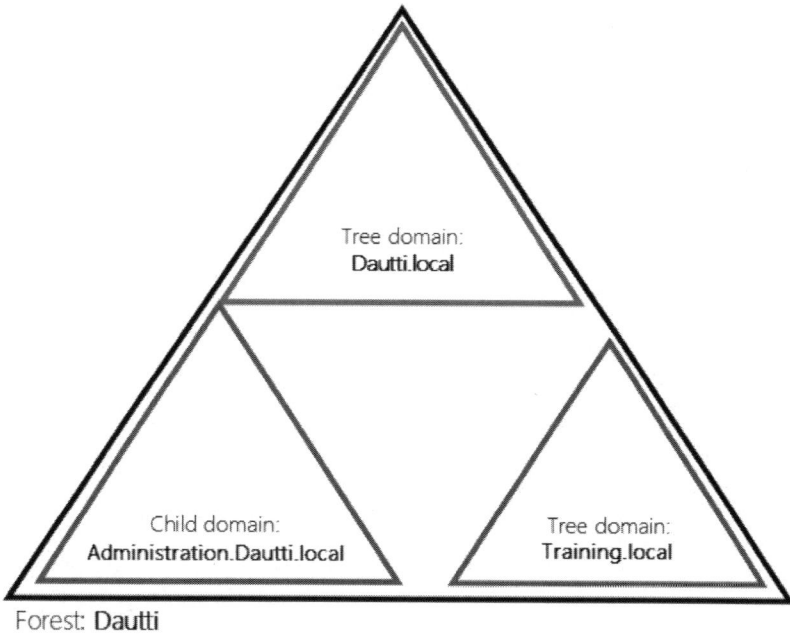

Figure 5.8 – The namespace concept in AD DS

To better understand namespaces, try associating them with the **Uniform Resource Locator (URL)**, a unique web address that helps users locate the websites on a web server. Next, let's learn about **sites** in domains, which represent physical or logical locations in a network.

Sites

There are physical and logical topologies even in AD, as with computer networks. Hence, a *domain* represents the logical topology, whereas a *site* depicts the physical topology in an AD infrastructure. From that, a site represents the physical or logical location of the computer network in a particular domain.

Next, let's learn about replication, which helps update AD objects between domains.

Replication

In an AD infrastructure, replication is the process of synchronizing the standard directory partition among all DCs in a forest. Additionally, replication topology is a set of communication paths through which the DC's replication data travels.

Next, let's learn about the **schema** that defines objects and attributes in AD.

Schema

Generally, **objects**, **classes**, and **attributes** define a schema object in AD. While the *class* represents the object type, the *attribute* represents the characteristics of the object. Thus, AD stores the objects identified by their classes and attributes. The schema defines the model containing the rules of the types of objects stored in AD. Furthermore, replication synchronizes the *schema* among all DCs in a forest.

Next, let's learn about Microsoft Passport, which provides a passwordless sign-in method.

Microsoft Passport

In today's technological trends, where users need to sign into different environments to access applications, websites, and other services, the traditional authentication format such as username and password is neither viable nor secure. The same applies to Windows OSs and services. Therefore, Microsoft has introduced Microsoft Passport, enabling users to authenticate without a password. Microsoft Passport is based on the **Fast ID Online** (**FIDO**) **Alliance** standard and consists of two components: a single sign-in service and a wallet service. From that, we can deduce that Microsoft Passport is a two-factor authentication mechanism with which users can prove who they are by providing something they uniquely possess.

In the preceding sections, you learned about the various components of the AD infrastructure. In the next section, you will delve deeper into the concept of DNS and its components.

Understanding DNS

The DNS concept dates back to the 1960s, also known as the ARPANET era. At that time, scientists engaged in the ARPANET project were trying to find a way of memorizing names instead of IP addresses. Decades later, at the beginning of the 1980s, the first specification documents were published about DNS in **Requests for Comments (RFC)** documents.

DNS has a tree structure (hierarchical), where each branch represents the root zone, and each leaf has zero or more resource records. Each zone can represent a root domain or multiple domains and sub-domains. A domain name consists of one or more components, known as labels, and is separated by a period; for example, *packtpub.com*. DNS is maintained by a database that uses a distributed clients/server architecture where network hosts represent the servers' names. Thus, the following steps illustrate how DNS works:

1. If you enter www.packtpub.com in your browser's address bar and press *Enter*, your computer's browser will request the internet to access the www.packtpub.com website.
2. The first server the browser runs into is the recursive resolver, which, in most cases, may be provided by your **internet service provider (ISP)**.
3. The *recursive resolver* will then contact the root servers scattered all over the globe and contain information about top-level domains; in our example, this is .com.
4. The top-level domains will provide the DNS information to the recursive resolver.
5. Once the recursive resolver receives information from the top-level domains, it will contact the domain name server, *packtpub.com*. Then, using the DNS's local DNS, the recursive resolver will try to figure out the IP address.
6. Once it learns the web server's IP address, the recursive resolver provides that IP to your computer's browser to access the web server's content through its newly accustomed IP address.

Now that you have learned how DNS works, let's learn how to add the DNS role. You need one in your organization's network to enable communications over friendly domain names.

Adding the DNS role

Similar to the AD DS role, DNS in Windows Server 2022, too, is a role that can be added using Server Manager, as shown in the following screenshot:

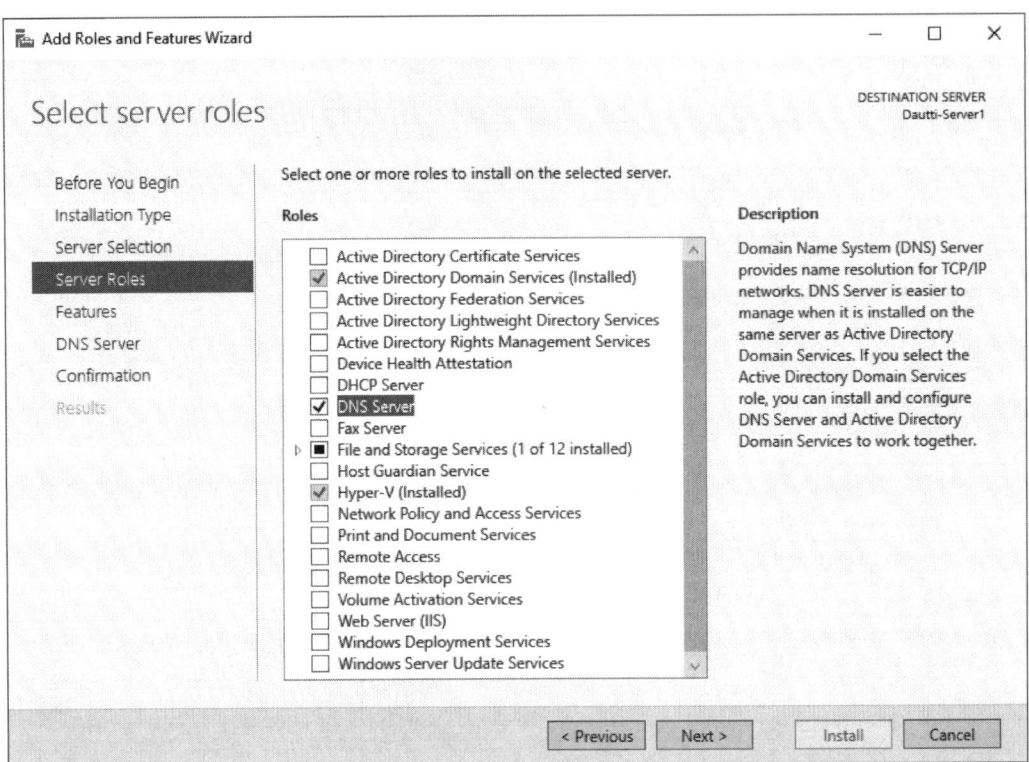

Figure 5.9 – Adding the DNS Server role in Windows Server 2022

Note that, when installing the DNS role, it can be installed either as a separate role, as shown in the preceding screenshot, or alongside AD DS, as shown here:

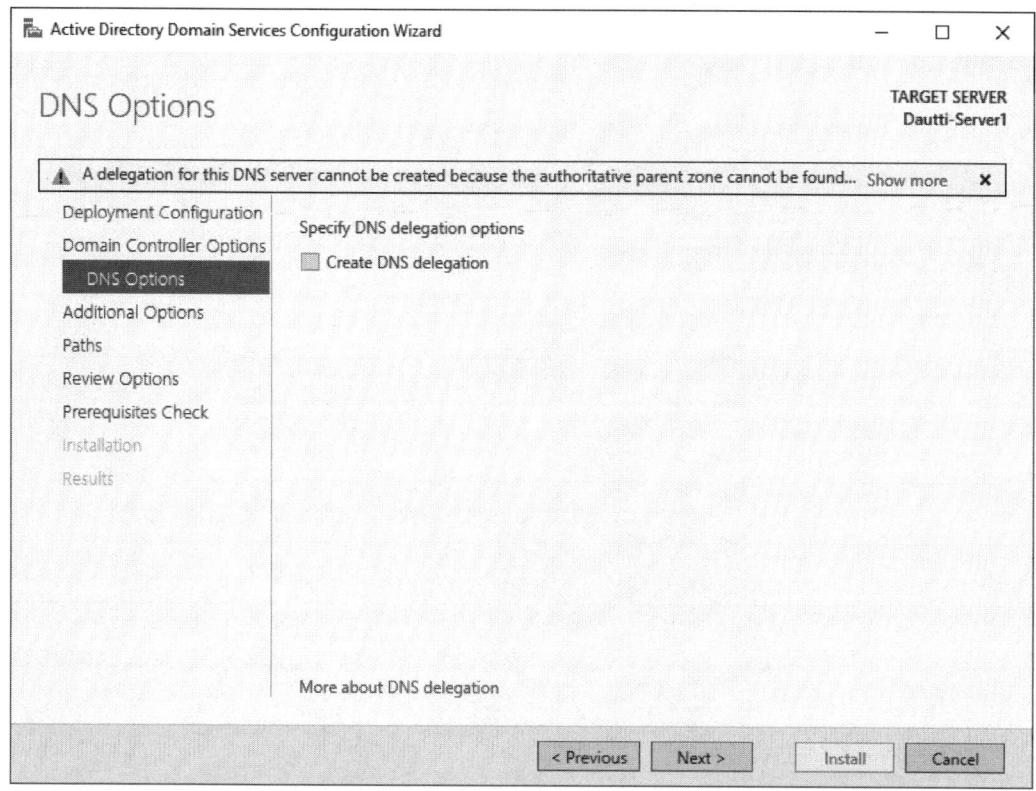

Figure 5.10 – Adding DNS alongside AD DS in Windows Server 2022

As you can tell, installing DNS is a step in adding the AD DS role itself. So, next, let's learn about the **hosts** and **lmhosts** files.

The hosts and lmhosts files

The hosts and **LAN manager hosts** (**lmhosts**) files are used for name resolution and are stored in the `C:\Windows\system32\drivers\etc` directory, as shown in the following screenshot. The host files contain the mapping of IP addresses to hostnames and are used for DNS name resolution. Unlike `hosts`, the `lmhosts` file contains the mapping of IP addresses to computer names and is used for NetBIOS name resolution. In both files, entries are inserted manually, and each entry should be kept on a separate line:

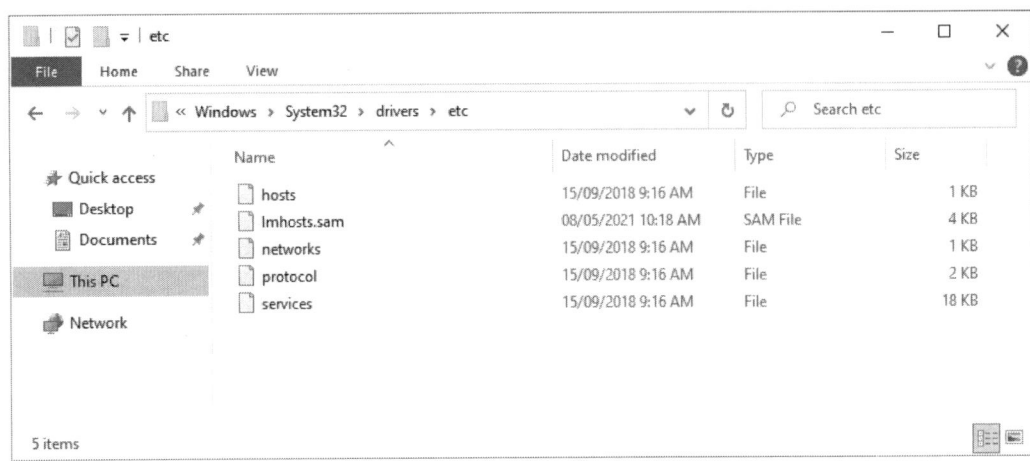

Figure 5.11 – The hosts and lmhosts files in Windows Server 2022

The following table presents examples of `hosts` and `lmhosts` entries:

Entries	Description
hosts entry	IP address FQDN hostname #Comment
lmhosts entry	IP address FQDN hostname Extension <tag>#Comment

Table 5.2 – Examples of hosts and lmhosts entries

Next, let's learn about hostnames, which represent local computer local domains, as explained earlier in this chapter.

Hostnames

A hostname represents a logical element that's been assigned to a device, as shown in the following screenshot. It is unique and used to identify the host in a network. Other than that, it is also used for communication on a LAN. Often, it is called a domain name too.

Understanding DNS

Figure 5.12 – Assigning a hostname in Windows Server 2022

Next, let's learn about **DNS zones**, administrative spaces that host the DNS records of specific domains.

DNS zones

A DNS zone is a hierarchical structure that enables the existence of DNS zones. For example, the **AD namespace** is very much related to the **DNS namespace**. Therefore, the DNS namespace can be divided into zones that can store information about domains. In this way, DNS provides three types of zones:

- The primary zone stores the direct copy of the DNS database and maintains all DNS zone records.
- The secondary zone is a backup of the primary zone, meaning that whenever the primary zone is unavailable, it resolves DNS queries.
- The stub zone, in principle, represents the secondary zone with no editable primary copy of the database and contains sufficient information to identify the authoritative DNS.

Authoritative DNS can be configured manually by a system administrator or dynamically by other DNSs. It represents the DNS server, which contains the DNS records of the actual domain. In contrast, **non-authoritative DNS** contains the cached information that previous DNS lookups have constituted.

Next, we'll learn about **Windows Internet Name Service (WINS)**, a legacy name resolution service for NetBIOS names.

WINS

To automate **NetBIOS name resolution**, you can use Microsoft's WINS server. A **WINS server** maps the IP addresses to NetBIOS names when your computer connects to a shared folder or printer. WINS is a feature in Windows Server 2022 and can be added through Server Manager using the **Add Roles and Features Wizard** area, as shown here:

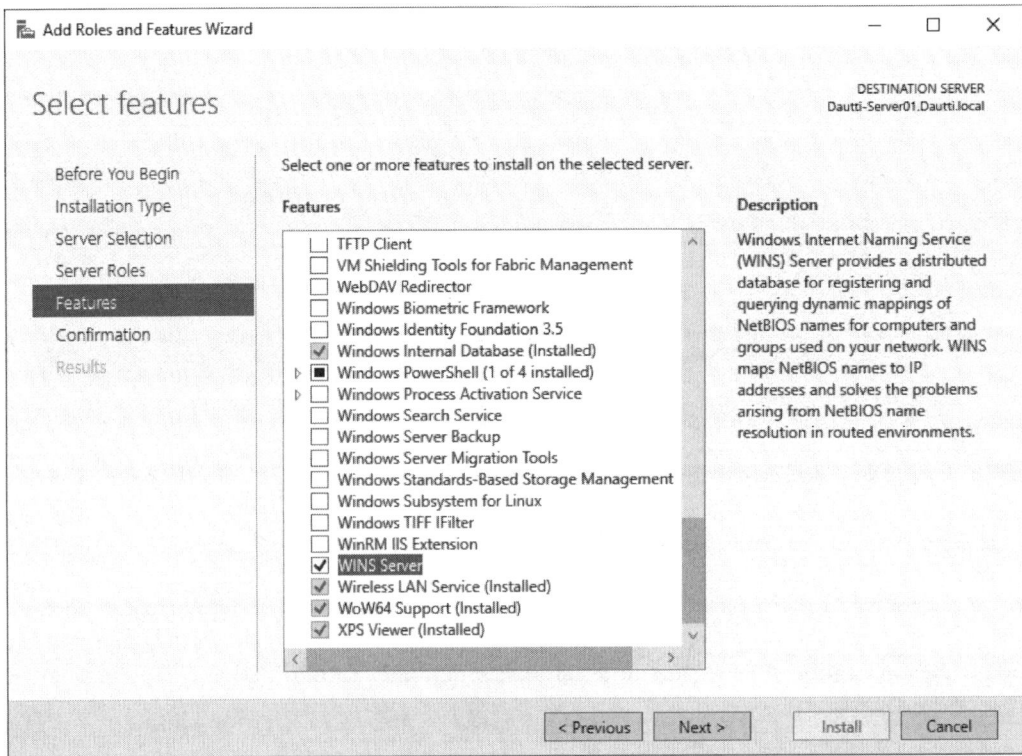

Figure 5.13 – Setting up the WINS feature in Windows Server 2022

Next, let's learn about the **Universal Naming Convention (UNC)**, which is just a standard for identifying a shared resource in a network.

UNC

Used in Unix, the UNC is a standard for identifying a computer network share. Its format, as shown in the following screenshot, uses double backslashes to precede the name of the server; for example, `\\servername\folder`.

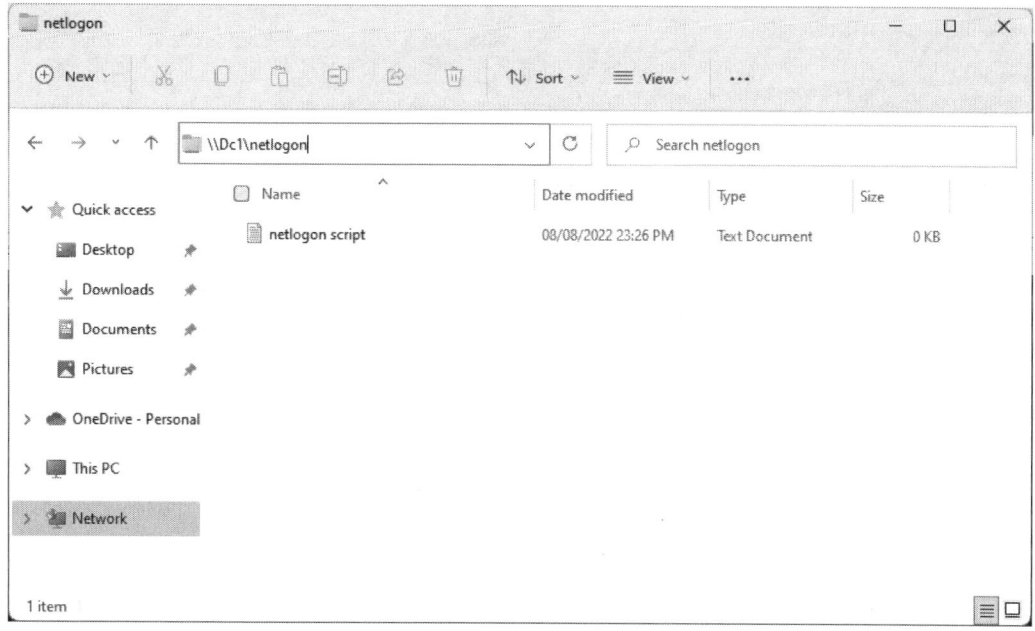

Figure 5.14 – A UNC path in Windows Server 2022

In this section, you learned about DNS and its components. In the next section, you will learn about OUs and containers. While OUs can contain additional containers and linked GPOs, containers can contain AD objects but not linked GPOs.

Understanding OUs and containers

The AD Users and Computers console provides OUs and default containers to ease the administration of objects. In the following sections, you will get acquainted with OUs and default containers.

What are OUs?

As mentioned earlier, to ease the administration of its objects, AD uses OUs. Users, groups, computers, and other OUs are generally placed within OUs. Usually, organizations create OUs to mirror their organizational business structures. Regardless of the number of tree domains in a forest, each domain can have its own OU hierarchy, as shown here:

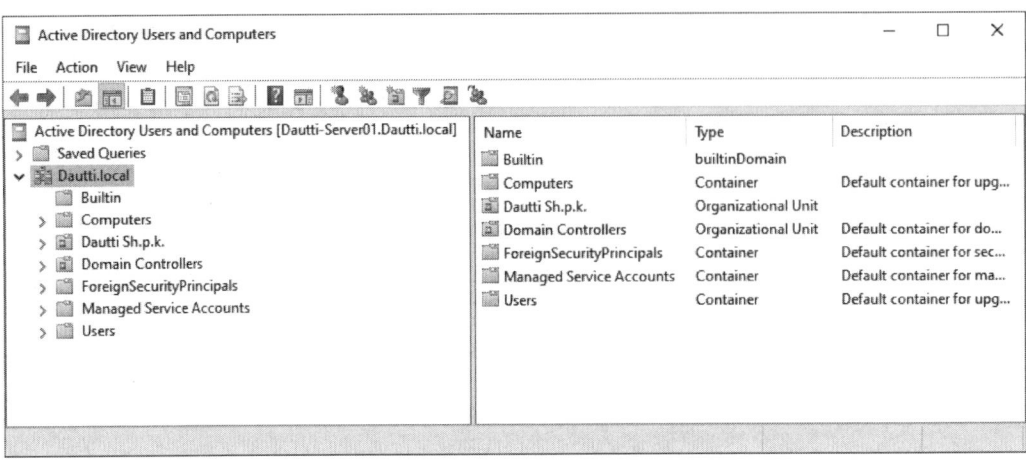

Figure 5.15 – An example of an OU hierarchy in Windows Server 2022

Try associating the OUs with folders in File Explorer to understand them better. Next, we'll learn about default, pre-built containers in AD that are used to store users, computers, and other objects.

Default containers

Once the server gets promoted to a DC, several default containers are created, as shown in the following screenshot. These default containers are unique because they cannot be renamed, deleted, created, or associated with a **Group Policy object** (**GPO**).

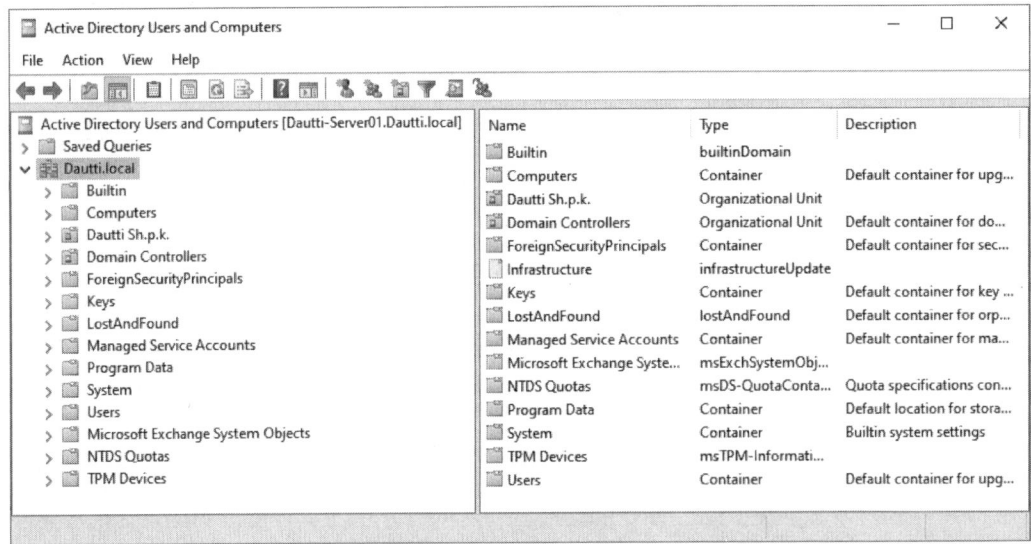

Figure 5.16 – Default containers in Windows Server 2022

Next, let's learn about hidden default containers.

Hidden default containers

Not every default container is needed for a system admin's day-to-day job! Because of that, by default, there are **hidden containers** too. These containers are hidden to stop the AD Users and Computers console from looking messy. However, it may seem that the security part is the biggest reason there are hidden containers in AD.

To unhide these hidden default containers, enable the **Advanced Features** option from the **View** menu, as shown in the following screenshot:

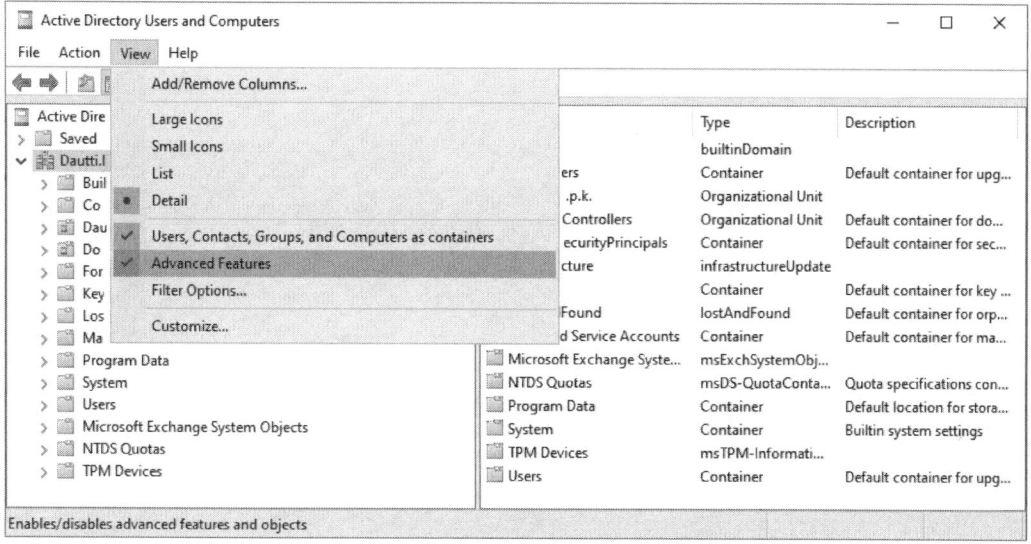

Figure 5.17 – Default hidden containers in Windows Server 2022

Now that you have learned about default containers, let's learn about their uses.

Uses of default containers

The following is a list of simple uses for some of the default containers in Windows Server 2022:

- **Computers** is a default container for upgraded computer accounts.
- **Domain Controllers** is a default container for DCs.
- **ForeignSecurityPrincipals** is a default container for **SIDs**.
- **Keys** is a default container for critical objects.

- **LostandFound** is a default container for orphaned objects.
- **Managed Service Accounts** is a default container for managed service accounts.
- **Users** is a default container for upgraded user accounts.

Next, let's learn about delegating control to an OU, which is a way to grant administrative control to objects to non-administrators.

Delegating control to an OU

Knowing that OUs facilitate the organization of AD objects, whenever you want to grant permissions to a particular user or a group of users in AD, you can do so via *delegating control* to an OU. However, before you assign permissions to a user or group of users, they need to be moved into an OU, as shown here:

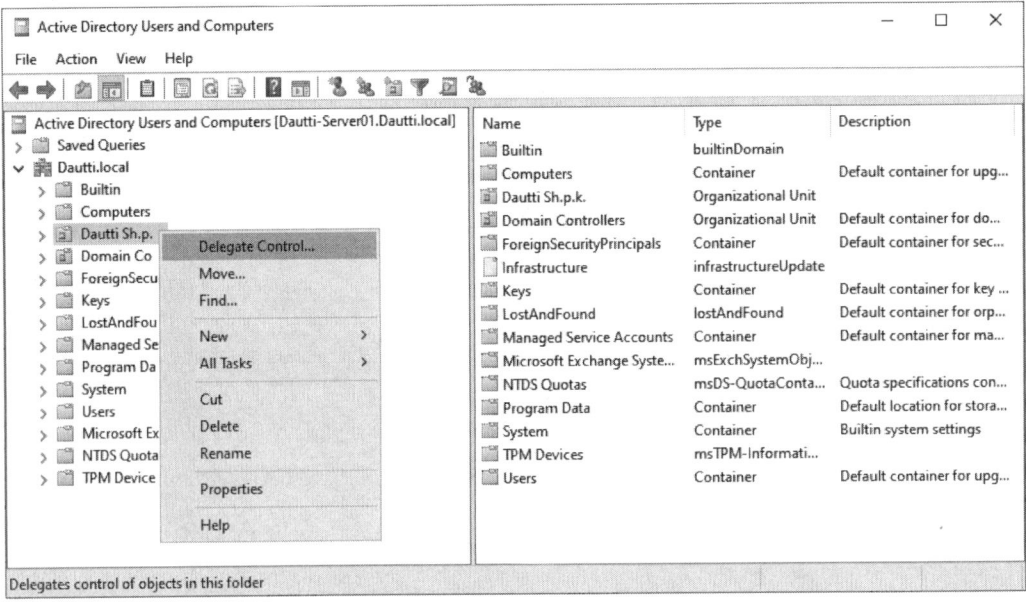

Figure 5.18 – Delegating control to an OU in Windows Server 2022

This section taught you the various aspects of OUs and containers. In the next section, you will learn about user and computer accounts and groups.

Understanding accounts and groups

User and computer accounts are used to access network services. Moreover, user and computer accounts reside in an AD in a Windows-based domain network. In such a centralized environment, groups are used to facilitate assigning rights and permissions. In the following sections, you will learn about different accounts and groups. So, let's dive in.

Domain accounts

Technically, the **domain account** is part of an AD, and as such, it is authenticated by the same entity – that is, AD. As a result, the domain account is authorized to access local and network services. It does that based on the access granted to the account or a group that an account belongs to.

To create a domain account in Windows Server 2022, follow these steps:

1. From **Windows Administrative Tools**, open the **Active Directory Users and Computers** console.
2. Right-click the **Users** container and select **New | User**.
3. Enter the user's required information, as shown in the following screenshot, and click **Next**:

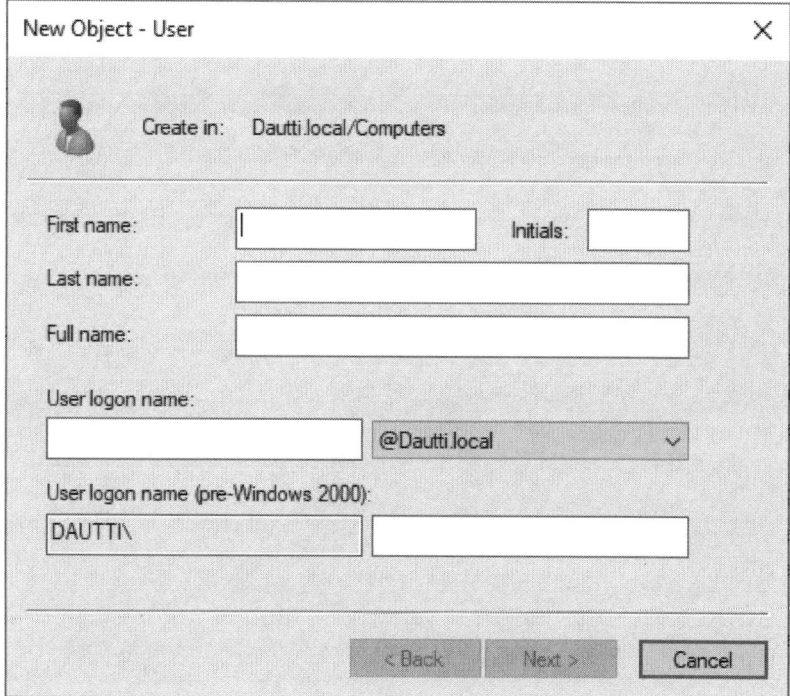

Figure 5.19 – Creating a domain account in Windows Server 2022

4. Provide a temporary password, confirm your choice, and click **Next**.
5. Click **Finish** to close the **New Object - User** window.

The next type of account we will look at is local accounts.

Local accounts

Unlike domain accounts, local accounts are the part of computers where that account has been created, and as such, the account is authenticated by Windows SAM. The local account is authorized to access local services based on the access granted to the account.

Additionally, the local account can access shared resources in a P2P network without permission.

To create a local account in Windows Server 2022, follow these steps:

1. From **Windows Administrative Tools**, open the **Computer Management** console.
2. Expand **System Tools | Local Users and Groups**, right-click the **Users** container, and select **New | User**.
3. Enter the user's required information, as shown in the following screenshot, and click **Create**:

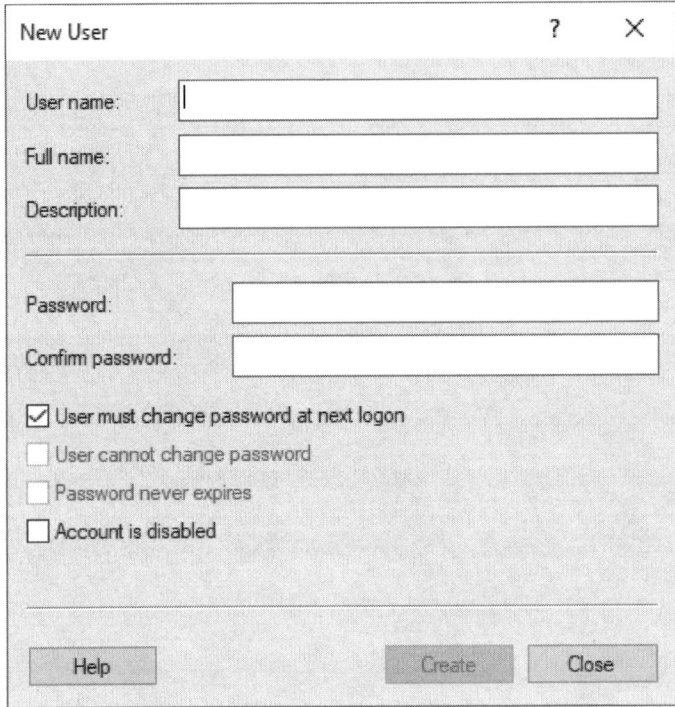

Figure 5.20 – Creating a local account in Windows Server 2022

> **Important note**
> The server must not be a DC if you wish to set up a local account in Windows Server 2022. Thus, there's no need to add the AD DS role.

The local account is an account that is stored on the local computer and authenticated locally by Windows SAM. Next, let's look at the various user profiles, which provide pieces of important information that identify a user.

User profiles

In general, in Windows Server-based networks, the following user profiles are used:

- The local user profile is created when the user logs on to a computer for the first time and is stored on a local computer (see *Figure 5.21*).
- The roaming user profile is a local profile that's copied and stored in a network share.
- The mandatory user profile is a roaming profile where a user logs off. Here, no changes in a profile are saved:

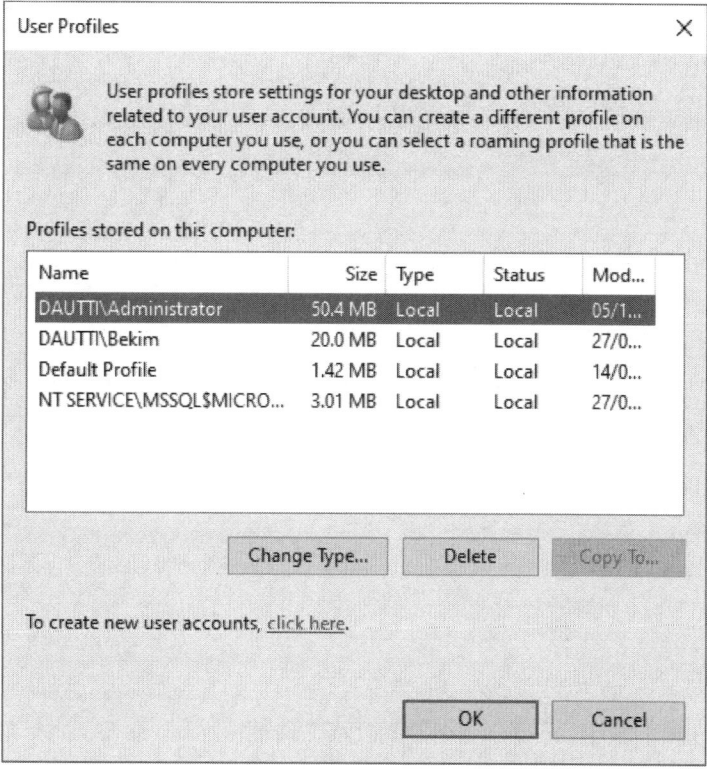

Figure 5.21 – User profiles in Windows Server 2022

From the preceding explanation, it can be concluded that the local user profile is stored on the local computer, the roaming user profile is stored in the network share, and the mandatory user profile is not saved. Therefore, the mandatory user profile also uses the fixed profile from the network share.

Next, let's learn about computer accounts, which, as the name suggests, identify computers in local or centralized domains.

Computer accounts

In AD, a **computer account** identifies a computer in a domain. Before joining a computer to a domain, its hostname must be unique in a network. Once a computer joins a domain, it continues to use its computer name to communicate with other computers and servers in a network. Computer accounts are managed through the **Active Directory Users and Computers** console, as shown in the following screenshot:

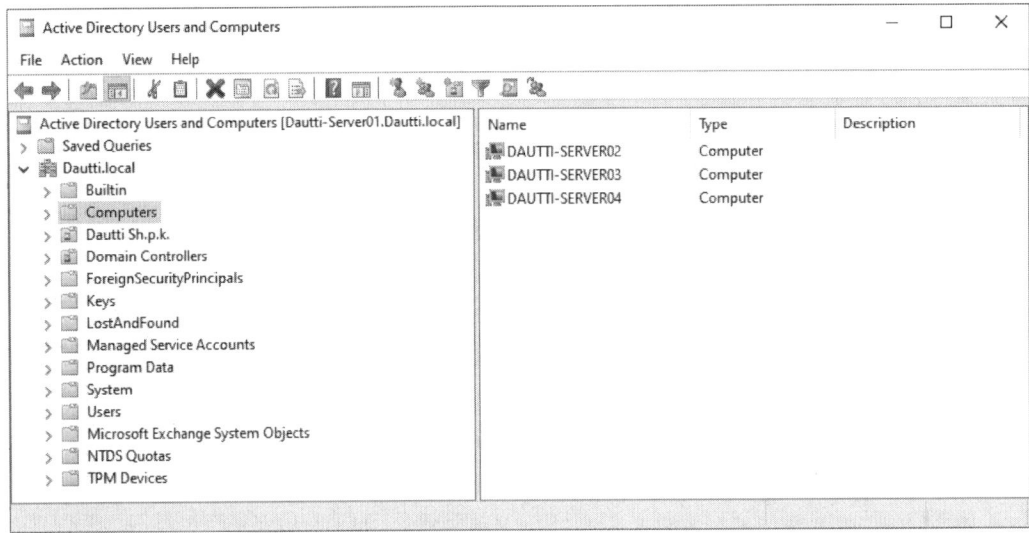

Figure 5.22 – Computer accounts in Windows Server 2022

So far in this section, we have learned about various user and computer accounts. Next, let's learn about groups.

Group types

In AD, a group represents a collection of AD objects. Groups are used to provide more structured administration instead of permissions and rights being assigned to each AD object individually. Note that a group is an object, which can also be moved to an OU.

As shown in the following screenshot, groups are managed through the **Active Directory Users and Computers** console. In AD, there are two types of groups:

- **Security groups**: Explicitly used to assign permissions to shared resources on a network
- **Distribution groups**: Mainly used to distribute email lists in an organization's network

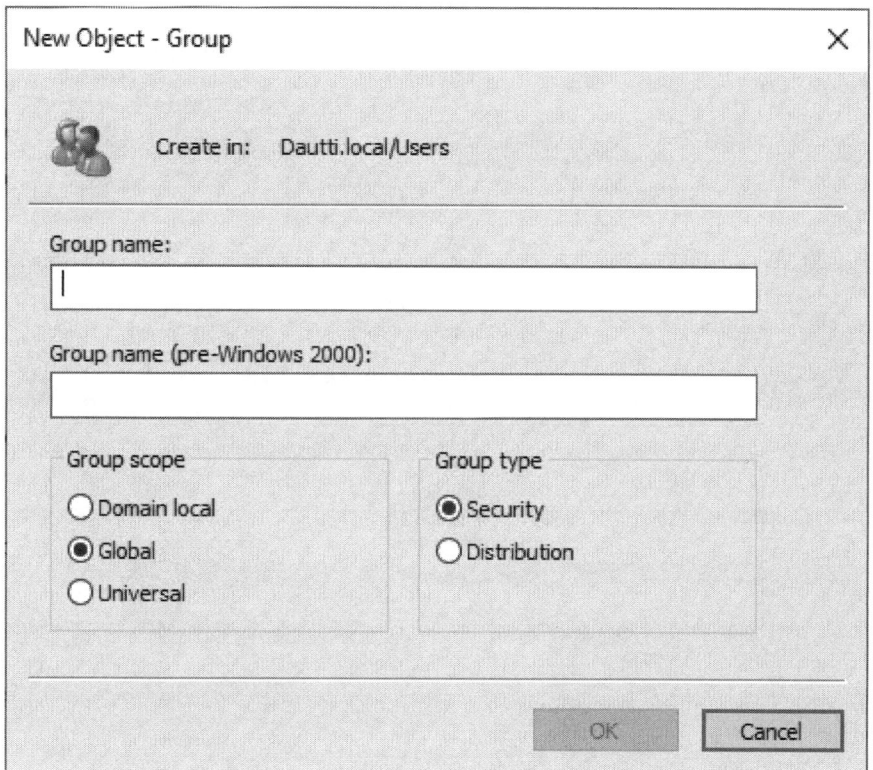

Figure 5.23 – Group types in Windows Server 2022

Next, let's learn about default groups and how to create them. They help control access to resources and delegate roles.

Default groups

As discussed in the previous section, groups are used to facilitate the administration of AD objects. Once the server gets promoted to a DC, a significant number of default groups are created, as shown in the following screenshot:

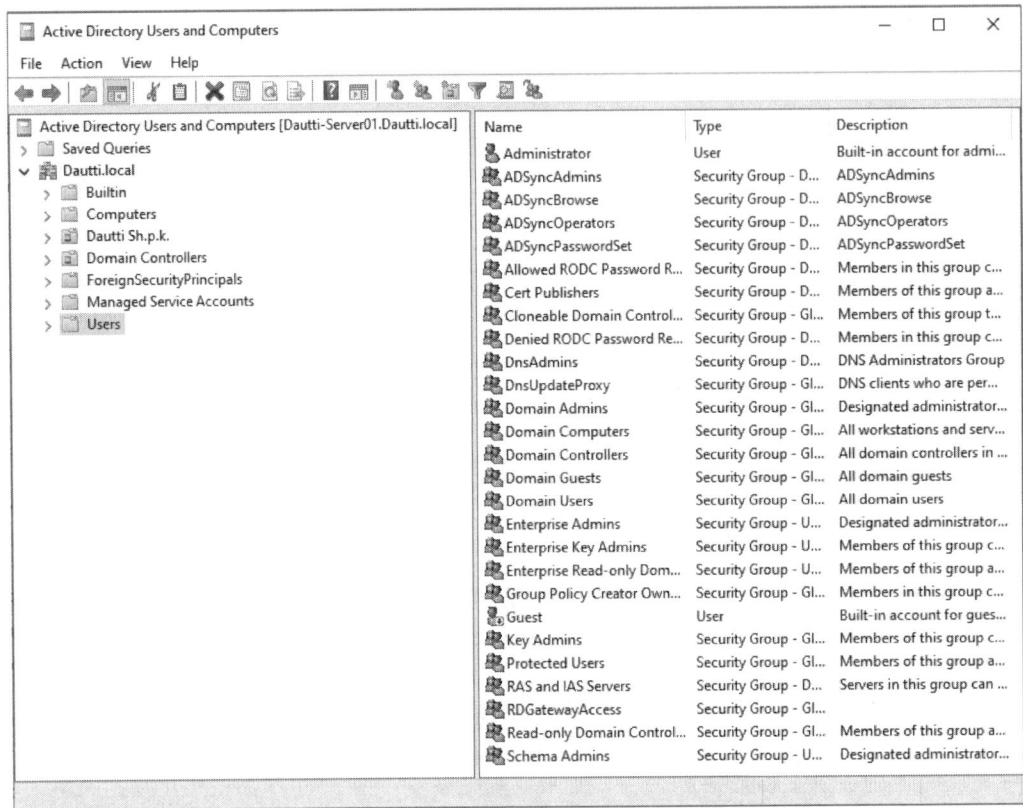

Figure 5.24 – Default groups in Windows Server 2022

Next, let's learn about group scopes and their types.

Group scopes

Whether it's a security or universal group, try to understand the group scope as an extension option in a forest, tree, or child domain. For example, in AD, there are three group scopes, as shown in *Figure 5.25*:

- **Domain local group**: This includes accounts, local domain groups, global groups, and universal groups from the parent's local group domain.
- **Global group**: This includes accounts and global groups from the parent's global group domain.
- **Universal group**: This includes accounts, global groups, and universal groups from any domain in the forest where a universal group belongs:

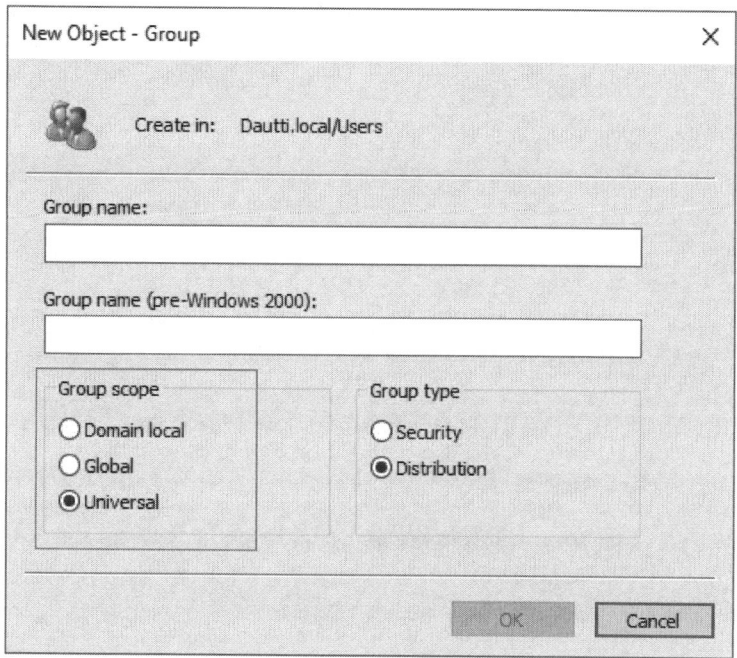

Figure 5.25 – Group scopes in Windows Server 2022

Moving forward, let's look at group nesting.

Group nesting

As we have learned so far, AD groups are objects that can be organized in a group nesting. Adding groups to other groups minimizes the number of individually assigned permissions to users or groups.

AGDLP and AGUDLP

Microsoft's recommendations for effectively using group nesting when assigning permissions are **Accounts, Global, Domain Local, Permissions (AGDLP)** and **Accounts, Global, Universal, Domain Local, Permissions (AGUDLP)**.

Follow these steps to assign permissions using AGDLP group nesting:

1. Add the accounts to the global group.
2. Add the global group scope to the domain local group.
3. Assign permissions to the domain local group.

Follow these steps to assign permissions using AGUDLP group nesting:

1. Add the accounts to the global group.
2. Add the global group to the universal group.
3. Add the universal group to the domain local group.
4. Assign permissions to the domain local group.

So far, you have learned about the AD infrastructure, DNS, and their components. You also learned what OUs and containers are. Then, you learned about computer accounts and groups and their types. Now, it is time to install the AD DS and DNS roles.

Chapter exercise – installing the AD DS and DNS roles and promoting the server to a DC

This chapter's exercise will teach you how to install the AD DS and DNS roles and promote the server to a DC. To get started, follow these steps:

1. Click the **Start** button. Then, from the **Start** menu, click the **Server Manager** tile.
2. In the **Server Manager** window, in the **WELCOME TO SERVER MANAGER** section, click **Add roles and features**, as shown in the following screenshot:

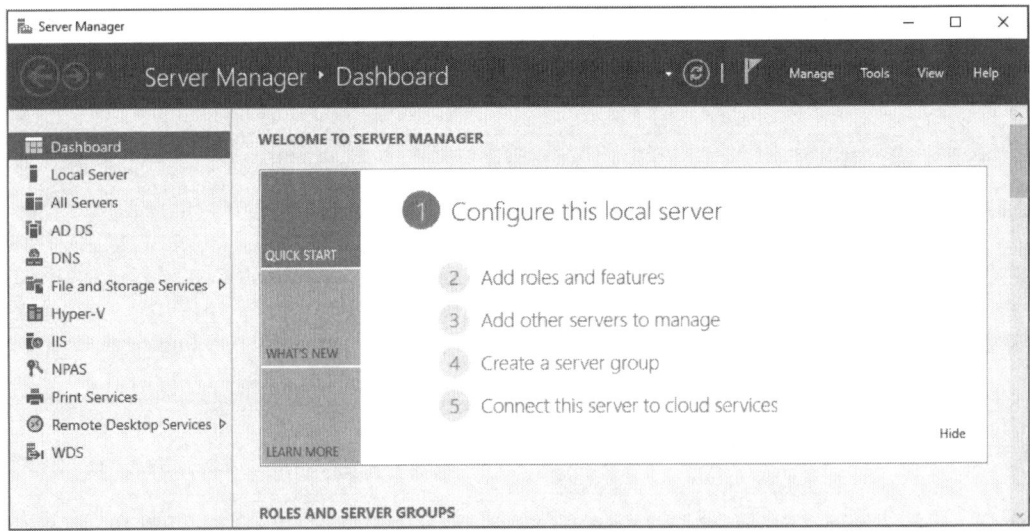

Figure 5.26 – Adding roles and features to the server using Server Manager

3. With the **Add Roles and Features Wizard** area open, click **Next**.
4. Select the **Role-based or feature-based installation** option and click **Next**.
5. Select a server from the **server pool** option and click **Next**.
6. Select the **Active Directory Domain Services** role, as shown in the following screenshot, and click **Next**:

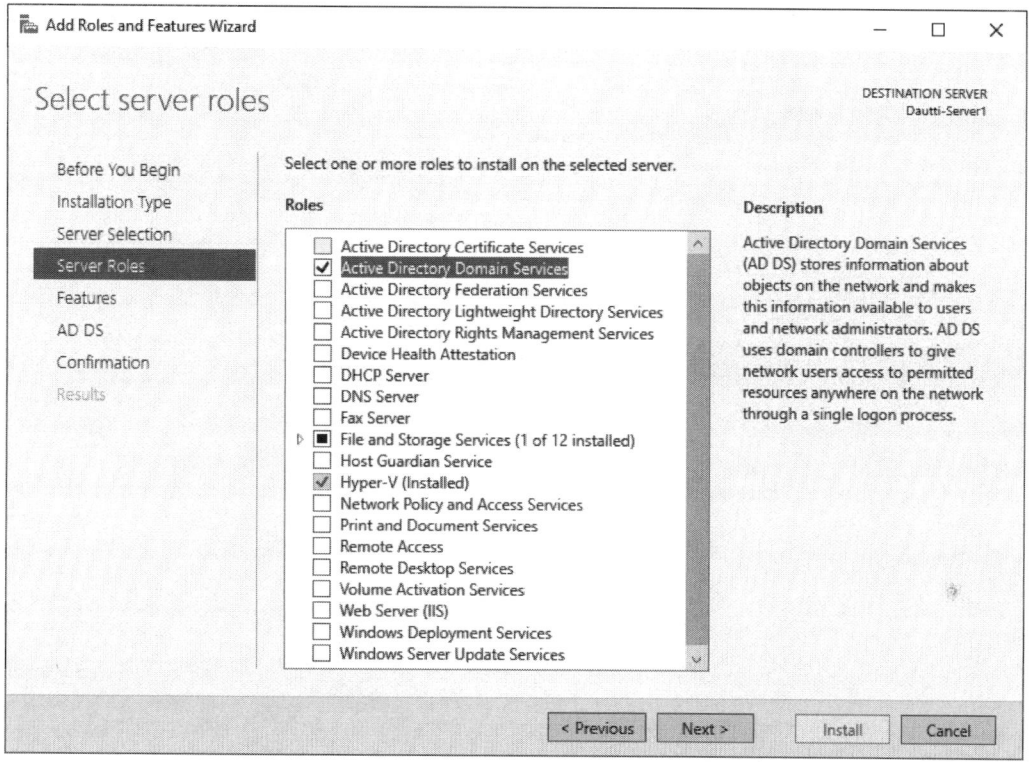

Figure 5.27 – The Add Roles and Features Wizard in Windows Server 2022

7. Click the **Add Features** button when adding features required for **Active Directory Domain Services**. Then, click **Next**.
8. Accept the default settings in the **Select features** step and click **Next**.
9. Take your time and read the AD DS description and the things to note regarding AD DS installation. Then, click **Next**.
10. Confirm your installation selections for the AD DS role and click the **Install** button.
11. Either hit **Close** or wait until the installation progress reaches its end.
12. Click **Close** to close the **Add Roles and Features Wizard** area.

13. Under **Notifications**, click **Promote this server to a domain controller**.
14. In the **Active Directory Domain Services Configuration Wizard** area, select the **Add a new forest** option, as shown in the following screenshot, and enter a **root domain name**. Then, click **Next**.

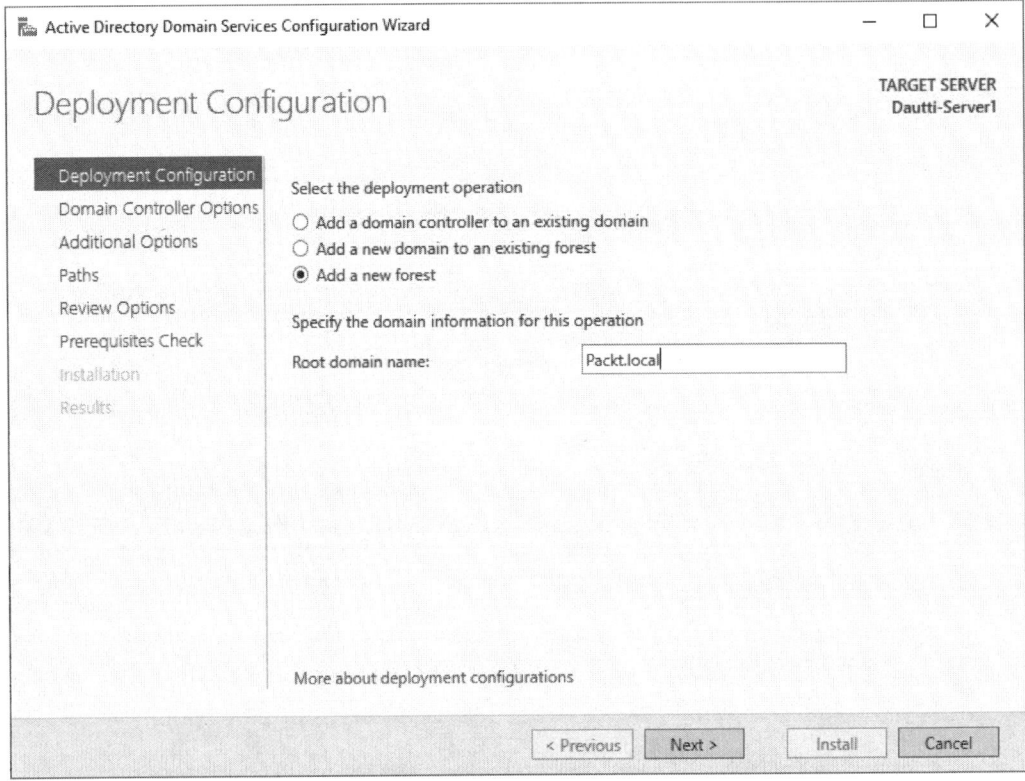

Figure 5.28 – Active Directory Domain Services Configuration Wizard

15. Accept the defaults for the forest and domain functional levels, and enter the **Directory Services Restore Mode (DSRM)** password. Then, click **Next**.
16. If you have an existing DNS on your network, manually create a delegation for that DNS server to enable reliable name resolution from outside of your domain. Otherwise, no action is required. Click **Next**.
17. Either accept the default NetBIOS entry or change it accordingly. Click **Next**.
18. Either accept the default paths or change them accordingly. Click **Next**.
19. Review your options and click **Next**.
20. Since the prerequisites have been met, click **Install**.

Now that you have taken all the necessary steps, the server will restart to complete promoting itself to a DC.

This exercise taught you how to install the AD DS and DNS roles and promote the server to a DC. Now, let's summarize this chapter

Summary

In conclusion, this chapter taught you about the various directory services and naming resolution concepts. First, you became familiar with the AD DS role and got acquainted with the DNS role. Now, you know how to set up a DC to authenticate users and computers and the naming resolution server in your organization's infrastructure. Furthermore, you became familiar with OUs and delegated control to OUs, which will help you easily organize and manage AD objects. Then, you were introduced to accounts and groups, which will help you place the right users in the respective groups of the organization.

Finally, this chapter provided a chapter exercise that explained how to install an AD DS role and promote the server to a DC.

In the next chapter, you will learn how to add roles to Windows Server 2022.

Questions

Answer the following questions to test your knowledge of this chapter:

1. An AD is a distributed database that stores objects in a hierarchical, structured, and secure format. (True | False)
2. _____ minimizes the number of individually assigned permissions to users or groups.
3. Which of the following user profiles is used mainly in Windows-based domain networks?

    A. Domain user profile

    B. Security user profile

    C. Roaming user profile

    D. Mandatory user profile

4. The WINS server maps the IP addresses to BIOS names. (True | False)
5. _____ is a set of communication paths through which the DC's replication data travels.

6. Which of the following are AD group scopes? (Choose two)

    A. OU
    B. Security group
    C. Global group
    D. Universal group

7. UNC is a standard for identifying a share in a computer network. (True | False)

8. _____ is a server that's responsible for securely authenticating requests to access your organization's domain resources.

9. Which **Microsoft Management Console (MMC)** snap-ins are used to manage AD? (Choose two)

    A. Active Directory Administrative Center
    B. Active Directory Users and Computers
    C. UNC
    D. OU

10. The best example of a domain is a client/server network where a dedicated server on the network is used to provide services. (True | False)

11. _____ stores the primary copy of the DNS database and maintains all DNS zone records.

12. Which of the following are forest-wide operations master roles? (Choose two)

    A. Master schema
    B. Domain naming master
    C. LAN manager hosts
    D. Default containers

13. Discuss the AD DS and DNS roles and their implementations.

14. Discuss Microsoft's recommendations, AGDLP and AGUDLP, for assigning permissions.

Further reading

To learn more about the topics that were covered in this chapter, take a look at the following resources:

- *AD DS Deployment*: https://docs.microsoft.com/en-us/windows-server/identity/ad-ds/deploy/ad-ds-deployment
- *What's New in DNS Server in Windows Server*: https://docs.microsoft.com/en-us/windows-server/networking/dns/what-s-new-in-dns-server

- *Creating an Organizational Unit Design*: `https://docs.microsoft.com/en-us/windows-server/identity/ad-ds/plan/creating-an-organizational-unit-design`
- *Managing Groups*: `https://docs.microsoft.com/en-us/windows/win32/ad/managing-groups`

6
Adding Roles to Windows Server 2022

Now that you have had a chance to get to know the two most crucial roles of Windows Server 2022, AD DS and DNS, and learned how to install them, it is time to get acquainted with the other roles and features of Windows Server 2022. Adding roles to Windows Server will help you define its function in an organization's network.

This chapter will explain the role and the importance of roles in determining how Windows Server functions when providing network services. First, you will get to know most of the roles and features that Windows Server 2022 supports. You will also learn how to add roles and set them up correctly when required. Then, you will understand application servers, web services, remote access, and file and print services. In addition to learning about roles, features will also be explained, along with the steps to add them to a server.

Finally, this chapter includes an exercise where you will install the **Internet Information Services** (**IIS**) and **Print and Document Services** (**PDS**) roles.

In a nutshell, the following topics will be covered in this chapter:

- Understanding server roles and features
- Understanding application servers
- Understanding web services
- Understanding Remote Access
- Understanding file and print services
- Understanding user rights, NTFS permissions, and share permissions
- Chapter exercise – installing the Web Server (IIS) and PDS roles

Technical requirements

To complete the exercises in this chapter, you will need the following equipment:

- A PC with Windows 11 Pro with at least 16 GB of RAM, 1 TB of HDD, and access to the internet
- Virtual machine 1 (file server) with Windows Server 2022 Standard (Desktop Experience) with at least 4 GB of RAM, 100 GB of HDD, and access to the internet
- Virtual machine 2 (web server) with Windows Server 2022 Standard (Desktop Experience) with at least 4 GB of RAM, 100 GB of HDD, and access to the internet
- Virtual machine 3 (print server) with Windows Server 2022 Standard (Desktop Experience) with at least 4 GB of RAM, 100 GB of HDD, and access to the internet

Understanding server roles and features

First, you should determine the function of a server within an organization's IT infrastructure before adding a proper role to it. The following sections will familiarize you with Windows Server 2022 roles, role services, and features.

Server roles

When adding a role in Windows Server 2022, as shown in *Figure 6.3*, usually, you need to determine the function of the server by the role (that is, network service) it is running in your organization's infrastructure. That way, the server role is the server's primary function. In the best-case scenario, the server should only have one role. However, the server can have multiple roles too. So, always try to understand the exact function the server needs to perform when choosing the required hardware.

Role services

Other than adding roles to the server and determining the server's function, there are situations when you need to add role services. This raises questions such as, *"What are role services?"*

First, let's try to understand role services with an example. Let's assume that you want to have an internet print server so that employees can print from outside the company's network (that is, via an extranet). First, you add the PDS role to the server, and then you add **Internet Printing** as a role service, as shown in *Figure 6.21*. That way, you augment the functionality of the role.

Server features

As we've discussed, other than roles and role services, features (see *Figure 6.2*) can be added to the server to support a given function. There are times when the following must be done:

- Installing .NET Framework 3.5 is required to help the role that is being added.

- Installing an **IP Address Management (IPAM)** feature is required to support the DHCP or DNS roles in the organization's network infrastructure.
- Installing the WINS server, alongside DNS, is required to solve problems that arise from NetBIOS name resolution in routed environments.

These situations are just some instances where feature installation is required.

Server Manager

Server Manager is commonly used to add roles in Windows Server 2022. Introduced with Windows Server 2008, Server Manager is a Windows administrative tool that administrators use to add, set up, and manage server roles. Its user interface is simple and easy to navigate. Usually, the **Scope** pane lists the roles that have been installed, whereas the **Details** pane displays the details of an established role. Therefore, it can be concluded that Server Manager (see *Figure 6.1*) is the one-stop administrative console for installing roles, configuring services, managing resources, and managing tasks in Windows Server 2022, both locally and on remote servers:

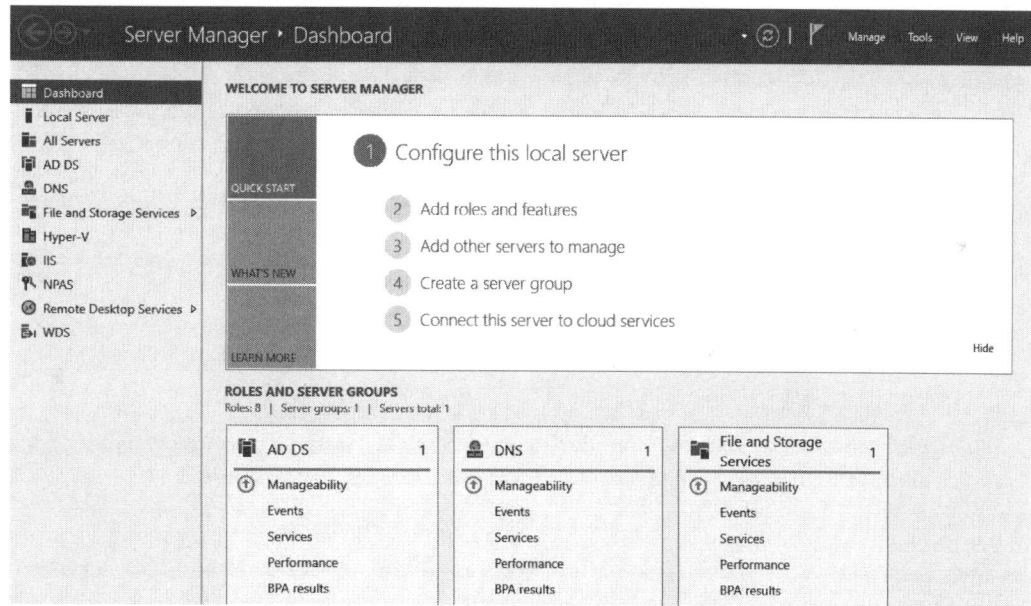

Figure 6.1 – Server Manager's user interface in Windows Server 2022

So far, you have learned what server roles and features are. The next section will walk you through the concept of application servers and their types.

Understanding application servers

When searching for the meaning of the word *application* in the Merriam-Webster dictionary, one of the definitions presented is also *an act of putting something to use*. From that, it can be understood that an application server is a server that provides usable services on a particular network. You will get acquainted with some well-known application servers in the following sections.

Let's begin by understanding what mail servers are and how to set them up.

Email server

In its simplest terms, an **email server**, often called a mail server, is a server that sends and receives emails. For a server to function as an email server, it must have email server software installed. Therefore, the Exchange Server client/server application will turn a server into an email server in Windows-based servers. Exchange Server then enables the system administrator to manage and create email accounts on the server. To send and receive emails, Exchange Server utilizes network protocols. The main features and communication protocols that are used by an email server are as follows:

- The **Mail Transport Agent** (**MTA**) is responsible for transporting the mail between mail servers.
- The **Mail Delivery Agent** (**MDA**) is responsible for delivering the mail from the server to a user's inbox.
- The **Mail User Agent** (**MUA**) is responsible for providing a platform for composing and reading emails.
- The **Simple Mail Transfer Protocol** (**SMTP**) uses port 25 and powers the MTA in transferring the mail between servers.
- The **Post Office Protocol** (**POP**) uses port 110 and is responsible for downloading emails from the server to the user's local computer.
- The **Internet Message Access Protocol** (**IMAP**) uses port 143 and is responsible for retrieving emails from the mail server and sending them to a user's mail application.

While Exchange Server is considered an advanced email server, in simple terms, you can set up an email service that only sends and forwards the emails in Windows Server 2022 by adding the SMTP Server feature. To add the SMTP Server feature to Windows Server 2022, as shown in *Figure 6.2*, follow these steps:

1. Click on **Add roles and features** in Server Manager's **WELCOME TO SERVER MANAGER** section.
2. On the **Before you begin** page, click **Next**.
3. Click **Next** on the **Installation Type** page.

4. On the **Server Selection** page, click **Next**.
5. There is no role to add, so click **Next** on the **Server Roles** page.
6. Select **SMTP Server** from the **Features** list:

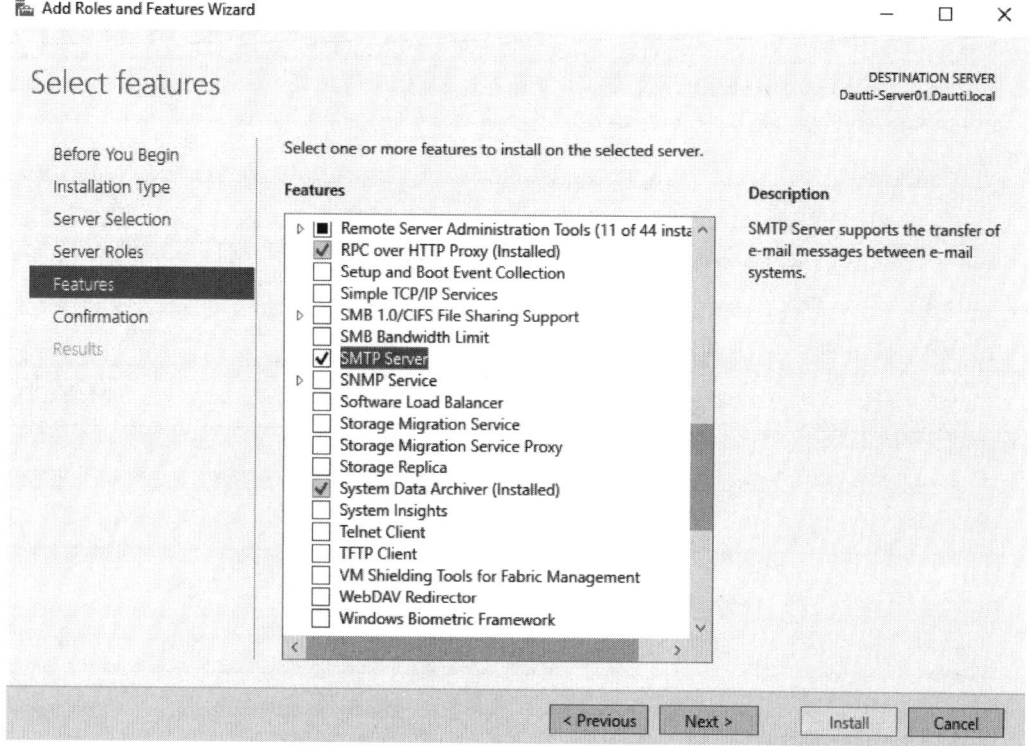

Figure 6.2 – Adding the SMTP Server feature to Windows Server 2022

7. There is no role service to add, so click **Next**.
8. On the **Confirmation** page, click **Install**.

> **Important note**
> Exchange Server 2022 is Microsoft's application for setting up a mail server in an organization's network. To do that, you should install and configure Exchange Server 2022.

Next, let's learn about the database server and its access protocols.

Database server

Simply put, a database server is considered a high-power computer that provides authorized users with database services and data access. Moreover, a database server stores data in a central location that can be backed up, enabling applications and users to access data across the network. Therefore, the SQL Server client/server application will turn a server into a database server in Windows-based servers. **Microsoft** (**MS**) SQL Server enables the system administrator to manage and create tables on the server. Furthermore, to provide data access, MS SQL Server requires protocols. The main features and communication protocols that are utilized by a database server are as follows:

- **Data** is the raw material of the database and, as such, is the main component of a database server. Without data, there is no database.
- A **database application** is an application through which the user interacts with the database server.
- **Users** are the people who use the database.
- **Open Database Connectivity** (**ODBC**) is a protocol that enables applications to access data in a database server.
- **Java Database Connectivity** (**JDBC**) is Sun Microsystem's protocol that enables Java applications to access data in a database server.
- **Object Linking and Embedding Database** (**OLEDB**) is Microsoft's protocol that enables applications to access data in a database server.

> **Important note**
> SQL Server 2022 is Microsoft's application for setting up a database server in an organization's network. To do that, you need to install and configure SQL Server 2022.

Next, let's learn what collaboration servers are.

Collaboration server

As its name suggests, a collaboration server provides centralized tools and resources to facilitate communication and interaction in the virtual workspace. This means that users from the same organization can exchange collaborative documents, instant messages, personal and group calendars, video meetings, and other services. Therefore, the SharePoint Server client/server application will turn a server into a collaboration server for Windows-based servers. SharePoint Server enables the system administrator to manage and create sites, libraries, documents, and more on the server. Furthermore, SharePoint Server utilizes networking protocols to provide access to sites, libraries, documents, and other resources.

> **Important note**
>
> SharePoint Server 2022 is Microsoft's application for setting up a collaboration server in an organization's network. To do that, you need to install and configure SharePoint Server 2022.

Next, let's learn what monitoring platforms are.

Monitoring server

Simply put, a monitoring platform can involve anything from monitoring the health of servers to monitoring their performance. However, in a broader context, a monitoring platform can manage the entire IT environment and collect data on issues across networks by monitoring servers on-premise, in the cloud, or by monitoring a hybrid environment from a central console. These platforms are designed to track the status of critical client/server applications, network services, IT infrastructures, websites, and other services. Moreover, monitoring platforms employ configurable alerts to quickly detect problems and notify system administrators of critical issues that must be resolved. Therefore, the **System Center Operations Manager** (**SCOM**) client/server application will turn a server into a monitoring server for Windows-based servers.

SCOM Server enables the system administrator to manage and monitor devices and services in an enterprise environment. From this, it is evident that server monitoring is an ongoing and complex process that, in addition to monitoring key server components such as CPU, memory, disk usage, and network interface, monitors network applications and services on the servers.

> **Important note**
>
> System Center 2022 is Microsoft's application for setting up a monitoring server in an organization's network. To do that, you must install and configure System Center 2022, including SCOM.

Next, let's learn what threat management servers are.

Data protection server

The data protection server establishes a platform that supports **business continuity and disaster recovery** (**BCDR**) strategies by facilitating an organization's data backup and recovery. Therefore, the System Center **Data Protection Manager** (**DPM**) client/server application will turn a server into a data protection server for Windows-based servers. In addition, DPM enables the system administrator to run the following:

- Application-aware backups for Exchange Server, SQL Server, and SharePoint Server
- Files, folders, and volumes backups

- System state backups
- **Virtual machine** (**VM**) backups on Hyper-V for both Windows and Linux

> **Important note**
> System Center 2022 is Microsoft's application for setting up a data protection server in a corporate network. To do that, you must install and configure System Center 2022, including DPM.

This section has taught you about some well-known application servers that are used today in on-premises environments. Aside from helping you get to know some of the most used client/server applications nowadays, it has helped you learn about their features, components, protocols, and capabilities. The following section will teach you about the various available web services.

Understanding web services

A web service is usually a network service but in web technology. Moreover, a web service represents communication between the browser and the web server based on the request/response paradigm. Commonly, it takes place over the internet using the **Hypertext Transfer Protocol** (**HTTP**) communication protocol. Therefore, to better understand web services, you must familiarize yourself with IIS, WWW, and FTP.

In the next section, you'll learn what **IIS** is.

IIS

IIS is Microsoft's web server, which provides reliable, manageable, and scalable web applications. IIS supports communication protocols such as HTTP, HTTPS, FTP, FTPS, SMTP, and NNTP for communication between the browser and the web server. In addition, for dynamic content on the server side, Microsoft has developed a scripting technology called **Active Server Pages** (**ASP**).

In IIS version 10, Microsoft has significantly increased security by providing support for scripts that take a long time to execute, including HTTP/2. Additionally, they added Microsoft Edge based on Chromium, a new browser released in January 2020. Other key features introduced with IIS 10 in Windows Server 2022 include upgraded HTTP/3 server-side cipher suite negotiation, IIS administration PowerShell cmdlets, wildcard host headers, and more. All these improvements have increased IIS's overall performance and security.

To set up a web server in Windows Server 2022, you need to add IIS as a role to the server, as shown in the following screenshot:

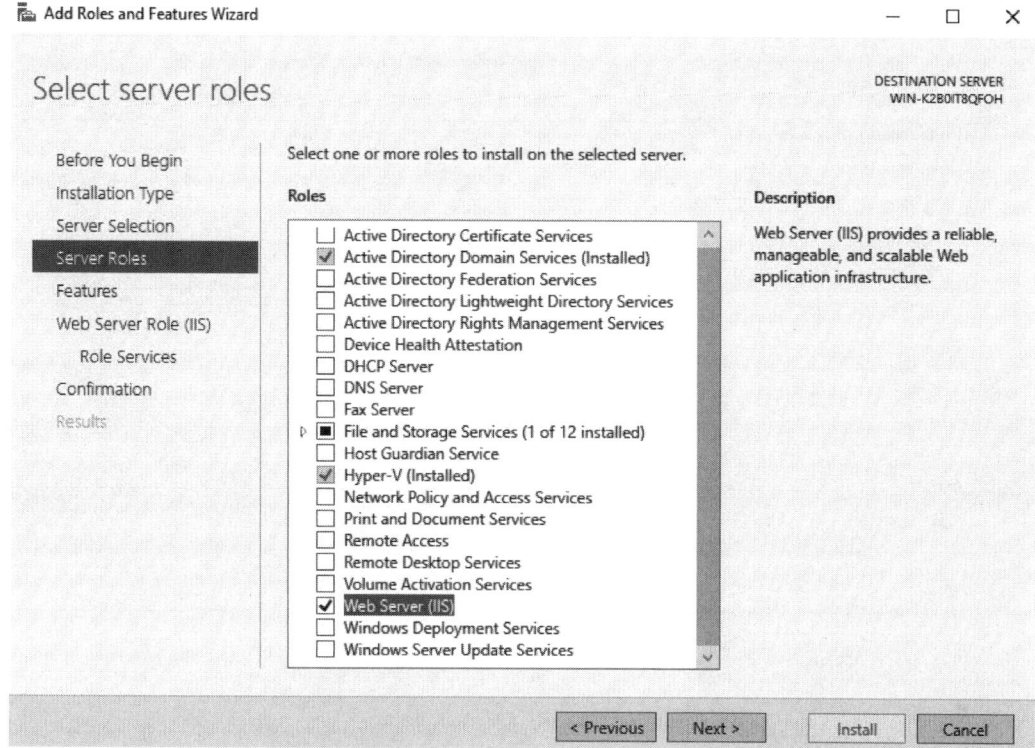

Figure 6.3 – Adding the Web Server (IIS) role in Windows Server 2022

IIS Manager is an administrator console that's used to manage the web server in Windows Server 2022, as shown in the following screenshot. IIS Manager can be accessed from Server Manager, Windows Administrative Tools, and the **Run** dialog box by running the `inetmgr` command.

Figure 6.4 – IIS Manager in Windows Server 2022

Next, let's learn about the **World Wide Web** (**WWW**), which is among the most used services on the internet.

WWW

Often, people confuse the internet with the WWW. Perhaps using the WWW while connected to the internet means they feel that the WWW is the internet! Who knows? However, like many other internet services, the WWW is also an internet service that's accessed through the HTTP protocol. It consists of electronic documents compiled with **Hypertext Markup Language** (**HTML**), as shown in the following screenshot:

Understanding web services 171

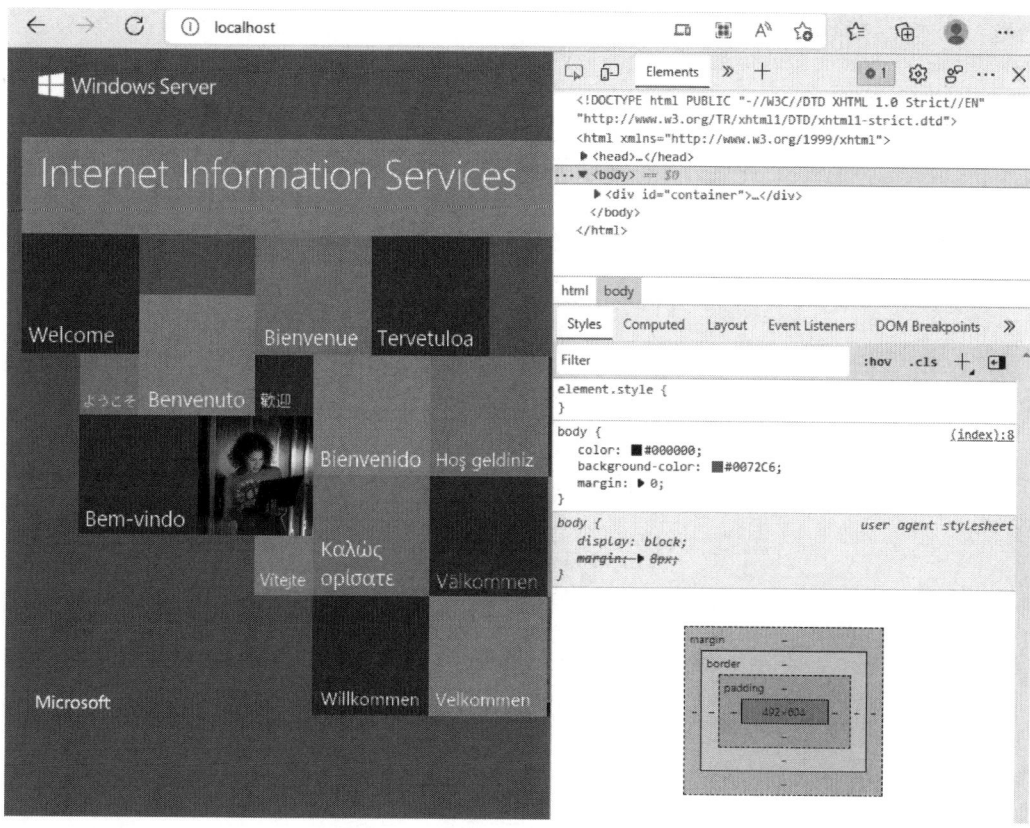

Figure 6.5 – A website and its source code

Next, let's learn about the FTP communication protocol, which is used to share files over the internet, and how to set it up.

FTP

File Transfer Protocol (**FTP**), as the name suggests, transfers files between computers over the internet. Moreover, it sends and receives corporate data in a corporate network. However, on websites and web servers, it is used to upload and download files. It is built on a client/server network architecture, so it uses port 21 to establish a session and port 20 to transfer data.

To set up an FTP server in Windows Server 2022, first, you need to add **Web Server (IIS)** as a role and then **FTP Server** as a role service, as shown in the following screenshot:

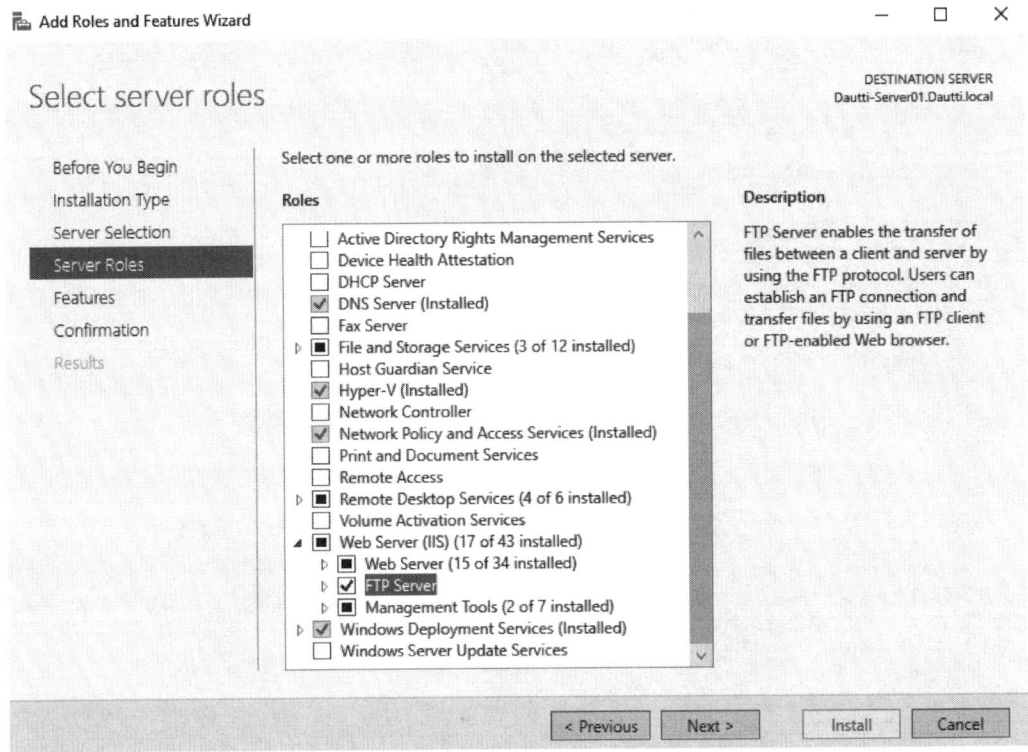

Figure 6.6 – Adding FTP Server as a server role in Windows Server 2022

Next, we will learn about worker processes, which help requests reach out to the web server application pool, and how to access them.

Separate worker processes

From an IIS perspective, a web directory represents a website with an application pool. As we are talking about a collection of applications, it is evident that there is more than one application. Furthermore, the same worker process supports each application in the application pool. This means that a worker process that serves an application pool is separated by another worker process that helps another application pool too. Hence, if a specific web application does not work, it does not affect the applications running in other application pools.

To access an application's pool worker processes, select **Application Pools**; then, from the **Actions** pane on the right-hand side of the **IIS Manager** administrative console, select **Advanced Settings...**, as shown here:

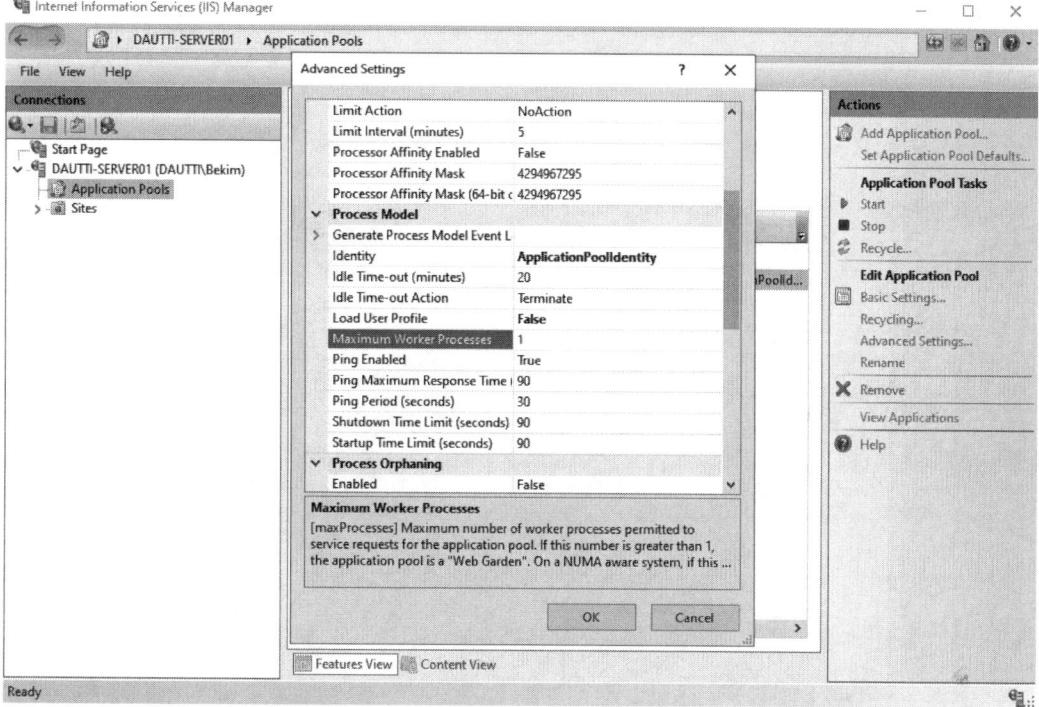

Figure 6.7 – An application's pool worker process in IIS

Now, let's learn how to add additional components to Windows Server's IIS.

Adding components to the IIS

Upon adding the **Web Server (IIS)** role, you will encounter the **Role Services** step (see *Figure 6.6*) in the **What is FTP?** section, where you will add the required components for IIS. You can add the necessary features even after you have added **Web Server (IIS)** to the server. First, however, you must use the Add Roles and Features Wizard from Server Manager to add additional components to Windows Server's IIS.

Now, let's learn about a site or website, a group of HTML documents on a web server.

Sites

A **site**, often referred to as a **website**, is a collection of web pages grouped to represent the content on the intranet or internet via web services. Commonly, HTML is issued to compile web pages and design a website. However, various scripting languages add dynamic content to a specific website. An example of a single web page website can be considered by adding **Web Server (IIS)** as a role in Windows Server 2022. Then, the default website is automatically created, as shown in the following screenshot:

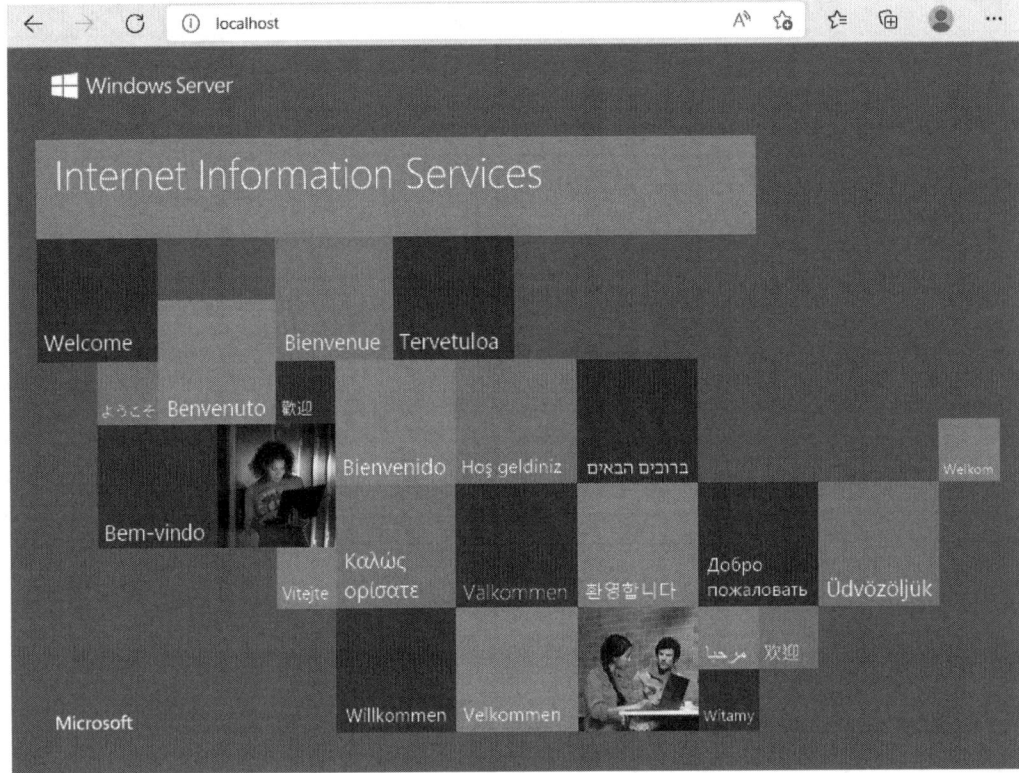

Figure 6.8 – Localhost in Windows Server 2022 powered by IIS

However, if you right-click **Sites** and select **Add Website...**, as shown in the following screenshot, you can add additional websites to Windows Server's IIS:

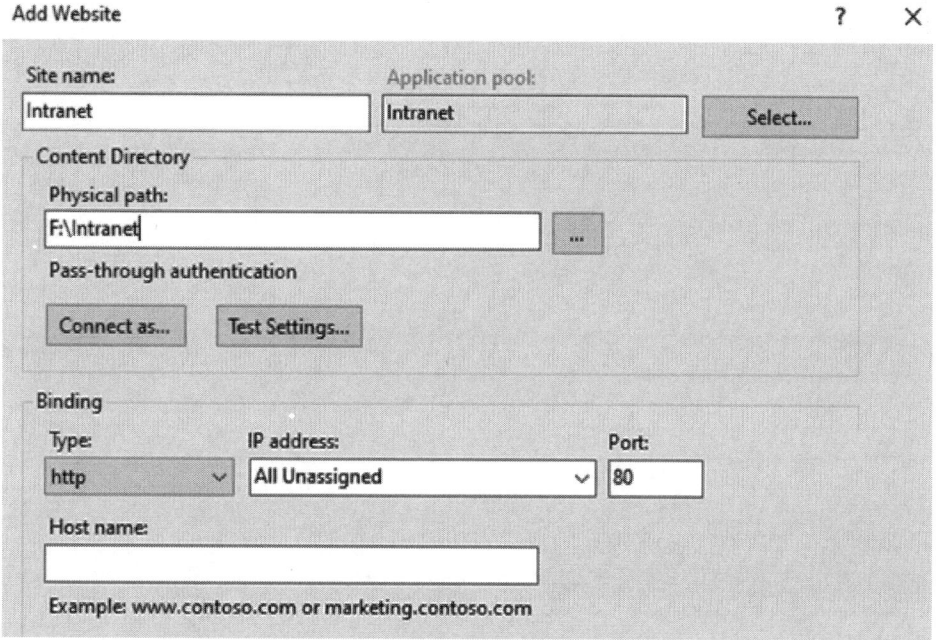

Figure 6.9 - Adding a website via IIS Manager in Windows Server 2022

Now, let's learn about the software ports that are used by client/server applications.

Ports

As you may know, there are hardware and software ports. A hardware port is any physical interface in a computer, peripheral device, or network device that allows interconnection for communication and management. By contrast, a *software port* (often known as an *application port*) is any logical endpoint where applications from the server communicate with other applications on LAN, WAN, and the internet. For example, a web server uses ports 80 and 443 for the HTTP and HTTPS protocols, respectively.

The following table lists the well-known application ports:

Protocol	Port	Transportation Protocol
FTP	21	TCP
SSH	22	TCP
Telnet	23	TCP
SMTP	25	TCP
HTTP	80	TCP and UDP

Protocol	Port	Transportation Protocol
POP3	110	TCP
NNTP	119	TCP
NTP	123	TCP
IMAP4	143	TCP
HTTPS	443	TCP

Table 6.1 – Well-known application ports

Now, let's learn about the **Secure Sockets Layer** (**SSL**), which can add security to communication between browsers and web servers.

SSL

SSL is a communication technology that encrypts the communication channel between a website on a web server and a browser on a server, as shown in *Figure 6.10*. The browser connects to a secure website with SSL over the HTTPS protocol on port 443. In such a secure infrastructure, certificates play an important role in encrypting all transmitted data. Certificates are used mutually by the website and the browser to negotiate a secure session between browser-to-server or server-to-server communications.

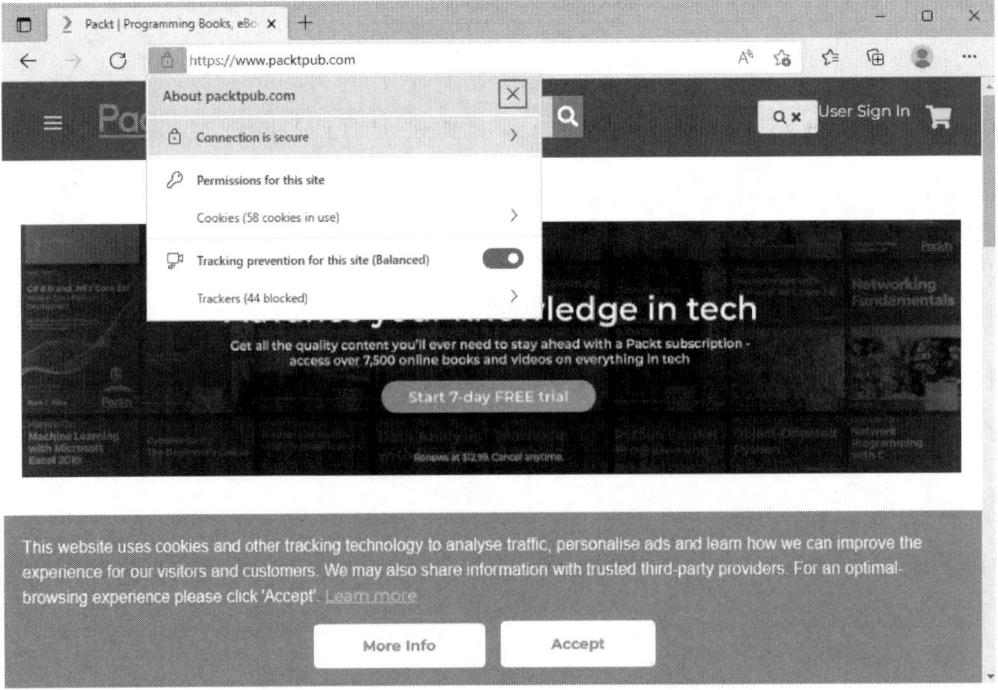

Figure 6.10 – Secured communication between a browser and a website

The preceding screenshot shows that Packt's website uses the HTTPS protocol instead of HTTP. As explained earlier, this website uses either **Secure Sockets Layer (SSL)** or **Transport Layer Security (TLS)** to secure communications between the browser on the server and Packt's website on a web server.

Certificates

As discussed in the previous section, a certificate is responsible for securing the communication channel between a website and a browser. The certificate, also known as a **digital certificate**, is an electronic document that ensures that entities can exchange data securely over the internet. Usually, certificates are issued by a secure entity known as a **Certificate Authority (CA)**, as shown in the following screenshot. In addition, the secure web infrastructure commonly utilizes a **public key infrastructure (PKI)**, which uses certificates to prove the ownership of the public key:

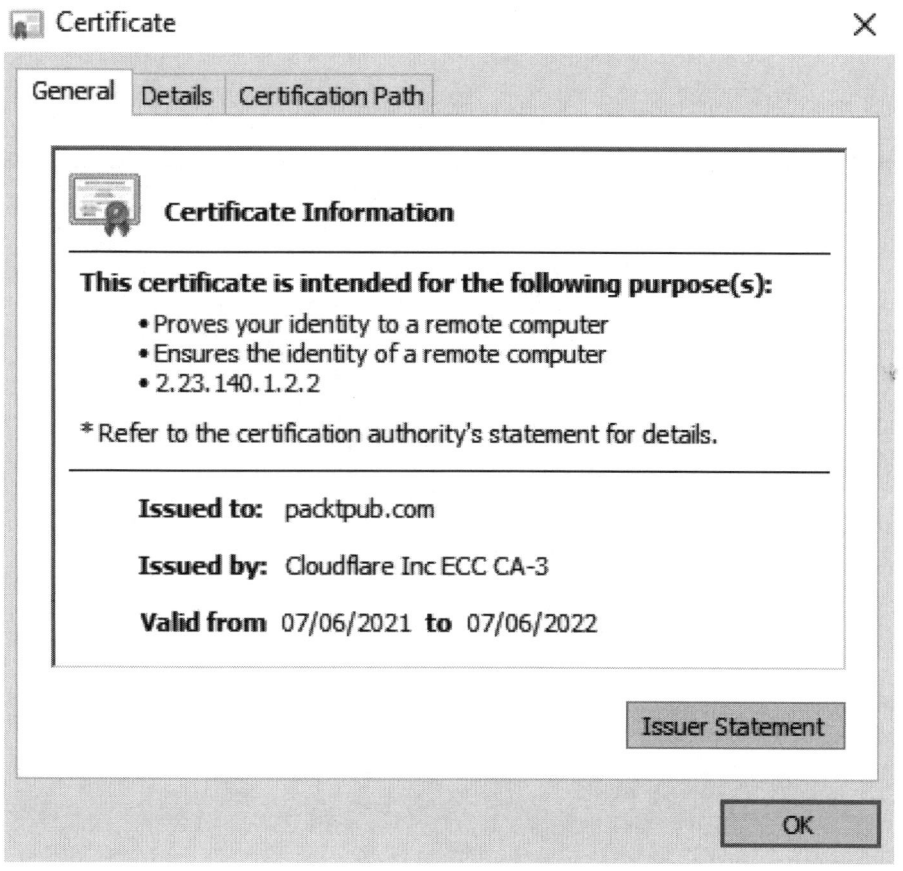

Figure 6.11 – Certificate issued by a CA

> **Important note**
> You can learn more about PKI at `https://docs.oracle.com/cd/B10501_01/network.920/a96582/pki.htm`.

This section has taught you about various web services and components. The next section will teach you about the Remote Access role in Windows Server 2022.

Understanding Remote Access

The **Remote Access** (**RA**) role in Windows Server 2022 enables remote access to resources within a corporate network. Moreover, remote access refers to the ability to monitor and control access to a computer or network anywhere at any time. It enables corporate users to work from a remote location by maintaining access to a corporate network. In Windows Server 2022, RA consists of a logical grouping of the following network access technologies:

- **DirectAccess**, introduced in Windows Server 2008 R2, uses IPsec to encrypt communication between the DirectAccess client and the DirectAccess server. In addition, it encapsulates IPv6 traffic over IPv4 to reach the intranet over the internet. As a result, access to a corporate intranet can be enabled without a **Virtual Private Network** (**VPN**).

- **Routing and Remote Access Service** (**RRAS**), the **Remote Access Service** (**RAS**) successor in Windows NT, was introduced in Windows 2000 and represented a combined service that establishes links between remote locations over VPN and dial-up traffic paths between the sub-networks.

- **Web Application Proxy** acts as a recursive proxy in Windows Server 2022. It uses **Active Directory Federation Services** (**AD FS**) to authenticate corporate users so that they can access web applications on the corporate intranet through an extranet.

To set up a Remote Access server in Windows Server 2022, you need to add **Remote Access** as a role to the server, as shown in the following screenshot:

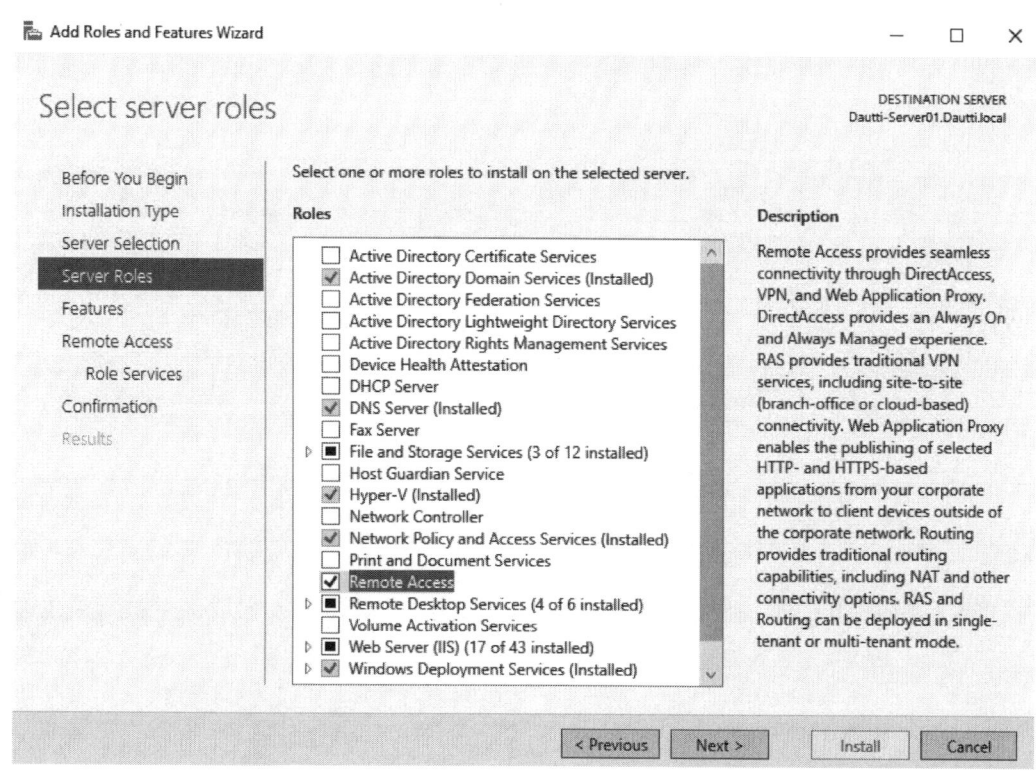

Figure 6.12 – Adding the Remote Access role in Windows Server 2022

Now, let's learn about the Remote Assistance feature and how to add it to the server.

Remote Assistance

Remote Assistance in Windows Server 2022 is a feature that enables a trusted helper to access the invitee's desktop remotely to assist in troubleshooting computer-related issues. In Help Desk terms, remote assistance refers to a computer activity where technical support is provided from a remote location over the internet. You must use the **Add Roles and Features Wizard** area to add the **Remote Assistance** feature to the server, as shown in the following screenshot:

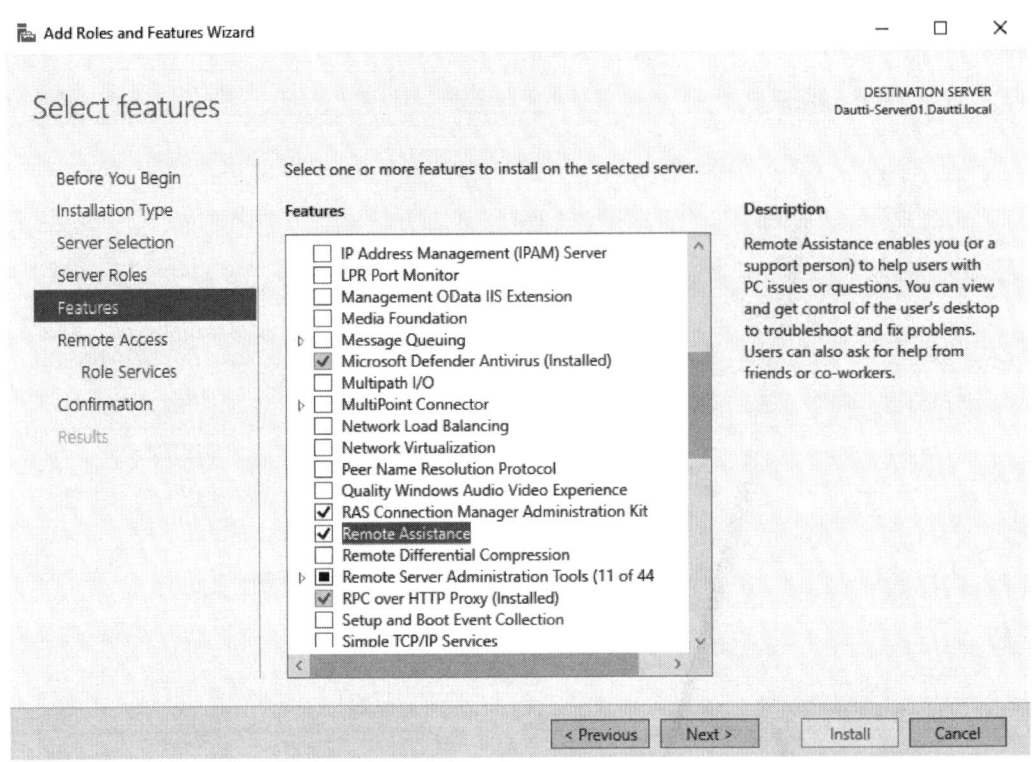

Figure 6.13 – Adding the Remote Assistance feature in Windows Server 2022

Now, let's look at the Remote Server Administration Tools feature, which allows you to administer servers remotely.

RSAT

Remote Server Administration Tools (**RSAT**) in Windows Server 2022 enables system administrators to manage the server roles and features of remote servers running Windows Server 2022 (both in GUI and CLI modes). RSAT is also available for client computers running Windows 10 and 11.

To enable **Remote Server Administration Tools** in Windows Server 2022, use the **Add Roles and Features Wizard** area, as shown in the following screenshot:

Understanding Remote Access

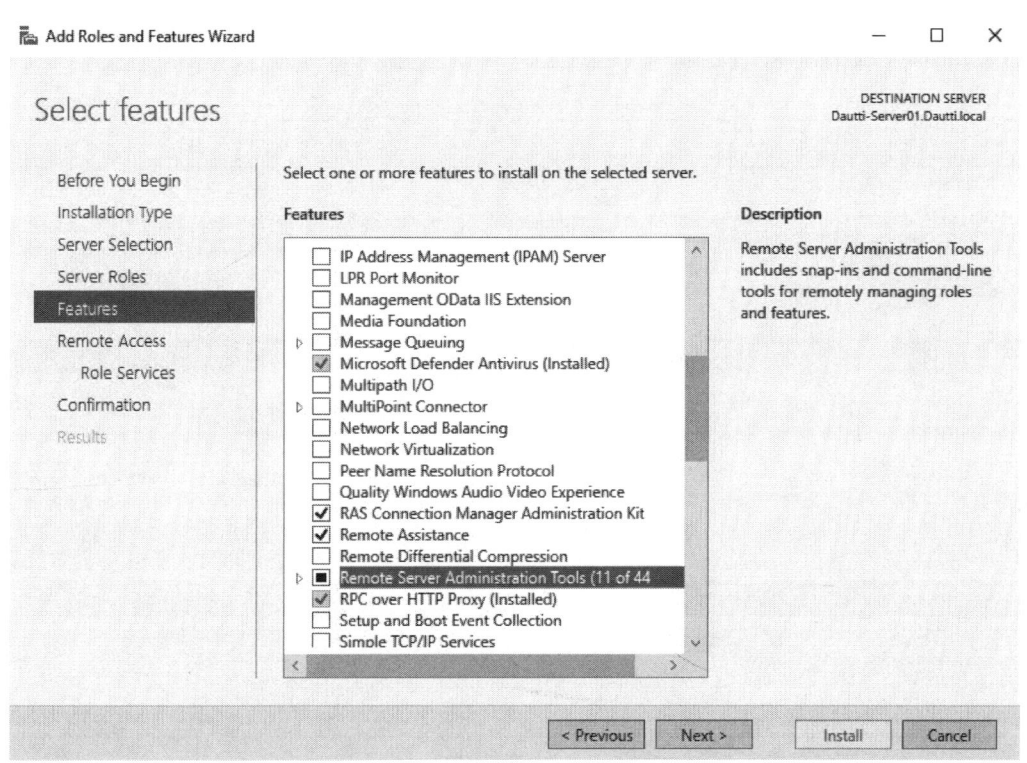

Figure 6.14 – Adding the RSAT feature in Windows Server 2022

Now, let's learn about **Remote Desktop Services** (**RDS**), a communication protocol that enables remote access to the server.

RDS

Known as **Terminal Services** (**TS**) until Windows Server 2008, RDS earned its name and identity with the release of Windows Server 2008 R2. This role allows you to set up a GUI with remote access to computers within an organization's network and over the internet. Additionally, RDS delivers individual, virtualized applications to users' desktops. To set up an RDS server in Windows Server 2022, you need to add the **Remote Desktop Services** role to the server, as shown in the following screenshot:

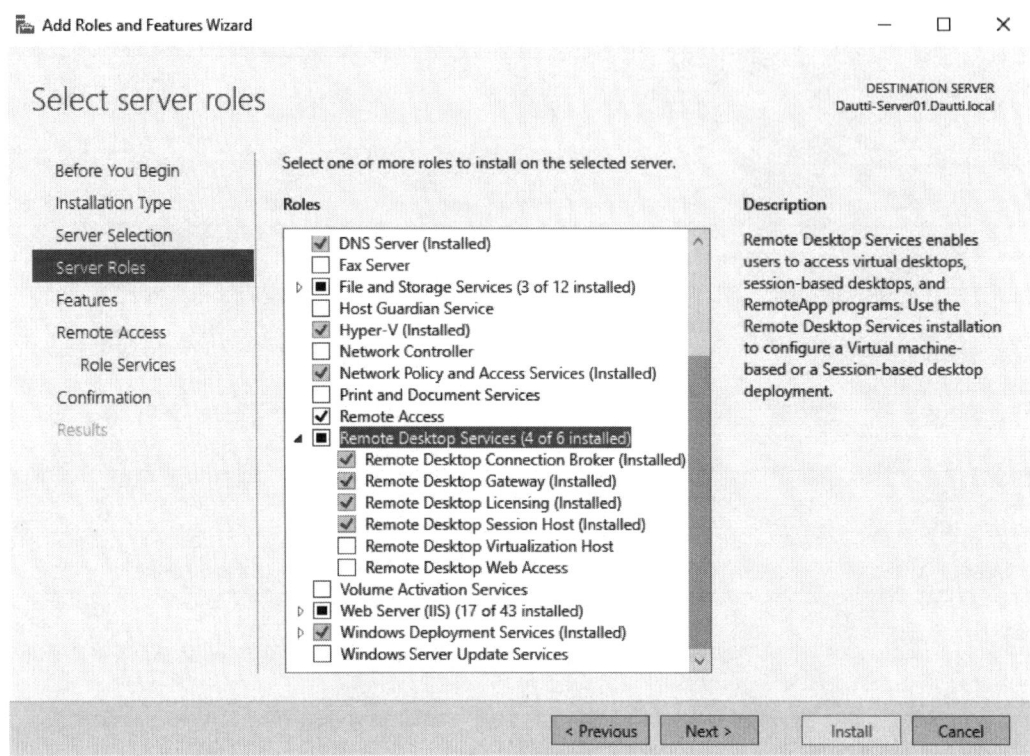

Figure 6.15 – Adding the RDS role in Windows Server 2022

Now, let's learn about the RDS Licensing server and how to set it up.

RDS Licensing

The **RDS Licensing** server manages RDS **Client Access Licenses (CALs)**. Users and computers use RDS CALs to access a **Remote Desktop Session Host (RDSH)** server. The RDS Licensing server provides two concurrent connections free of cost by default. If you need additional RDS CALs, then you need to purchase them.

To set up an RDS Licensing server in an organization's network with Windows Server 2022, first, you need to add the **Remote Desktop Services** role and then add the **Remote Desktop Licensing** role services, as shown in the following screenshot:

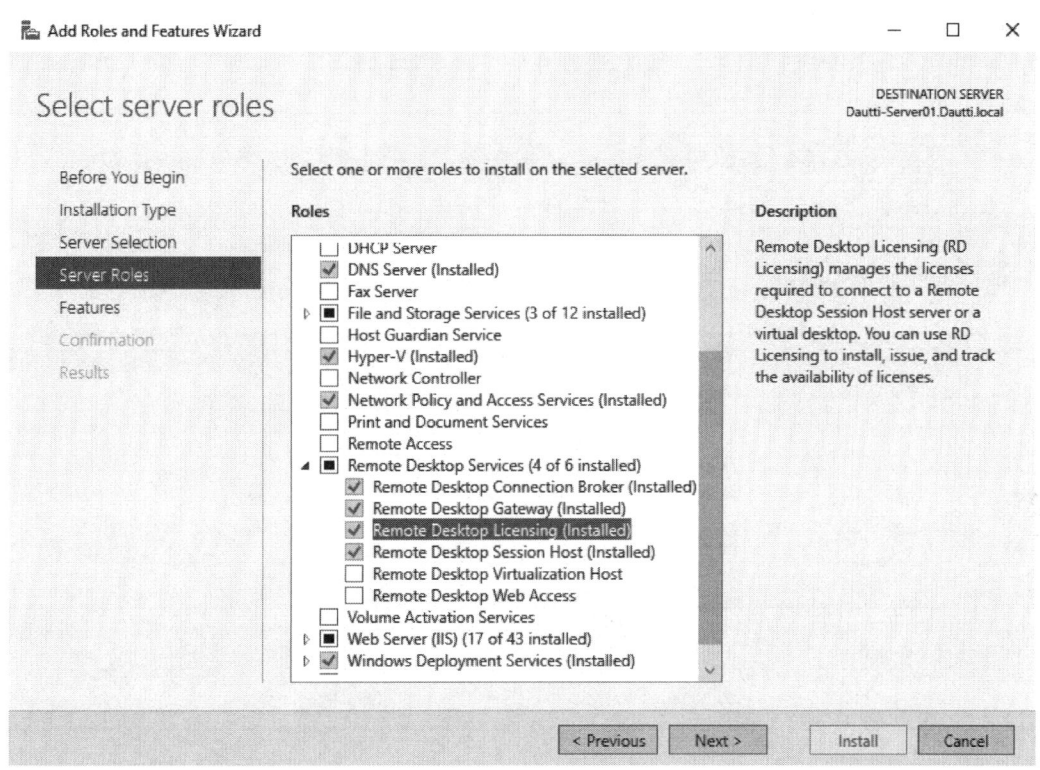

Figure 6.16 – Adding Remote Desktop Licensing role services in Windows Server 2022

Now, let's understand the **Remote Desktop Gateway** (**RDG**) server, which helps users access the company's intranet over the internet.

RDG

A **Remote Desktop Gateway** (**RDG**) server, which is part of the RDS role, is a role service in Windows Server 2022 that enables authorized users to connect to computers within an organization's network and over the internet using a **Remote Desktop Connection** (**RDC**) client. To set up an RDG server in your organization's network with Windows Server 2022, first, you need to add the **Remote Desktop Services** role and then add **Remote Desktop Gateway** role services, as shown in the following screenshot:

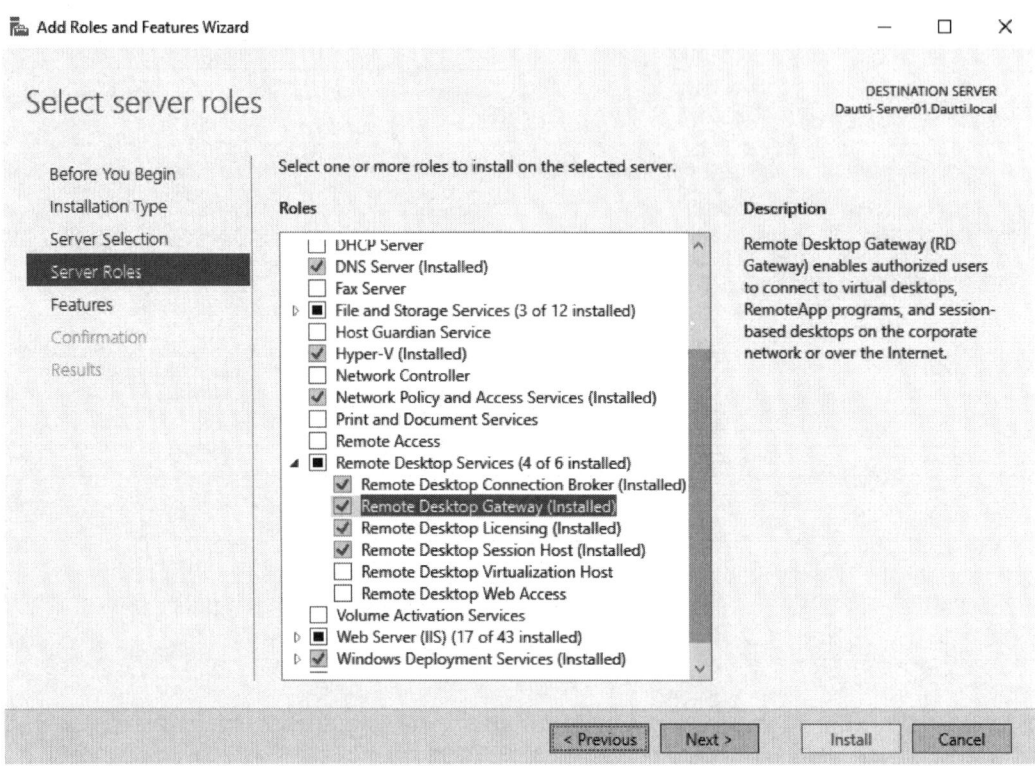

Figure 6.17 – Adding Remote Desktop Gateway role services in Windows Server 2022

Next, we'll turn our attention to **Virtual Private Networks** (**VPNs**), which allow us to set up secure communications. We'll also briefly look at how to deploy them.

VPN

As you may know, a **Virtual Private Network** (**VPN**) is a logical internet connection for securely transmitting data. As its name suggests, a VPN creates a virtual point-to-point link between two computers on the WAN and the internet. A VPN enables remote users to connect to a corporate network over the internet using tunneling protocols and data encryption algorithms. This kind of network is usually deployed in two ways:

- The *remote access VPN* connects remote users (telecommuters) with the server on their organization's private network.
- The *site-to-site VPN* enables organizations to connect two separate networks over the internet.

To set up a VPN server in Windows Server 2022, first, add the **Remote Access** role and then add the **DirectAccess and VPN (RAS)** role services, as shown in the following screenshot:

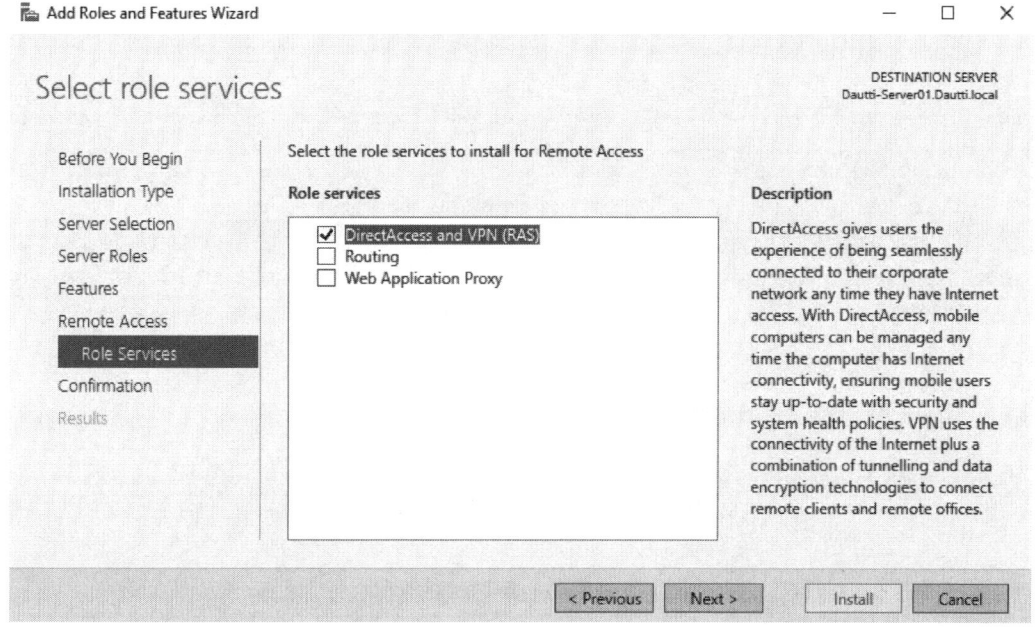

Figure 6.18 – Adding the DirectAccess and VPN (RAS) role services in Windows Server 2022

Now, let's learn about the **Application Virtualization** (**App-V**) service, which represents a virtualized application that runs within the simulated environment.

App-V

Microsoft App-V delivers virtualized applications to users. These virtualized applications are installed on a server and are provided to users in a service format. In addition, Microsoft App-V contains a centralized management system that administrators use to control how much access users have to each application. From a user's perspective, users interact with the virtualized applications the way it is supposed to run on a local machine. To set up an App-V server, download the **Microsoft Desktop Optimization Pack** (**MDOP**) from Microsoft's website.

> **Important note**
> You can learn more about MDOP at `https://technet.microsoft.com/en-us/windows/mdop.aspx?`.

Now, let's learn about the various ports that are used by client/server applications and communications protocols.

Multiple ports

As discussed earlier in the *Understanding RDS* section, port 3389 is used by RDS to send and receive data. However, that is only for accessing one computer at a time. So, what happens when you try to access more than one computer simultaneously through RDS?

While the first computer uses port 3389, sequential port numbers are assigned to other computers on the LAN, starting with 3390. Similarly, an IP socket is used to access multiple computers simultaneously from a remote location. An **IP socket** is a combination of an IP address and a port number that tells the application where to deliver the data:

- **Syntax**: `Public_IP_address:Port_number`
- **Example**: `192.168.2.10:8080`

This section taught you about the Remote Access role in Windows Server 2022 and its various role services and features. The next section will teach you about file and print services.

Understanding file and print services

It can be stated that *file and print services* are as old as computer networks themselves! That is because computer networks were born out of the need to share resources. Hence, file and print services were among the pioneering services in computer networks. These two services, which will be discussed in detail in the following sections, have been transformed into essential services, whether for home or corporate networks. These days, every **Network Operating System** (**NOS**), including Windows Server 2022, can provide file and print services.

File Services role

In Windows Server 2022, the **File and Storage Services** role is automatically added upon installing an operating system, as shown in the following screenshot. Does that surprise you? Maybe not, because you should remind yourself that you have just installed a NOS on the server!

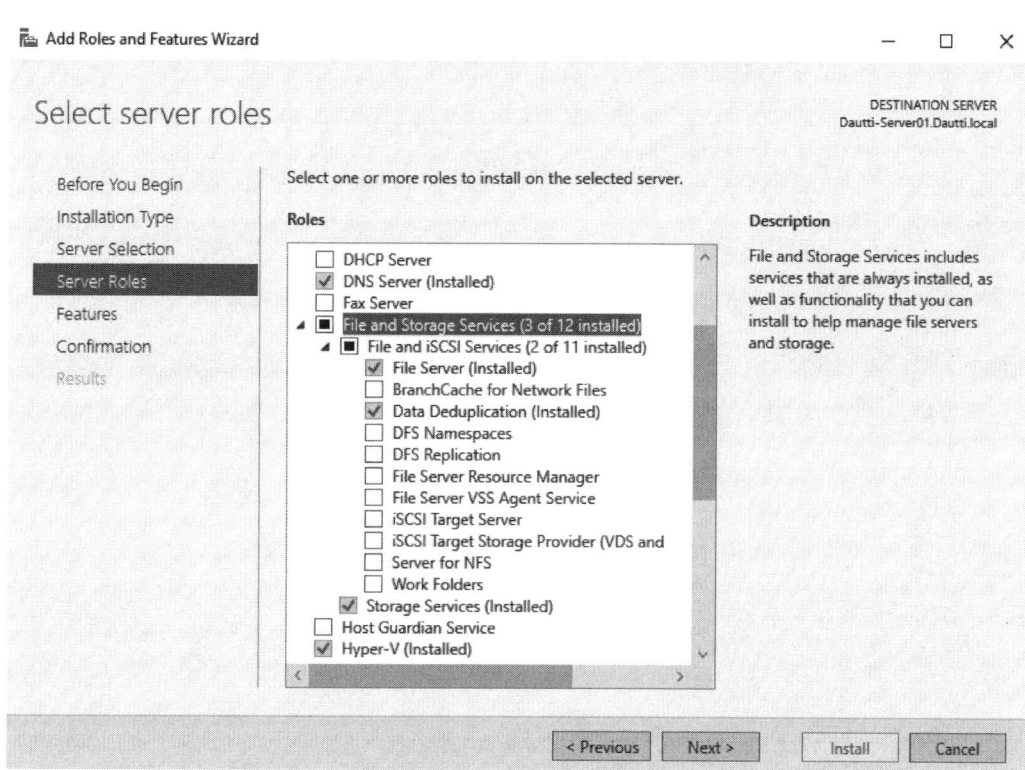

Figure 6.19 – The File and Storage Services role in Windows Server 2022

As mentioned earlier, file services have always been essential network services. From file sharing to work folders or DFS namespaces to BranchCache for network files, it is all about the data's availability and access from anywhere at any time.

PDS role

PDS is a service that enables centralized printing on a network. However, as its name suggests, PDS offers more than just a network printing service. It also provides a service for document scanning. With a scanning service, users receive scanned documents from the network scanner and send them to shared network resources. Usually, PDS is added as a role in Windows Server 2022, as shown in the following screenshot. Hence, to set up a print server in Windows Server 2022, select the **Print and Document Services** role to add **Print Server** role services:

Adding Roles to Windows Server 2022

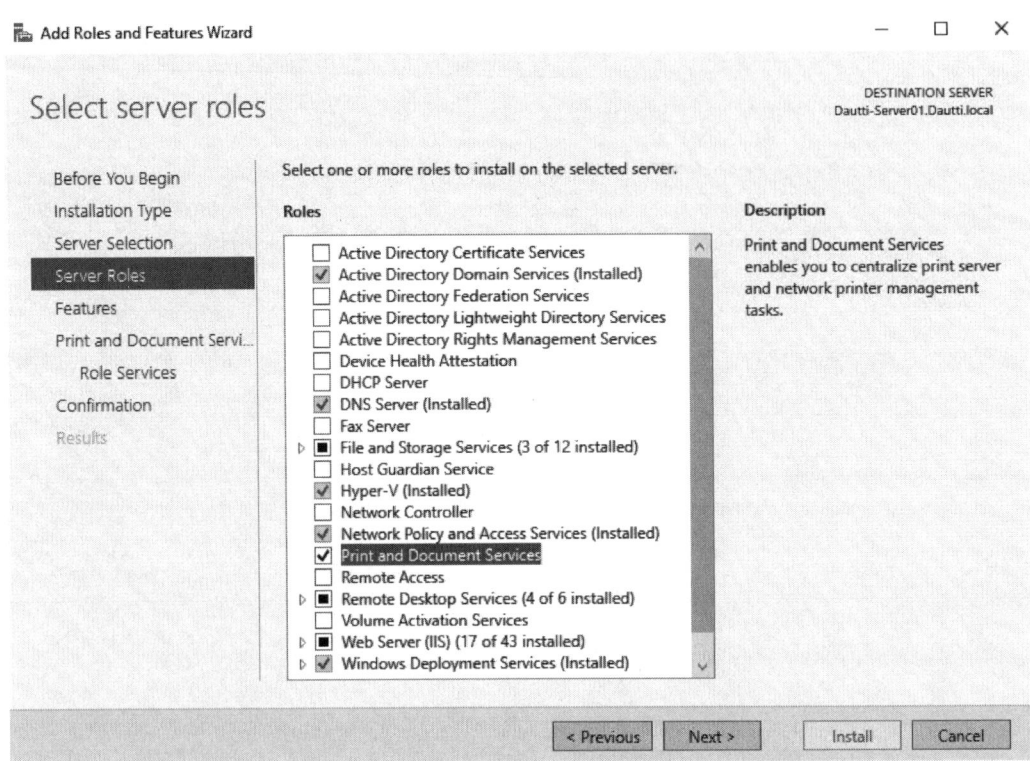

Figure 6.20 – Adding the Print and Document Services role in Windows Server 2022

The following role services can be installed as part of PDS, as shown in the following screenshot:

- **Print Server** allows you to manage printing queues and deploy and migrate print servers.

- **Internet Printing** allows you to set up a website that users can use to print over **Internet Client Printing** (**ICP**).

- **Line Printer Daemon** (**LPD**) **Service** enables Unix-based computers and non-Windows OSs to use **Line Printer Remote** (**LPR**) to print:

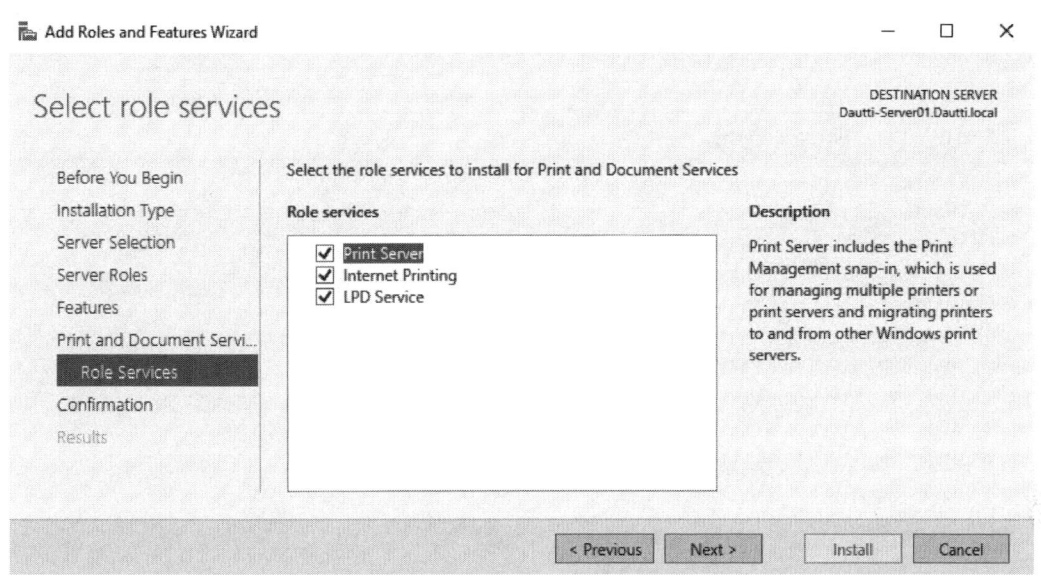

Figure 6.21 – Adding Print and Document Services role services in Windows Server 2022

In the following sections, we will look at various concepts related to a printer. We will begin with the local printer.

Local printer

As the name suggests, a **local printer** is a printer that is physically connected to a computer through either the parallel port (referred to as the printer port) or USB port. This printer primarily serves the computer that it is connected to. However, if a host computer shares the printer, it also serves other computers on the network.

Network printer

A **network printer**, as shown in the following screenshot, unlike a local printer, is a dedicated printer on the computer network that provides printing services. Moreover, a network printer uses either Ethernet or a Wi-Fi interface and can be accessed by multiple devices simultaneously on the same network. Also, the network printer can be accessed over the internet if such a setup exists:

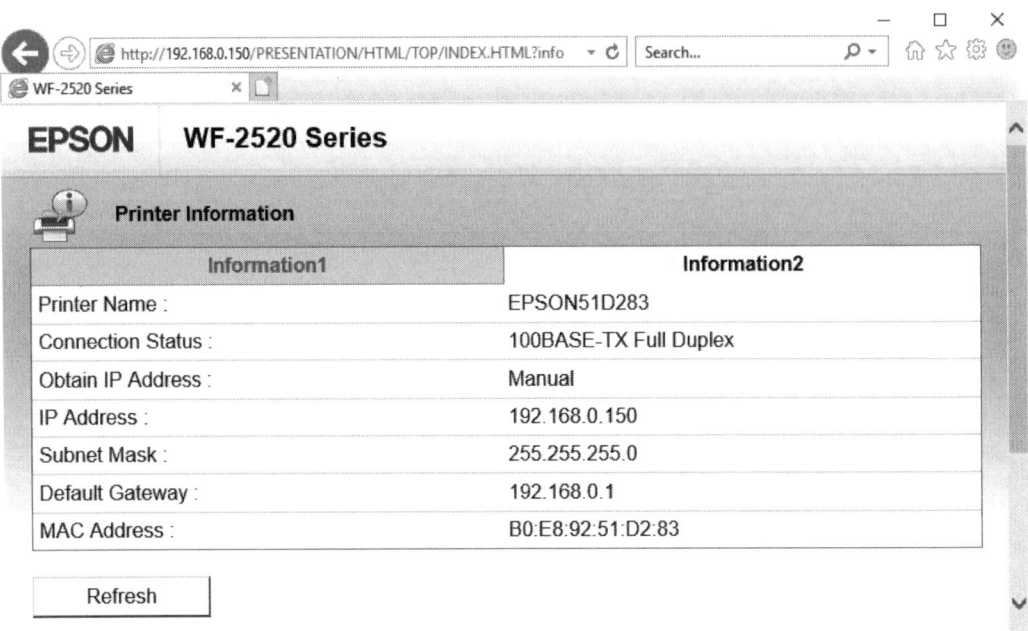

Figure 6.22 – Adding Print Server role services in Windows Server 2022

The preceding screenshot shows the information on a network printer. As you can see, accessing a network printer is very important regarding its IP address as it allows you to connect to the printer and print on a LAN or over the internet (if the printer uses a public IP address).

Printer pooling

Printer pooling in Windows Server 2022 is a feature that helps configure two or more physical printers into one logical printer. Printers installed on the print server must be almost identical or able to use the exact print driver. From the client's perspective (referred to as the frontend), though several physical network printers are available at the backend, it looks like a single printer. This logical connection of printers balances their load, hence increasing their usability and, at the same time, providing users with efficient printing:

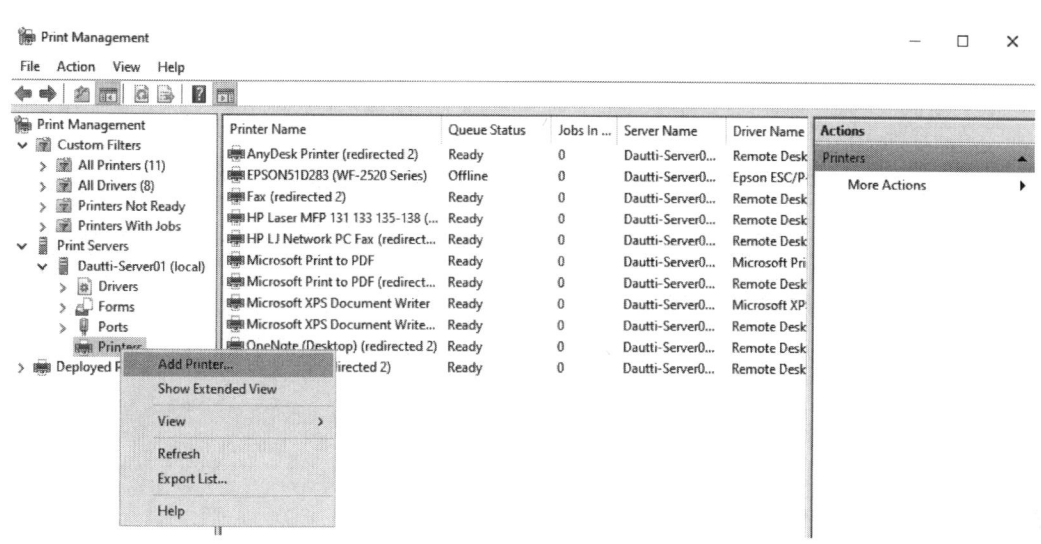

Figure 6.23 – Adding printers with printer pooling in Windows Server 2022

You can set up printer pooling in Windows Server 2022 by adding the **Print and Document Services** and **Print Server** role services. Then, you must install the necessary printers and configure printer pooling through the **Print Management** administrative console, as shown in the preceding screenshot.

Web printing

So far, we've learned that the network printer tries to relate **web printing** with printing over a web browser to understand it better and easier. That said, before setting up web printing in your organization's network, you must add the **Print and Document Services** role and **Internet Printing** as a role service. In addition, the **Web Server (IIS)** role is required:

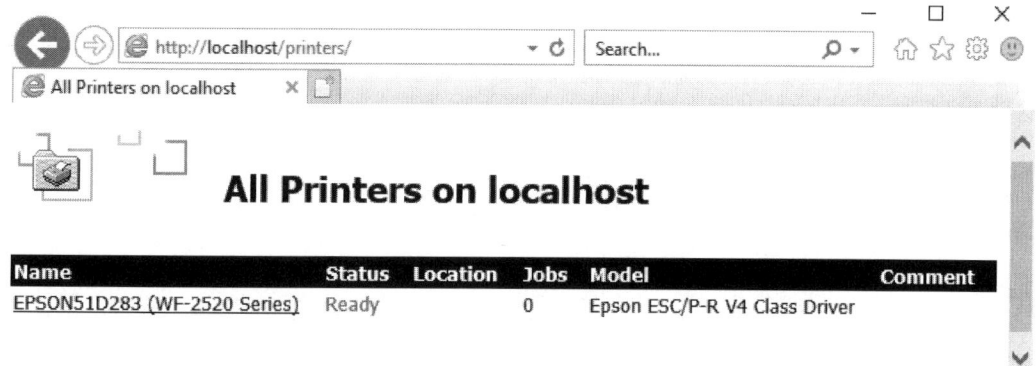

Figure 6.24 – Web printing in Windows Server 2022

Enter http://servername/printers in your browser's web address bar to access printers via a web browser. You should get something similar to the preceding screenshot.

Web management

Like local and network printing, web printing also has its way of managing print jobs. Hence, users can manage print jobs **with web printing management** jobs through the web browser similar to how they would when accessing a local printer or network printer. For example, to manage printers through a web interface, enter http://servername/printers in your browser's address bar and select the printer. The next page, as shown in the following screenshot, lists the print jobs that can be managed. First, however, you must add the **Internet Printing** role services to the server once you add the PDS role:

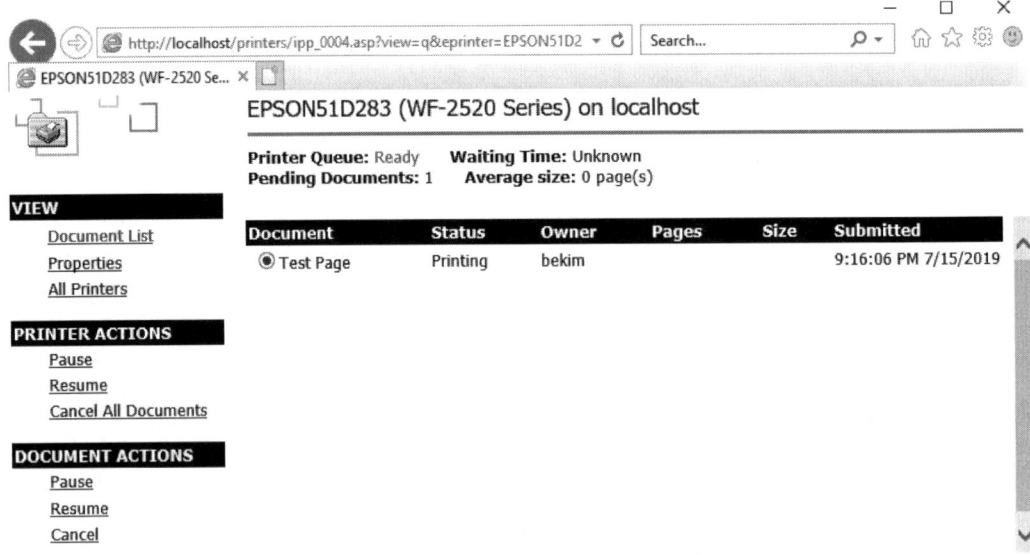

Figure 6.25 – Web printing management in Windows Server 2022

Now, let's understand printer driver deployment, which must be installed on a local or remote server to connect to the printer and run printing tasks.

Printer driver deployment

When managing printers from the **Print Management** administrative console, everything from driver deployment to adding printers can be accomplished, as shown in the following screenshot. That said, the printer driver deployment represents adding and updating a printer driver in the **Print Management** administrator console and installing the printer driver on one or more workplace computers:

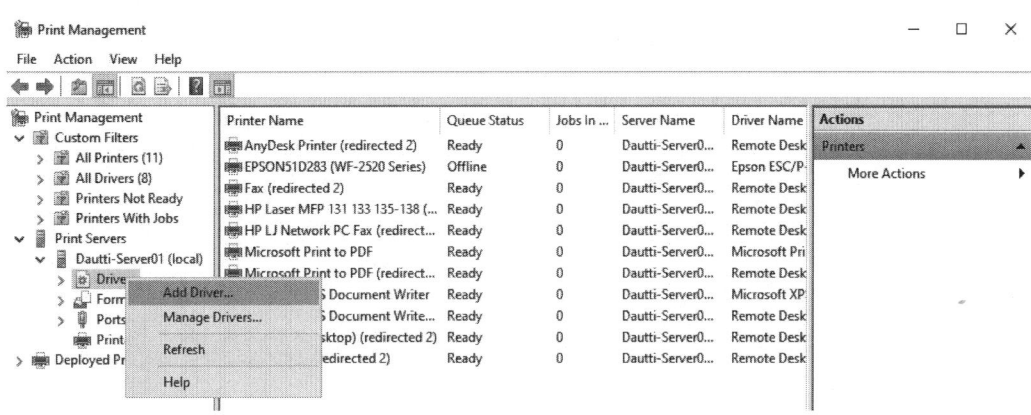

Figure 6.26 – Deploying print drivers and the Print Management console in Windows Server 2022

In this section, you learned about the PDS role, which can help you set up a print server in your organization. In the next section, we'll briefly look at user rights, NTFS permissions, and share permissions so that we can access files and folders from a local computer and network.

Understanding user rights, NTFS permissions, and share permissions

First things first, let's become familiar with user rights and permissions. If you open the **Properties** area of a folder in any Windows OS and then click on the **Security** tab, you will see the permissions for the `<user>` unit under the **group or user names** section. That section lists the following permission types:

- **Full control** allows you to read, write, modify, execute, change attributes and permissions, and delete files and subfolders.
- **Modify** allows you to view, modify, add, and delete files and sub-folders.
- **Read & execute** allows you to run and manage files.
- **List folder contents** allows you to view data files and a list of a folder's content.
- **Read** allows you to view files and file properties.
- **Write** allows you to write in a file.
- **Special permissions** provides access to additional advanced permissions.

However, note that each permission may contain either an allowed or denied setting, as shown in the following screenshot. Users are then allowed or denied access to the files and folders based on the assigned user rights. This means that every user's assigned allowed or denied setting contains specific permissions that determine the user's type of access to the objects:

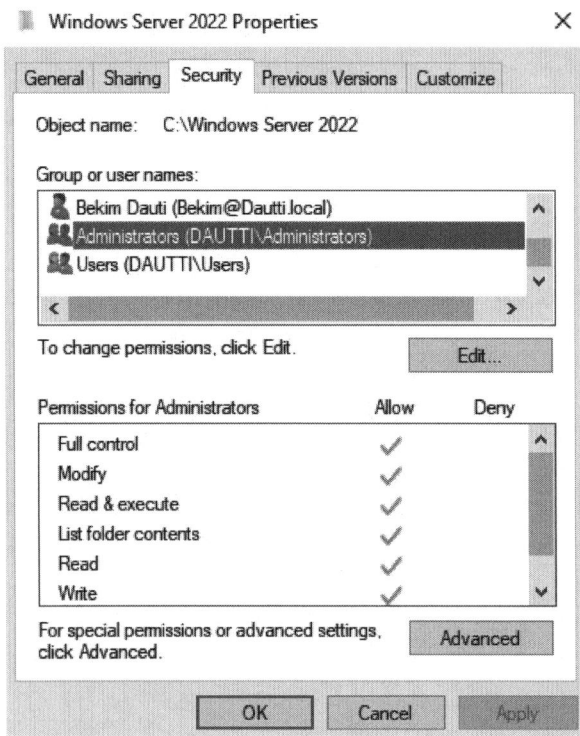

Figure 6.27 – NTFS permissions in Windows Server 2022

Another thing you must consider is comparing **New Technology File System** (**NTFS**) permissions with **share permissions**. Keeping in mind that NTFS is a native filesystem in Windows Server 2022, when we talk about NTFS permissions, we are dealing with file and folder access to the local server, on the server's storage device. In contrast, share permission has more to do with accessing files and folders shared over the network. Since we've already covered NTFS permissions, let's look at what shared permissions offer, as shown in the following screenshot:

- **Full Control** allows you to read, modify, and edit permissions and take ownership.
- **Change** allows you to read, execute, write, and delete files and subfolders.

- **Read** allows you to list and view the content:

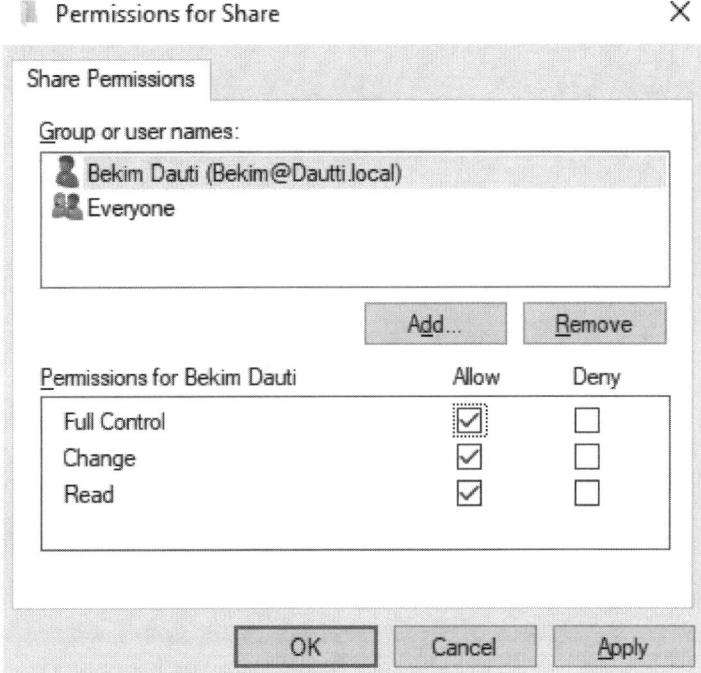

Figure 6.28 – Share Permissions in Windows Server 2022

Another perspective on user rights involves their assignment through **Local Group Policy Editor** (gpedit.msc), **Local Security Policy**, or **Default Domain Policy** by navigating to the Computer Configuration\Windows Settings\Security Settings\Policies\User Rights Assignment path. If the server is a domain member, then you will notice that some policies have already been configured, as shown in the following screenshot:

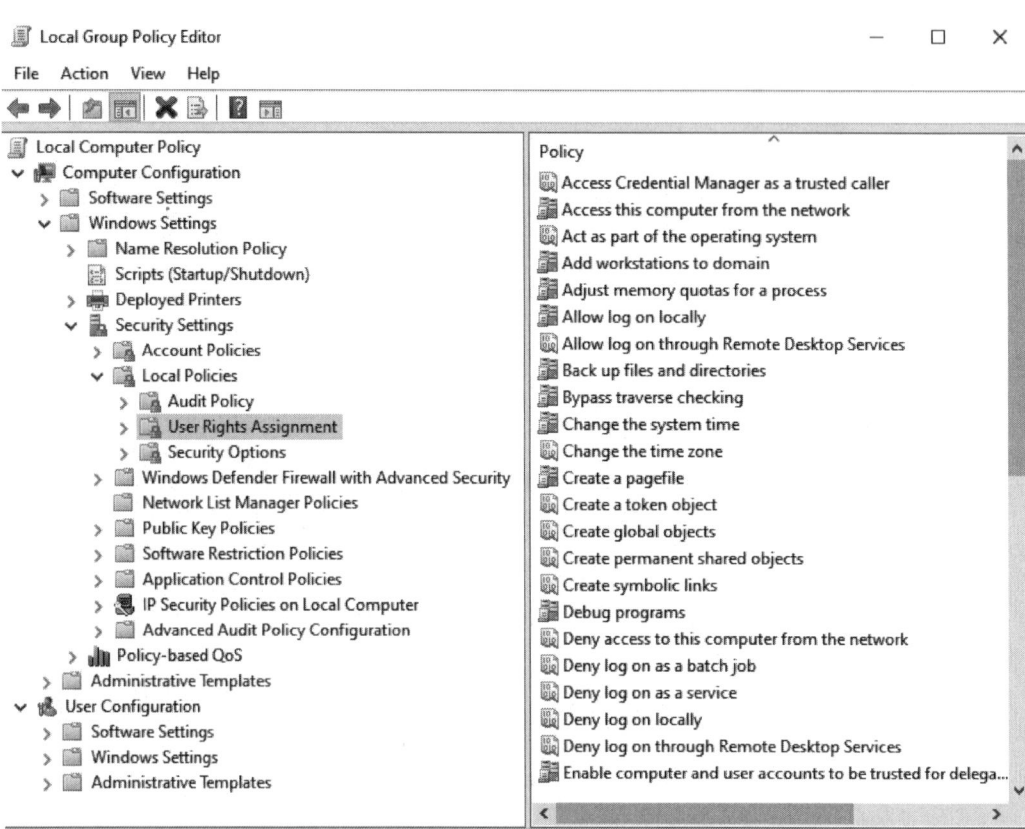

Figure 6.29 – User Rights Assignment in Windows Server 2022

In this section, we learned that there is a clear difference between user rights and permissions. While user rights have to do with user accounts, permissions have to do with access to objects (files and folders). Now, let's learn about file server auditing, a monitoring method that determines who modified the organization's data, when it was changed, and why.

Understanding file server auditing

Since a file server stores critical and sensitive data for an organization, auditing is significant in assessing management controls in its IT infrastructure. Thus, auditing the file server is necessary to record who has done what and when with the data.

To configure auditing in Windows Server 2022, open **Local Group Policy Editor** (`gpedit.msc`), **Local Security Policy**, or **Default Domain Policy** and navigate to the `Computer Configuration\Windows Settings\Security Settings\Local Policies\Audit Policy` path, as shown in the following screenshot:

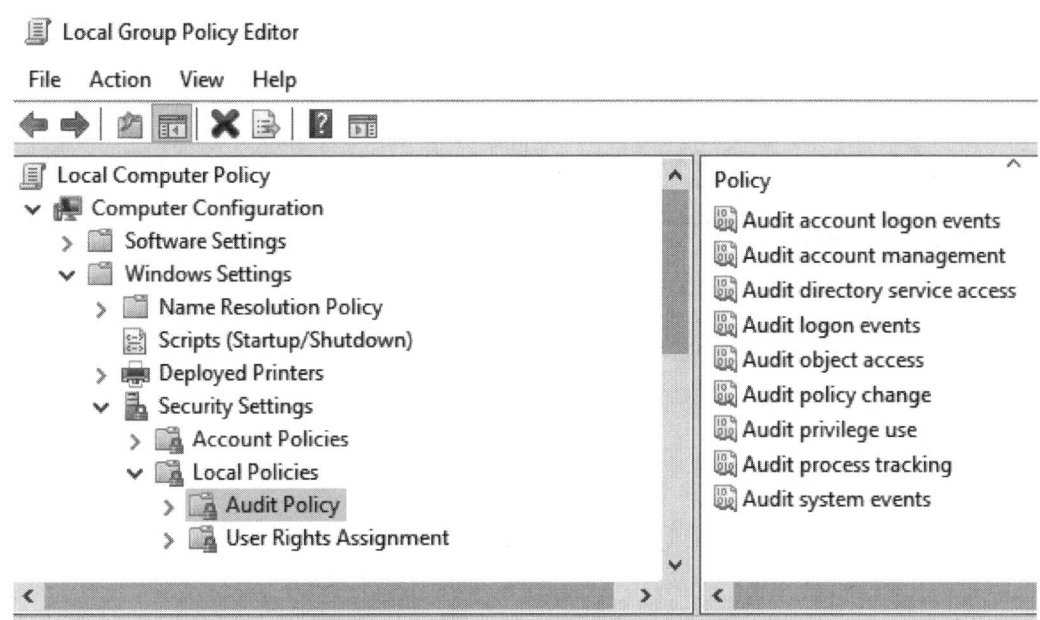

Figure 6.30 – Auditing in Windows Server 2022

In this section, you learned about the file and print services in Windows Server 2022 and their various features. Next, we will move on to this chapter's exercise, where we will look at installing the **Web Server (IIS)** and **Print and Document Services** roles.

Chapter exercise – installing the Web Server (IIS) and PDS roles

In this exercise, you will learn how to install the **Web Server (IIS)** and **Print and Document Services** roles.

Installing the Web Server (IIS) role

To install the **Web Server (IIS)** role in Windows Server 2022, follow these steps:

1. Click the **Start** button. Then, in the **Start** menu, click **Server Manager**.
2. Click on the **Add roles and features** hyperlink in the **Server Manager** window.
3. Shortly, the **Add Roles and Features Wizard** area will open, as shown in the following screenshot:

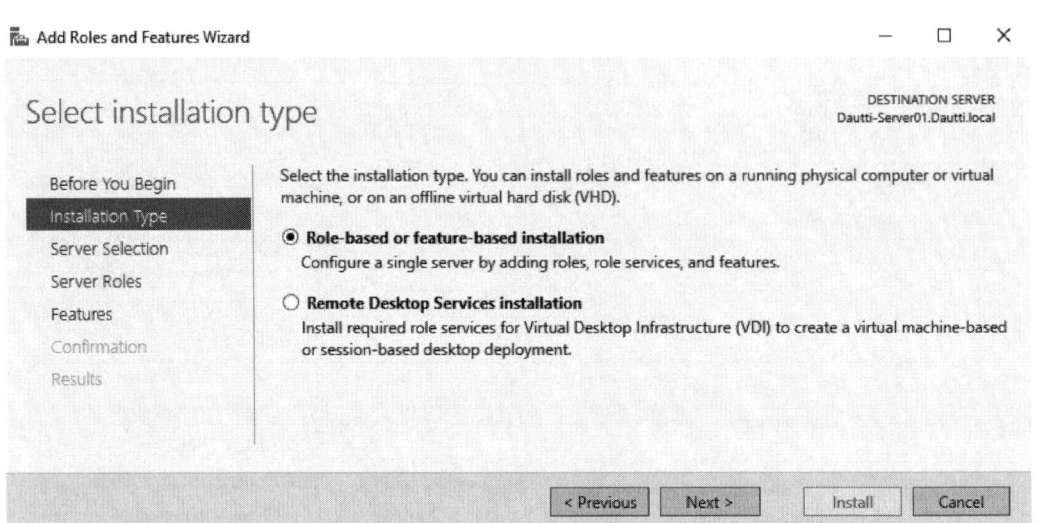

Figure 6.31 – The Add Roles and Features Wizard area in Windows Server 2022

4. Accept the **Role-based or feature-based installation** option and click **Next**.
5. Ensure that the correct server is highlighted from the server pool. Then, accept the **Select a server from the server pool** option and click **Next**.
6. Select the **Web Server (IIS)** role from the list of roles.
7. Click the **Add Features** button when the **Add features required for Web Server (IIS)** popup appears.
8. No feature is required to add the **Web Server (IIS)** role at this stage, so click **Next**.
9. In the **Web Server (IIS)** definition and the things to note regarding installing **Web Server (IIS)**, click **Next**.
10. Either accept the **Web Server (IIS)** role services or customize them to your needs.
11. Confirm your installation selections for the **Web Server (IIS)** role by clicking the **Install** button.
12. When the installation completes, click the **Close** button to close the **Add Roles and Features Wizard** area.
13. With that, the **Web Server (IIS)** role will be installed. A server restart is not required.

Installing a PDS role

To install a **Print and Documents Service** role in Windows Server 2022, follow these steps:

1. Click the **Start** button. Then, in the **Start** menu, click **Server Manager**.
2. Click on the **Add roles and features** hyperlink in the **Server Manager** window.

3. Shortly, the **Add Roles and Features Wizard** area will open.
4. Accept the **Role-based or feature-based installation** option and click **Next**.
5. Ensure that the correct server is highlighted from the server pool. Then, accept the **Select a server from the server pool** option and click **Next**, as shown in the following screenshot:

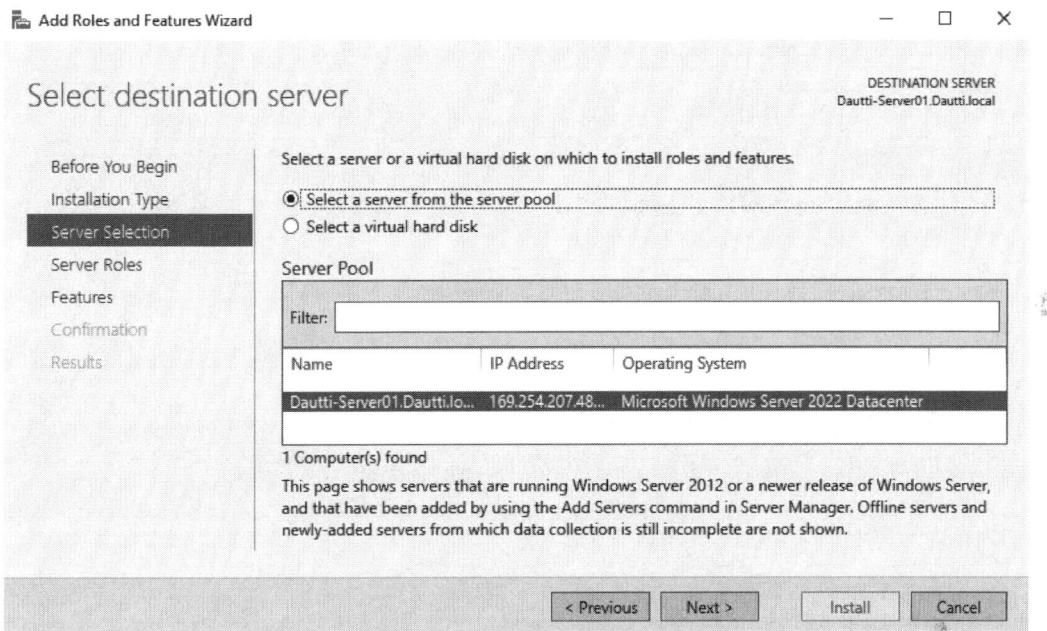

Figure 6.32 – Accepting the defaults

6. From the list of roles, select the **Print and Document Services** role.
7. Click the **Add Features** button when the **Add features that are required for Print and Document Services** dialog box pops up.
8. At this stage, no feature is required to add a PDS role, so click **Next**.
9. In the PDS definition and the things to note regarding PDS installation, click **Next**.
10. Either accept the PDS role services or customize them to your needs.
11. Confirm the installation selections for the PDS role by clicking the **Install** button.
12. When the installation completes, click the **Close** button to close the **Add Roles and Features Wizard** area.

With that, the PDS role will be installed. A server restart is not required.

Summary

In this chapter, you got acquainted with well-known client/server application servers such as the email server, database server, collaboration server, monitoring server, and data protection server.

Then, you learned about the **Web Server (IIS)** role, which can help you set up a web and FTP server and secure communication between the web server and a browser through SSL and digital certificates. Furthermore, you learned about Remote Access services, which help establish remote access to the organization's computers and servers. Then, you became familiar with user rights, NTFS permissions, and share permissions, which will help you understand the concepts of accessing and securing files locally and on the network.

Finally, this chapter concluded with an exercise that showed you how to install the **Web Server (IIS)** and **Print and Document Services** roles.

The next chapter will teach you about Group Policy in Windows Server 2022, which can be used to add more controls to user and computer accounts.

Questions

Answer the following questions to test your knowledge of this chapter:

1. A server role is a primary task that a server should perform. (True | False)
2. _____ transfers files from computer to computer, computer to a server, or vice versa, both on LAN and WAN.
3. Which of the following are NTFS permissions in Windows Server 2022? (Choose 3)

    A. Modify

    B. Write

    C. Change

    D. Read

4. A web service is a communication between two devices based on the request/response methodology that uses the FTP protocol. (True | False)
5. _____ is any logical endpoint where applications from your computer communicate with other applications on other computers, both on LAN and WAN.
6. Which of the following protocols are utilized by mail servers? (Choose 2)

    A. **File Transfer Protocol (FTP)**

    B. **Hypertext Transfer Protocol (HTTP)**

C. **Simple Mail Transfer Protocol (SMTP)**

   D. **Post Office Protocol (POP)**

7. Remote assistance is a feature that enables a helper to access the host's desktop remotely to assist with resolving issues. (True | False)

8. _____ is responsible for securing the communication channel between a website and a browser.

9. Which of the following ports is used by RDS?

   A. 25

   B. 110

   C. 443

   D. 3389

10. Web printing enables users to print files to network printers through Windows Explorer. (True | False)

11. _____ have to do with user access to shared folders and drives on the network.

12. Which of the following are share permissions? (Choose 2)

    A. Read

    B. Change

    C. Write

    D. Modify

13. Discuss the **Remote Access** and **Remote Desktop Services** roles.

14. Discuss user rights, NTFS permissions, and share permissions.

Further reading

To learn more about the topics that were covered in this chapter, take a look at the following resources:

- *IIS Web Server Overview*: `https://docs.microsoft.com/en-us/iis/get-started/introduction-to-iis/iis-web-server-overview`

- *DirectAccess*: `https://docs.microsoft.com/en-us/windows-server/remote/remote-access/directaccess/directaccess`

- *User Rights Assignment*: `https://docs.microsoft.com/en-us/windows/security/threat-protection/security-policy-settings/user-rights-assignment`

Part 3: Configuring Windows Server 2022

Part 3 covers **group policy** (**GP**) and virtualization. Upon completing this part, you will be able to configure GPOs and virtual machines. In addition, you will get to know the storage technologies and be able to configure them.

This part of the book comprises the following chapters:

- *Chapter 7, Group Policy in Windows Server 2022*
- *Chapter 8, Virtualization with Windows Server 2022*
- *Chapter 9, Storing Data in Windows Server 2022*

7
Group Policy in Windows Server 2022

So far in this book, you have learned how to set up Windows Server 2022, including installing it, adding post-installation tasks, and adding roles and features. This chapter will teach you about **Group Policy** (**GP**) in Windows Server 2022, an advanced configuration that restricts users from changing user and computer settings. After that, you will learn how GPs are managed on a local server and a domain controller. In addition, you will learn about **Group Policy Object** (**GPO**) configuration settings and how they are processed.

The second part of this chapter will help you learn about the **Local Group Policy Editor**, which allows you to manage GPOs on a local server. Moreover, you will learn how to update local GPOs. With that, you will be able to understand computer and user configurations.

Finally, we will cover a chapter exercise that provides a few examples of GPOs for system admins.

The following topics will be covered in this chapter:

- Understanding GP
- Types of GP editors
- Chapter exercise – examples of GPOs for system administrators

Technical requirements

To complete the exercises in this chapter, you will need the following equipment:

- A PC with Windows 11 Pro that has at least 16 GB of RAM, 1 TB of HDD space, and access to the internet
- A virtual machine with Windows Server 2022 Standard (Desktop Experience) that has at least 4 GB of RAM, 100 GB of HDD space, and access to the internet

Understanding GP

Sometimes, a system administrator may need to configure the home page of a company's website so that it opens when the browser is launched on the organization's computers. They may also need to deny access to all removable media drives on the organization's computers or block Microsoft accounts from being used on Windows 10 and 11 computers. These settings and many other advanced configurations can be accomplished via GP in a Windows Server domain-based network without third-party tools and utilities. Additionally, it enables the system administrator to run advanced configurations that restrict users from changing user and computer settings on local and domain-joined computers.

GP is a Windows Server feature that applies restrictions at the user and computer levels. In contrast, GPOs are administrative templates that enable system administrators to configure what users can and cannot do on computers, peripheral devices, and network applications across the organization's network. Furthermore, GPOs can be considered security tools as they can apply security settings to users and computers on a network managed by a domain controller.

By default, the configured GPOs are stored in the `C:\Windows\SYSVOL\sysvol\<domain>\Policies` path in the domain controller, as shown in the following screenshot:

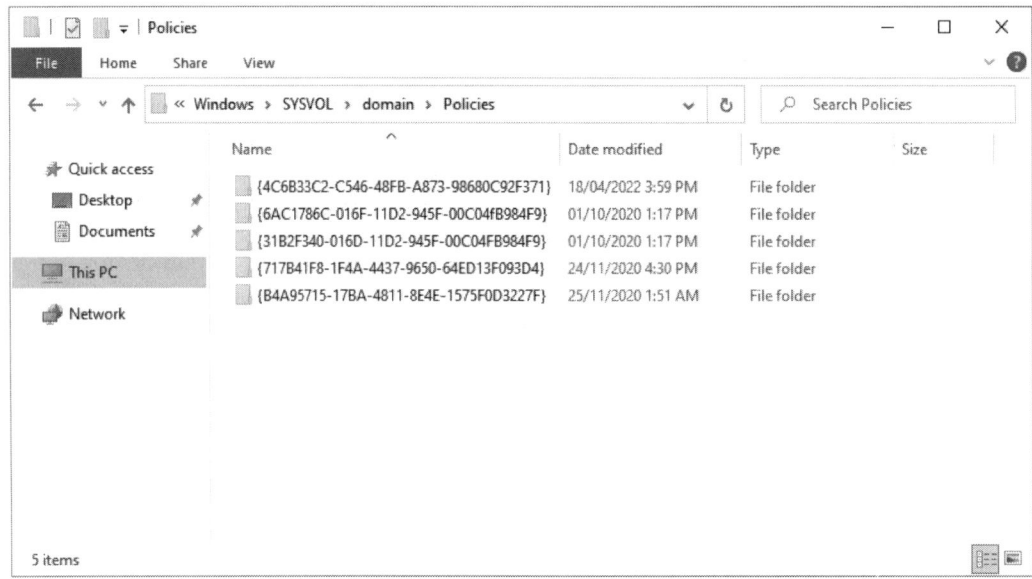

Figure 7.1 – The default location for GPOs in Windows Server 2022

Now that we've learned about GP and GPOs, let's learn how to manage them.

Managing GPOs

The **Group Policy Management** (**GPM**) snap-in, as shown in *Figure 7.2*, is a system administrator's favorite tool for managing GPOs. GPM is a console in the domain controller that allows you to configure and deploy GPOs across the organization's Windows Server domain-based network. It is a one-stop feature where system administrators can configure various Windows Server settings for every user and computer in the Windows Server IT environment. The GPM console consists mainly of the **Forest** pane and the **GPOs** pane. The **Forest** pane displays the hierarchical structure of the domain, whereas the **GPOs** pane contains the **Status**, **Linked Group Policy Objects**, **Group Policy inheritance**, and **Delegation** tabs.

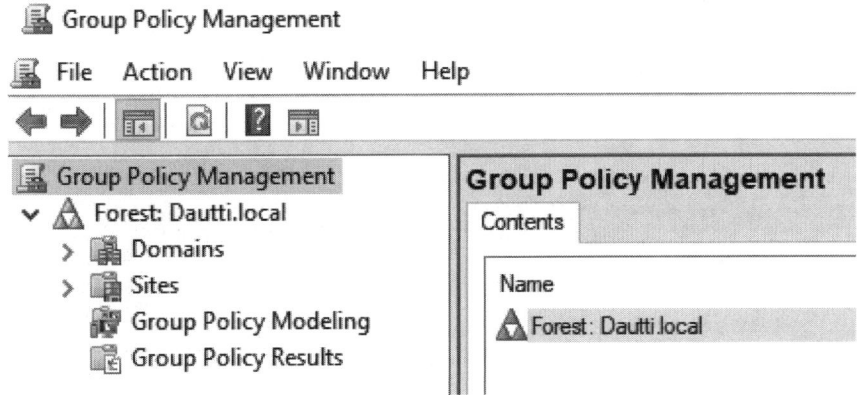

Figure 7.2 - The GPM console in the domain controller

The GPM console in Windows Server 2022 can be accessed in several ways. The following sections will teach you various ways of accessing the GPM console.

Accessing the GPM console from the Administrative Tools menu

To access the GPM console from the **Administrative Tools** menu in the **Start** menu, follow these steps:

1. Click the **Start** button.
2. From the **Start** menu, select **Windows Administrative Tools**.
3. In the **Administrative Tools** window, select **Group Policy Management**, as shown in the following screenshot:

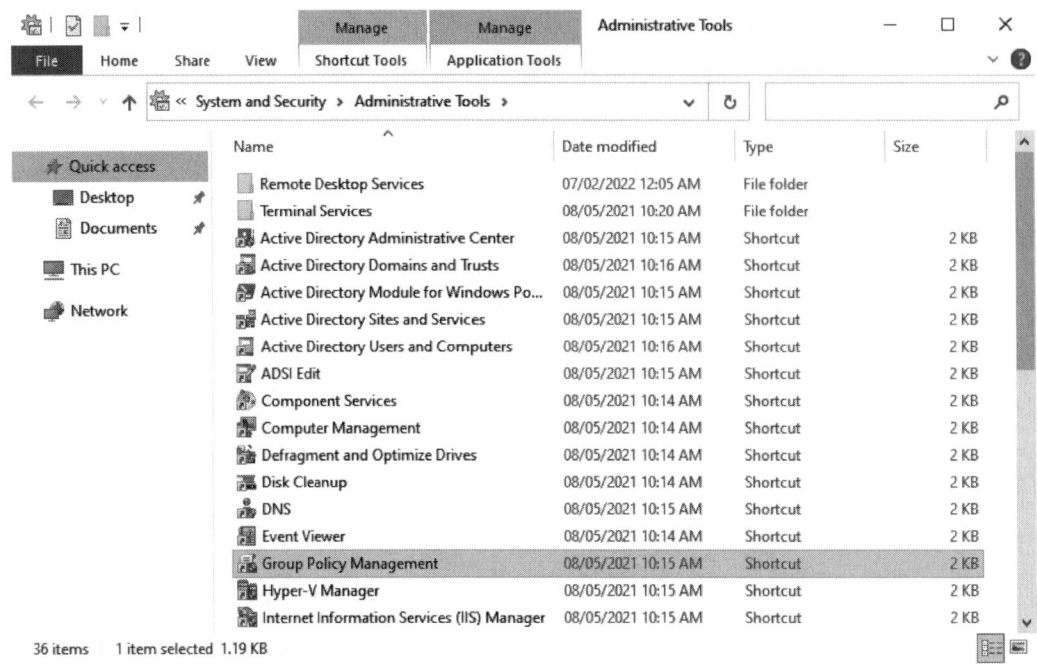

Figure 7.3 – Accessing the GPM console from the Administrative Tools menu

It can also be accessed from the **Run** dialog box.

Accessing the GPM console from the Run dialog box

To access the GPM console from the **Run** dialog box, follow these steps:

1. Press the Windows key + *R* to open the **Run** dialog box.
2. In the **Run** dialog box, enter gpmc.msc, as shown in the following screenshot, and click **OK**:

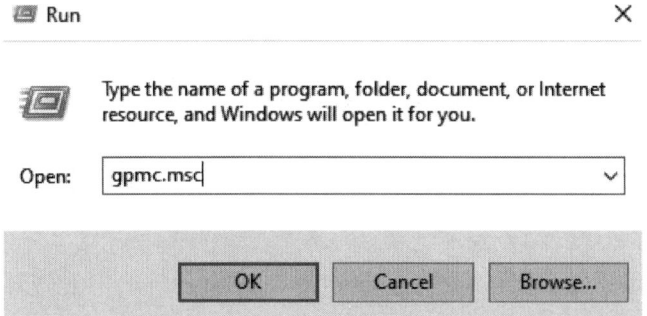

Figure 7.4 – Accessing the GPM console from the Run dialog box

Finally, let's learn how to access the GPM console from the **Server Manager** menu.

Accessing the GPM console from the Server Manager menu

To access the GPM console from the **Server Manager** menu in the **Start** menu, follow these steps:

1. Click the **Start** button.
2. From the **Start** menu, select **Server Manager**.
3. In the **Server Manager** window, click **Tools** and select **Group Policy Management** from the menu, as shown here:

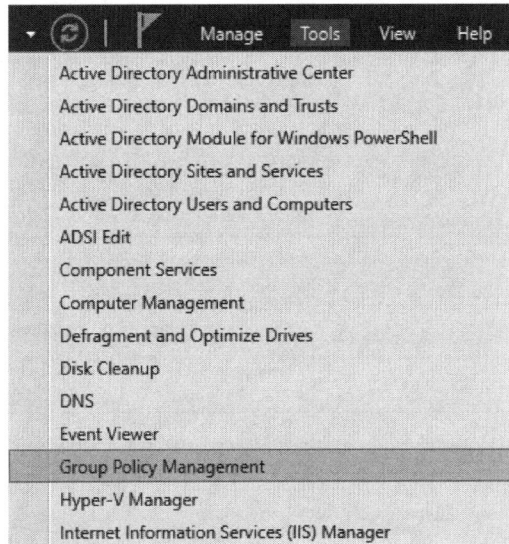

Figure 7.5 – Accessing the GPM console from the Server Manager menu

Now, let's learn about the various GPO configuration settings that can help you configure the GPOs.

GPO configuration settings

As mentioned earlier, GPOs are administrative templates that system administrators configure. They can have configuration settings that determine the applicability of the policy, equivalent to registry keys in the Registry Editor. Hence, a GPO's settings can be configured to specific matters that affect users and computers. Thus, the GPO settings contain the following three configurable settings, as shown in *Figure 7.6*:

- **Not Configured** is the default setting for GPOs, meaning that the registry value has not been established.

- **Enabled** is a configured setting indicating that a GPO is enabled, meaning that the registry value has been set to 0x1.
- **Disabled** is a configured setting indicating that a GPO is disabled, meaning that the registry value has been set to 0x0:

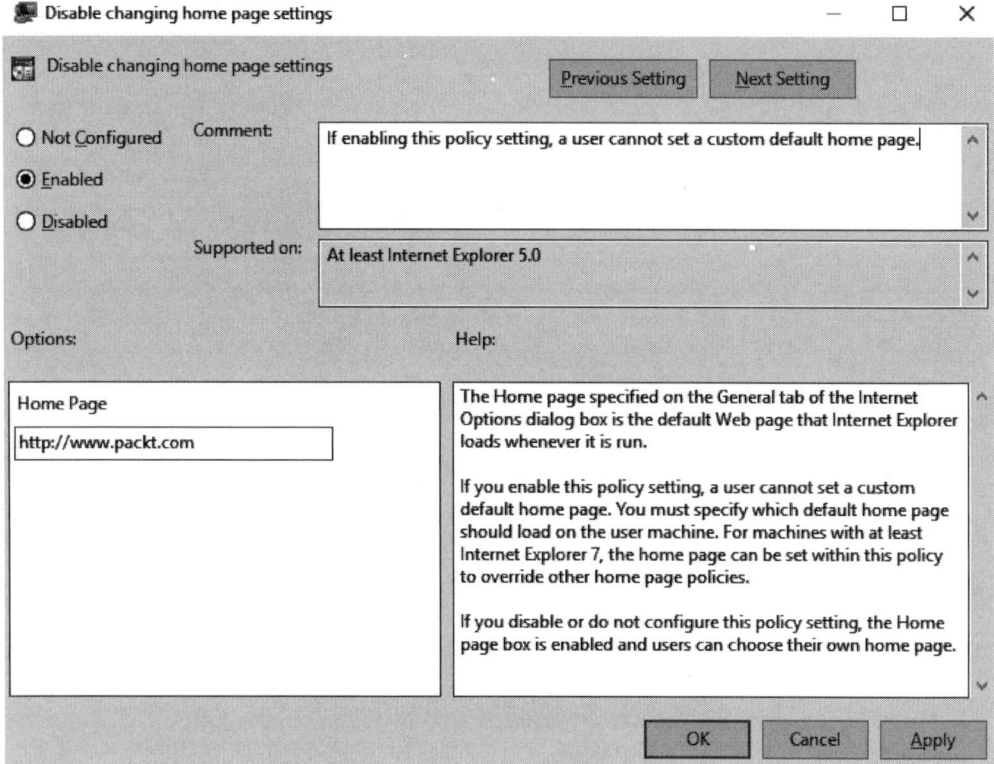

Figure 7.6 – GPO settings configuration values

Once the GPO has been configured, it must be deployed. Therefore, it is good to learn the order in which GPOs are processed. So, let's understand how **GPO processing** takes place.

Processing GPOs

Since GPOs are applied at the user and computer levels, their settings can be enforced, and users cannot change them. Hence, GPOs on a local computer can be configured using the **Local Group Policy Editor**, whereas, on a domain controller, GPOs can be configured through the GPM console. Therefore, GPO processing is carried out in the following order:

1. **Local** indicates that the GPOs have been applied to the computer's local policy for every user.
2. **Site** indicates that the GPOs have been applied at the site where the computer belongs.

3. **Domain** indicates that the GPOs have been applied to the domain of which the computer is a member.
4. **OUs** indicates that the GPOs have been applied to the organizational unit where the computer has been placed.

From a computer perspective, GPOs are configured in the following two ways:

- The **local computer** indicates that the GPOs have been configured on the local computer.
- The **domain computer** indicates that the GPOs have been configured on the domain controller.

Another important consideration with GP settings has to do with applicability. Remember that GPOs assigned to user accounts are applied when the user logs on to a computer, whereas GPOs assigned to computer accounts are used when the computer is turned on.

> **Important note**
> Microsoft has developed a new GP settings reference spreadsheet for Windows Server 2022. The computer and user configurations administrative template includes these new GP settings. It can be downloaded from https://www.microsoft.com/en-us/download/details.aspx?id=104005.

In this section, you got acquainted with the Windows Server 2022 Group Policy feature, including the GPM console for managing GPs, GPO configuration settings, and the order in which GPOs are processed. The next section will cover the **Local Group Policy Editor** and how to update the local GPOs and configuration settings for users and computers.

Types of GP editors

In addition to the **Group Policy Management Editor**, which is used in the domain controller, the **Local Group Policy Editor** is another **Microsoft Management Console** (**MMC**) snap-in that allows you to manage GPO settings on a local computer. With the **Local Group Policy Editor,** a system admin can configure local GPOs by enabling or disabling configurable settings in Local Group Policy at the user and computer levels.

Unlike the GPM console, which becomes available once you have installed **Active Directory Domain Services** (**AD DS**) on the server and set up a domain controller, the **Local Group Policy Editor** can be found in Windows Server 2022.

Local Group Policy Editor

While the **Group Policy Management** console deploys GPOs in domain environments, the **Local Group Policy Editor** console deploys GPOs on standalone servers, not a domain member.

To access the **Local Group Policy Editor** console in Windows Server 2022, follow these steps:

1. Press the Windows key + R to open the **Run** dialog box.
2. In the **Run** dialog box, enter gpedit.msc and press **OK**.

The **Local Group Policy Editor** console will open, as shown in the following screenshot:

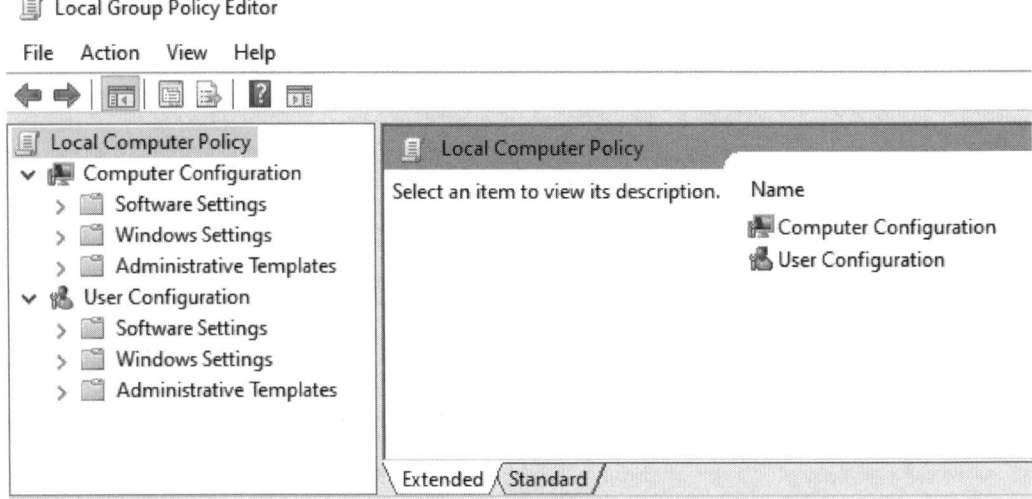

Figure 7.7 – The Local Group Policy Editor console in Windows Server 2022

Besides learning how to access the GPOs, you should also know how to update local GPOs. You'll learn how to do that next.

Updating local GPOs

Once you have configured the GPOs, you can enforce them on your local server. Follow these steps to do so:

1. Press the Windows key + R to open the **Run** dialog box.
2. In the **Run** dialog box, enter gpupdate /force, as shown in the following screenshot, and click **OK**:

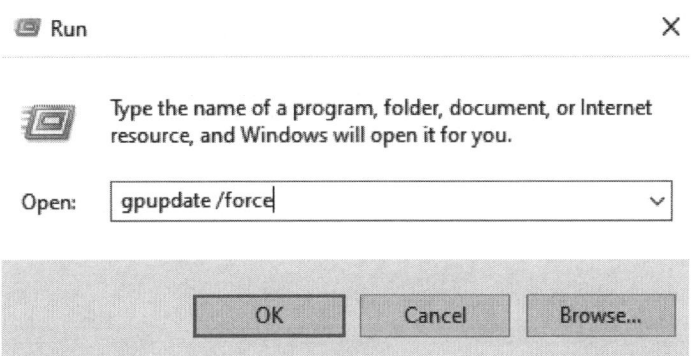

Figure 7.8 – Running the gpupdate /force command via the Run dialog box

3. Shortly after, the Command Prompt window will open up and display a message stating **Updating policy...**, as shown in the following screenshot:

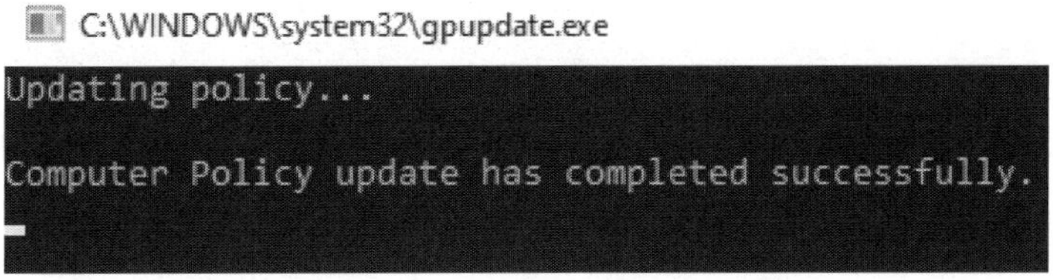

Figure 7.9 – The process of deploying the policy

4. Once the computer policy update has finished, the Command Prompt window will close automatically.

> **Tip**
> Typing `gpedit.msc` also works in Cortana/search box (the **Start** menu), Windows PowerShell, and the Command Prompt window (`cmd.exe`). Also, it is recommended that you don't use the `gpupdate /force` command when updating GPOs. Instead, you just need to use the `gpupdate` command. This is because the `gpupdate/force` command causes administrative overhead on clients and servers, and as such, it causes all GPOs to be reprocessed.

GPO configuration settings

There are two types of GPO configuration settings: computer configuration settings and user configuration settings. We'll explore these two types of GPO configuration settings in the following sections.

Computer configuration GPO settings

As explained in the *Processing GPOs* section, GPOs that have been assigned at the computer level are applied when computers are turned on. Therefore, computer configuration GPO settings are bound to computers, regardless of the user logged on to that computer. Hence, computer configuration GPO settings apply to both local and domain-joined computers.

To set up GPOs at the computer level, follow these steps:

1. Press the Windows key + R to open the **Run** dialog box.
2. In the **Run** dialog box, enter gpedit.msc and press **OK**.
3. In the **Forest** pane of the GPM console, right-click the domain and select **Create a GPO in this domain, and Link it here...**, as shown in the following screenshot:

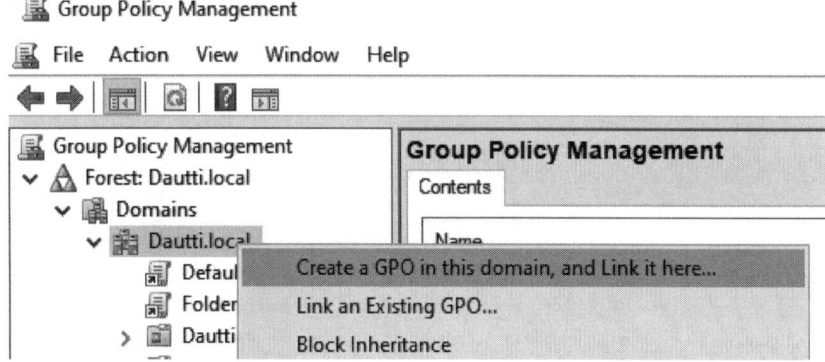

Figure 7.10 – Creating a GPO in the domain controller

4. In the **New GPO** window, enter a name for the new GPO, then click **OK**.
5. In the **Group Policy Management** pane, select the **Linked Group Policy Objects** tab.
6. Right-click the newly created GPO and select **Edit** from the context menu.
7. In the **Group Policy Management Editor** window, expand **Policies** under **Computer Configuration**, and then select the desired computer administrative template to configure, as shown here:

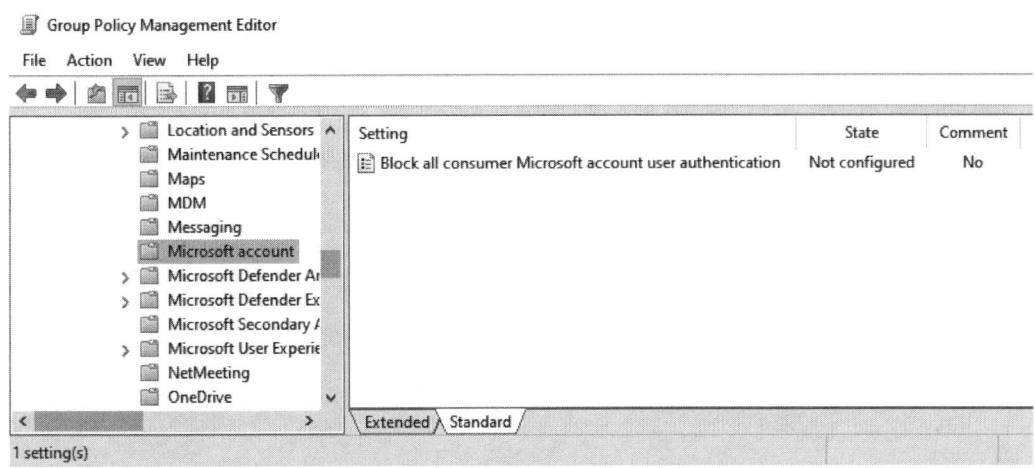

Figure 7.11 – Setting up the computer configuration policies

8. Close the **Group Policy Management Editor** window.
9. In the **Group Policy Management** pane, right-click the recently created GPO and select **Enforced**.
10. In the dialog box that appears, click **OK**.

User configuration GPO settings

Unlike computer configuration GPO settings, user configuration GPOs represent the GPOs assigned at the user level. Therefore, user configuration GPO settings are applied to the user account, regardless of the computer they have logged on to.

To set up GPOs at the user level, follow these steps:

1. Repeat *Steps 1 to 6* from the *Computer configuration GPO settings* section.
2. In the **Group Policy Management Editor** window, expand **Policies** under **User Configuration**, and then select the desired user administrative template to configure, as shown here:

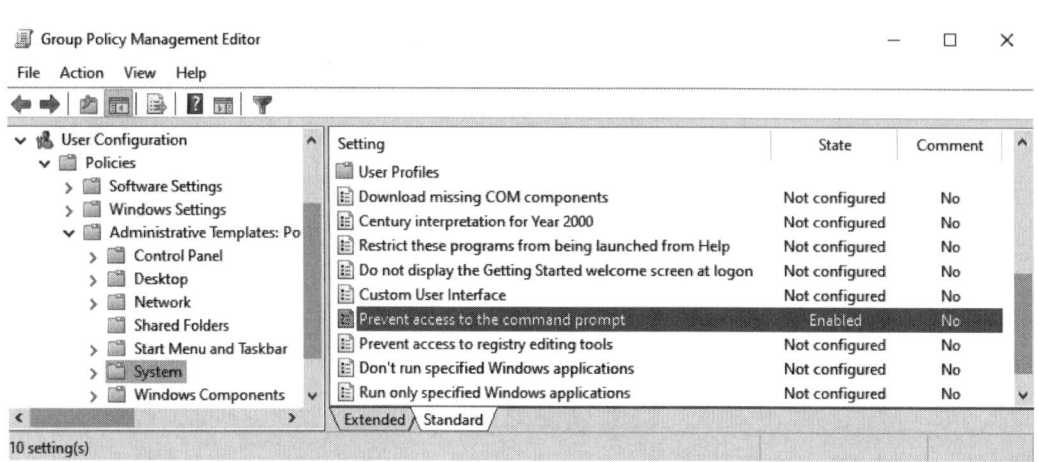

Figure 7.12 – Setting up the user configuration policies

3. Close the **Group Policy Management Editor** window.
4. In the **GPOs** pane, right-click the recently created GPO and select **Enforced**.
5. In the **Group Policy Management** dialog box, click **OK**.

In this section, you got acquainted with the **Local Group Policy Editor,** another way of managing GPOs. You also learned about the different configuration policies. The next section will provide several examples of GPOs for system administrators.

Chapter exercise – examples of GPOs for system administrators

In this chapter exercise, you will get acquainted with some GPOs that may interest system administrators. For example, renaming the administrator account, disabling the guest account, blocking Microsoft accounts, prohibiting access to the Control Panel and PC settings, and denying access to all removable media drives are some of the most used GPOs these days. In addition, you will learn how to implement the needed GPOs in your organization using GPMC.

Renaming the administrator account

To rename the administrator account using a GPO in Windows Server 2022, follow these steps:

1. Navigate to the `Computer Configuration\Policies\Windows Settings\SecuritySettings\Local Policies\Security Options` path to reach the GPO.
2. Double-click **Accounts: Rename the administrator account**.

3. In the **Properties** dialog box, check the **Define this policy setting** box and enter the new name you want to use for the administrator account.

4. Click **OK** to close the **Properties** dialog box.

This policy is being applied to computer configuration settings. Securing the network services in particular and the whole network of the organization, in general, are among the highest priorities. For that reason, it is recommended to rename the administrator account in Windows Server 2022 to stop it from being misused. Next, let's rename the guest account using a GPO.

Renaming the guest account

To rename the guest account using a GPO in Windows Server 2022, follow these steps:

1. Navigate to the `Computer Configuration\Policies\Windows Settings\SecuritySettings\Local Policies\Security Options` path to reach the GPO.

2. Double-click **Accounts: Rename guest account**.

3. In the **Properties** dialog box, check the **Define this policy setting** box and enter the new name that you want to use for the guest account.

4. Click **OK** to close the **Properties** dialog box.

This policy is being applied to computer configuration settings. For security reasons, to minimize the chance of the guest account being misused, you may want to change the name of the guest account. Now, let's block the Microsoft account using a GPO.

Blocking the Microsoft accounts

To block Microsoft accounts from using a GPO in Windows Server 2022, follow these steps:

1. Navigate to the `Computer Configuration\Policies\Windows Settings\SecuritySettings\Local Policies\Security Options` path to reach the GPO.

2. Double-click **Accounts: Block Microsoft accounts**.

3. In the **Properties** dialog box, check the **Define this policy setting** box and select **Users can't add or log on with Microsoft accounts** from the drop-down list.

4. Click **OK** to close the **Properties** dialog box.

This policy is being applied to user configuration settings. For security reasons, to prevent users from adding and logging into the organization's computers with their Microsoft accounts, system administrators can block the usage of Microsoft accounts. So, now, let's deny access to the Control Panel and PC settings using a GPO.

Prohibiting access to the Control Panel and PC settings

To restrict access to the Control Panel and PC settings using a GPO in Windows Server 2022, follow these steps:

1. Navigate to the `User Configuration\Policies\AdministrativeTemplates\Control Panel` path to reach the GPO.
2. Double-click **Prohibit access to the Control Panel and PC settings**.
3. Select the **Enable** option in the dialog box, then click **OK** to close the dialog box.

This policy is being applied to the user's configuration settings. If you wish for users in your organization's network to be unable to make changes to their computers, you can deny their access to the Control Panel and PC settings. Now, let's learn how to restrict access to all removable media drives.

Denying access to all removable storage classes

To deny access to all removable storage classes using a GPO in Windows Server 2022, follow these steps:

1. Navigate to the `User Configuration\Policies\AdministrativeTemplates\System\Removable Storage Access` path to reach the GPO.
2. Double-click **All the removable storage classes: Deny all access**.
3. Select the **Enable** option in the dialog box, then click **OK** to close the dialog box.

This policy is being applied to the user's configuration settings. System administrators can completely restrict the use of removable storage by enabling or **prohibiting access** to the Control Panel and PC settings.

This section presented a few GPO configuration examples from the long list of GPOs supported by Windows Server 2022. Although only a few, these GPOs are essential as they highlight some of the most commonly used GPOs in a production environment. With that, we have reached the end of this chapter's exercise.

Summary

In this chapter, you learned about GP in Windows Server 2022, through which you can restrict users from changing both user and computer settings.

You got acquainted with the Windows Server 2022 Group Policy feature, including the GPM console for managing GPs. Furthermore, you learned about the **Local Group Policy Editor,** another way of managing GPOs. You even had the opportunity to learn about computer and user configuration settings. Additionally, you learned how to update GPOs on a local server and domain controller.

With the skills you have acquired from this chapter, you can deploy GPOs on local and domain-joined computers.

Finally, this chapter concluded with a chapter exercise that provided several examples of GPOs for system admins.

In the next chapter, you will learn about virtualization in Windows Server 2022.

Questions

Answer the following questions to test your knowledge of this chapter:

1. GPOs are processed in the following order: Local, Site, Domain, and **OUs**. (True | False)
2. The _____ are administrative templates that enable system administrators to configure what users can and cannot do on computers, peripheral devices, and network applications across the organization's network.
3. Which of the following represent GPO configuration values? Choose two:

    A. Enabled

    B. Disabled

    C. Allow

    D. Deny

4. The GPM is a console in the domain controller that allows you to configure and deploy GPOs across an organization. (True | False)
5. The _____ displays the hierarchical structure of the domain, whereas the _____ contains the **Status**, **Linked GPOs**, **GP inheritance**, and **Delegation** tabs.
6. Which of the following commands is used to update GPOs?

    A. `gpupdate /enforce`

    B. `gpupdate /setup`

    C. `gpupdate /run`

    D. `gpupdate /force`

7. **Not Configured** is the default setting for GPOs, meaning that the registry value has not been manipulated. (True | False)
8. The _____ is another **Microsoft Management Console** (**MMC**) snap-in that enables you to manage GPO settings on a local computer.

9. GPOs assigned to the computer level are applied when computers are in which state?

   A. Turned on
   B. Turned off
   C. Hibernate mode
   D. Sleep mode

Further reading

To learn more about the topics that were covered in this chapter, check out the following resources:

- *Group Policy Objects*: https://docs.microsoft.com/en-us/previous-versions/windows/desktop/policy/group-policy-objects
- *Linking GPOs to Active Directory Containers*: https://docs.microsoft.com/en-us/previous-versions/windows/desktop/policy/linking-gpos-to-active-directory-containers
- *Applying Group Policy*: https://docs.microsoft.com/en-us/previous-versions/windows/desktop/policy/applying-group-policy
- *Group Policy Best Practices*: https://www.netwrix.com/group_policy_best_practices.html

8
Virtualization with Windows Server 2022

As you may know, the topic of cloud computing is very popular these days. This is because many businesses and organizations either access or offer services in/from the cloud. Regardless, cloud computing happens to be a complex infrastructure powered by virtualization. For example, many servers are grouped to form a cluster in a data center that provides cloud services. On top of that cluster, many **virtual machines** (**VMs**) are running to make up the cloud's infrastructure.

In this chapter, you will learn about virtualization and become familiar with the Hyper-V software, a Microsoft product. This helps you enable virtualization on Windows clients and servers. Then, you will learn the steps to add the **Hyper-V role** in Windows Server 2022, know **Hyper-V Manager**, and learn how to create VMs. By doing so, you will understand what virtualization is, how you can enable the Hyper-V role, and how to create VMs.

Finally, this chapter concludes with an exercise where you will learn how to install the Hyper-V role on Windows Server 2022.

To sum up, the following topics will be covered in this chapter:

- Understanding server virtualization
- Getting to know Hyper-V Manager
- Chapter exercise – installing Hyper-V on Windows Server 2022

Technical requirements

To complete the lab for this chapter, you will need the following equipment:

- A PC with Windows Server 2022 Standard with at least 8 GB of RAM, 500 GB of HDD, and access to the internet

Understanding server virtualization

Let's try to understand virtualization from a technical point of view while considering technology that allows you to set up a VM. By doing this, we can execute the OS, application, and other favorite tools on that VM. Additionally, virtualization allows you to set up storage devices and network resources. Thus, instead of running ten physical servers, all those servers can be organized to run in one physical server in a virtualized mode.

Windows Server 2022 comes equipped with a Hyper-V feature that enables virtualization. As a descendant of Windows Virtual PC, Microsoft's Hyper-V was introduced in Windows Server 2008. Since then, Hyper-V has managed to attract the interest of system administrators, thereby positioning itself firmly in competition with VMware in terms of virtualization platform market shares. In technical terms, Hyper-V provides the services that are used to create and manage VMs and their resources.

In the next few sections, you will learn about various aspects of server virtualization, beginning with virtualization modes.

Virtualization modes

Generally, the following three virtualized modes are the most commonly used in today's virtualized environments:

1. The **fully virtualized mode** allows you to securely execute one or more OSs in isolation in a single physical server where guest OSs cannot be recompiled. Instead, it utilizes the host's OS resources, as shown in the following screenshot:

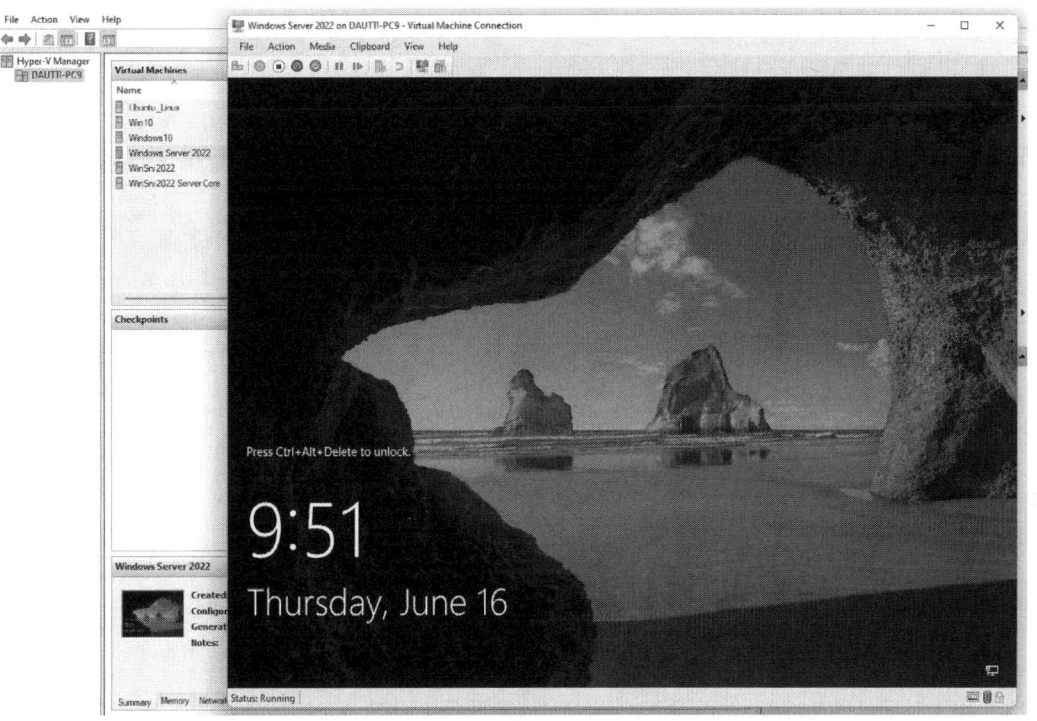

Figure 8.1 – Windows Server 2022 running in an isolated and secure virtual environment

2. **Paravirtualized mode** can be understood as a computer inside a computer with an installed OS that does not simulate the hardware. Instead, it offers an **application programming interface** (**API**) that allows you to recompile the guest OS.

3. **Containerization mode** provides a container that represents a standalone package containing the runtime environment, system tools, and settings needed to run the application code.

> **Important note**
> A host OS is the OS on a physical server, whereas a guest OS is the OS on a VM. For example, I am running Windows 11 Pro on a laptop as the host OS in my setup, and I am running Windows Server 2022 Standard as the guest OS.

Now, let's learn what the Hyper-V architecture is to understand virtualization as a concept.

The Hyper-V architecture

The Hyper-V architecture is based on a hierarchical format where the first level represents the hypervisor as the main element that constitutes the Hyper-V virtual platform. Thus, a hypervisor is accommodated at the root and directly accesses the underlying hardware. Then, the root component creates branch OSs that represent isolated executable environments.

Specifically, the branched OS represents a logical isolation unit without access to the underlying hardware. This way, it allows you to run guest OSs on these parts. In addition, components such as the **virtualization service provider** (**VSP**) and **virtualization service consumer** (**VSC**), through logical channels for communication known as the **virtual machine bus** (**VMBus**), enable communication between the root portion and the branch OSs, as shown in the following diagram:

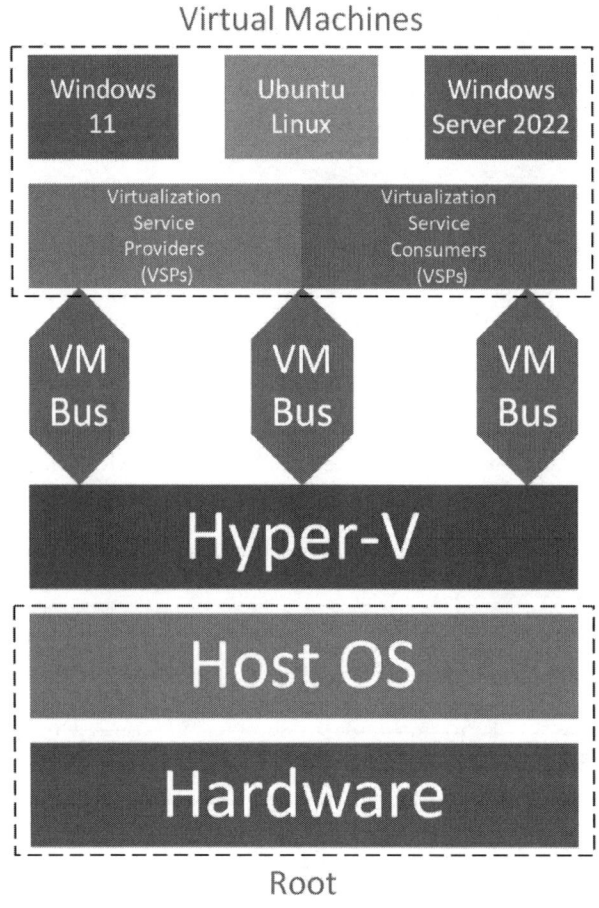

Figure 8.2 – Hyper-V architecture

Now that you've completed the *Understanding server virtualization* and *The Hyper-V architecture* sections, you should understand that the preceding diagram presents an example of a fully virtualized mode. Now that you're adequately acquainted with the Hyper-V architecture, let's learn about Hyper-V's installation requirements.

Hyper-V's installation requirements

To enable Hyper-V, the server's processors should support virtualization. That said, the processor on that physical server should support virtualization technology. Therefore, the hosting machine must be powered by an Intel or AMD processor with Intel **Virtualization Technology** (**VT**) or **AMD Virtualization** enabled.

Next, let's understand what is meant by **nested virtualization**.

Nested virtualization

Nested virtualization helps run a VM inside another VM. In other words, the host machine hardware allows you to run off Hyper-V from within a VM, thus enabling you to set up additional VMs. Although it's an exciting concept, it is unusual. However, since Windows Server 2016, Microsoft has supported nested virtualization. So, to understand it better and make it more manageable, consider running Hyper-V from the guest OS in the same way that it runs from the host OS. Therefore, the idea of a VM inside a VM will help you nest one Hyper-V within another effectively.

To set up nested virtualization in Windows Server 2022 using Windows PowerShell, follow these steps:

1. Right-click the **Start** button and select **Windows PowerShell (Administrator)** from the admin menu.

2. In the Windows PowerShell window, run the following two commands:

   ```
   Set-VMProcessor  -VMName <VMname>
   -ExposeVirtualizationExtensions
   $true
   Get-VMNetworkAdapter  -VMName <VMname> |
   Set-VMNetworkAdapter  - MacAddressSpoofing On
   ```

3. Finally, install Hyper-V (please refer to the *Chapter exercise – installing Hyper-V on Windows Server 2022* section later in this chapter).

With that, you've learned about virtualization modes, the Hyper-V architecture, Hyper-V's installation requirements, and nested virtualization. Now, it's time to become familiar with Hyper-V Manager.

Getting to know Hyper-V Manager

Hyper-V Manager is an administrative tool that is used to manage VMs. The following operations can be carried out with Hyper-V Manager:

- Creating, importing, and deleting VMs
- Creating a virtual switch
- Creating SAN Manager
- Inspecting and editing disks
- Stopping services

Hyper-V Manager's user interface, as shown in the following screenshot, consists of a server pane, a VM pane, a checkpoint pane, selected VM details, and an **Actions** pane:

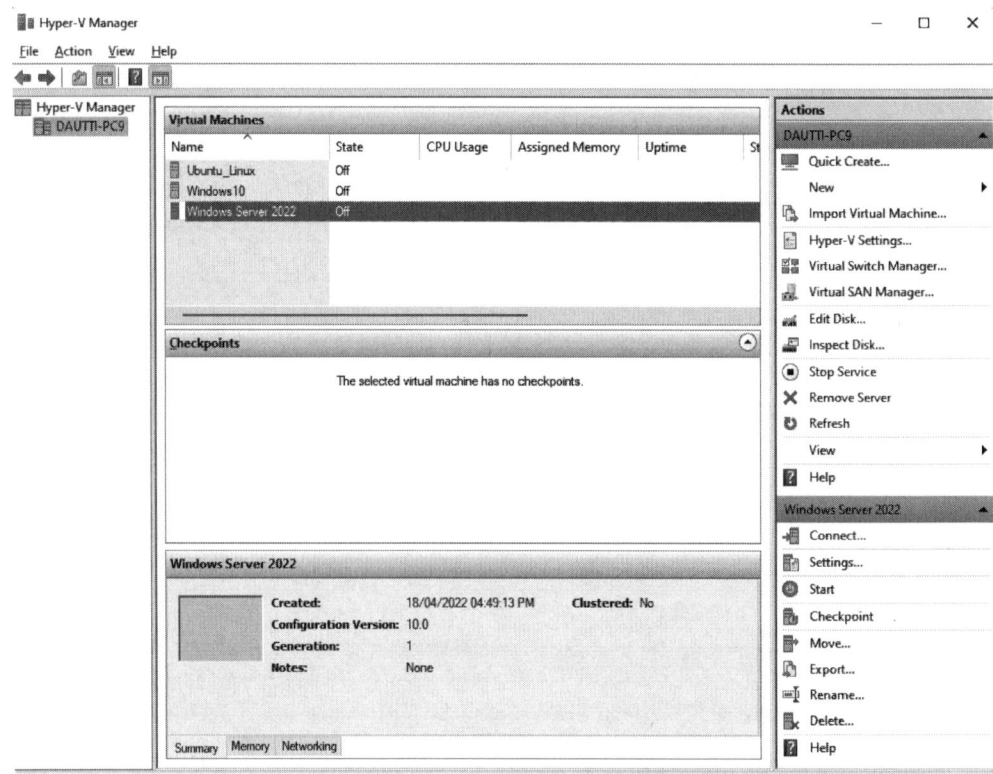

Figure 8.3 – Hyper-V Manager in Windows Server 2022

Next, you'll learn how to configure Hyper-V settings, which will help you apply them to the VMs you will build in Hyper-V.

Configuration settings in Hyper-V

Once you install the Hyper-V role in the server, it is recommended that you spend a little time getting to know the Hyper-V settings. You can set up the Hyper-V settings by clicking on **Hyper-V Settings...** in the **Actions** pane. The settings that can be set up are shown in the following screenshot:

- **Virtual Hard Disks** specifies the location on your server for storing virtual hard disk files.
- **Virtual Machines** specifies the location on your server for storing VM configuration files.
- **NUMA Spanning** provides VMs with additional computing resources, allowing you to run more VMs simultaneously.
- **Storage Migrations** specifies how many storage migrations can be performed simultaneously on the server.
- **Enhanced Session Mode Policy** allows you to redirect local devices and resources from servers running a **virtual machine connection** (**VMConnect**):

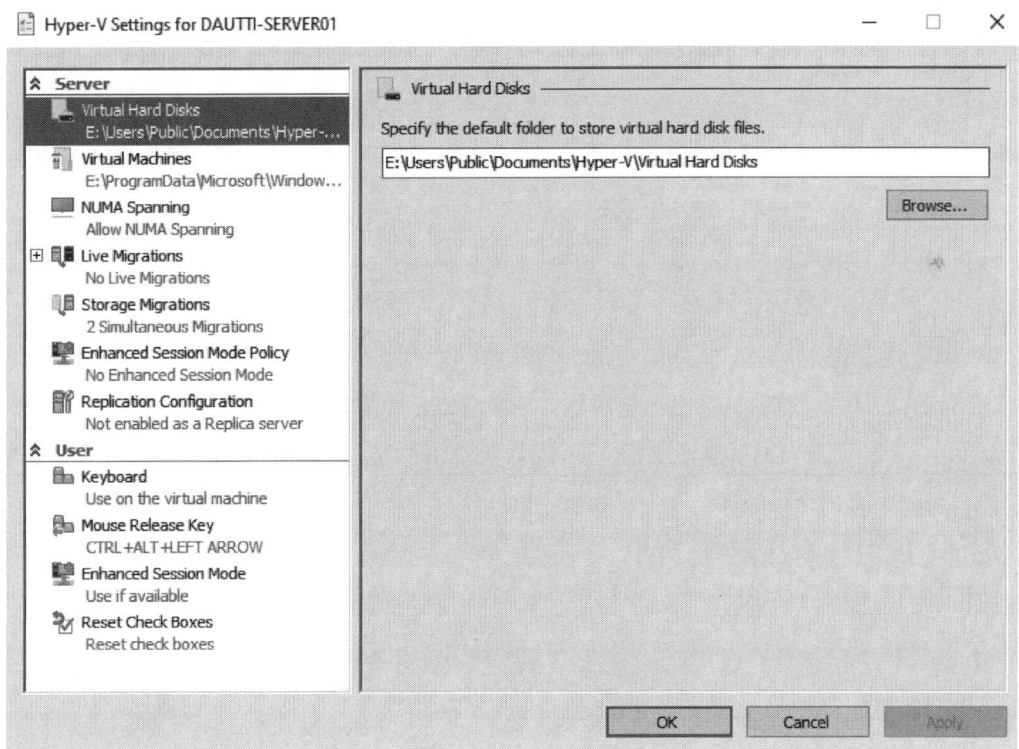

Figure 8.4 – Hyper-V Settings in Windows Server 2022

Now that you are familiar with the various Hyper-V settings, let's learn how to create and configure **virtual hard disks** (**VHDs**).

Creating and configuring VHDs

To create a VHD in Windows Server 2022 using Hyper-V Manager, follow these steps:

1. Click the **Start** button. Then, choose **Windows Administrative Tools** in the **Start** menu.
2. In the **Windows Administrative Tools** window, click **Hyper-V Manager**.
3. In the **Hyper-V Manager** window, click **New**, and then click **Hard Disk…** from the **Actions** pane drop-down, as shown in the following screenshot:

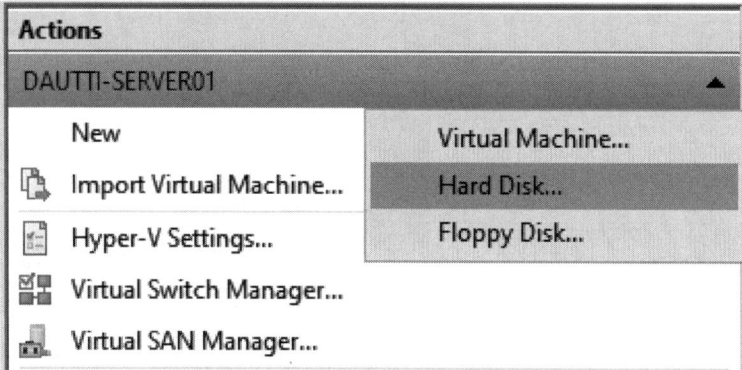

Figure 8.5 – Creating a virtual hard disk

4. Click **Next** on the **Before You Begin** page of the **New Virtual Hard Disk Wizard** area.
5. Select the format you want to use for the virtual hard disk and click **Next**.
6. Select the type of virtual hard disk that you want to create and click **Next**.
7. Specify the name and location of the virtual hard disk file and click **Next**.
8. Create a blank virtual hard disk or copy the contents of an existing physical disk and click **Next**.
9. Click **Finish** to create the virtual hard disk and close the **New Virtual Hard Disk Wizard** area.

Now, let's learn how to manage the VM's memory, which will act as the RAM of the VM.

Managing a VM's virtual memory

You can use Hyper-V Manager to manage a VM's memory. Therefore, you must turn the VMs off before setting up their memory. To operate a VM's memory in Windows Server 2022 using Hyper-V Manager, follow these steps:

1. Click the **Start** button. Then, choose **Windows Administrative Tools** in the **Start** menu.
2. In the **Windows Administrative Tools** window, click **Hyper-V Manager**.
3. In the **Hyper-V Manager** window, right-click any of the VMs with an off state and select **Settings…**, as shown in the following screenshot:

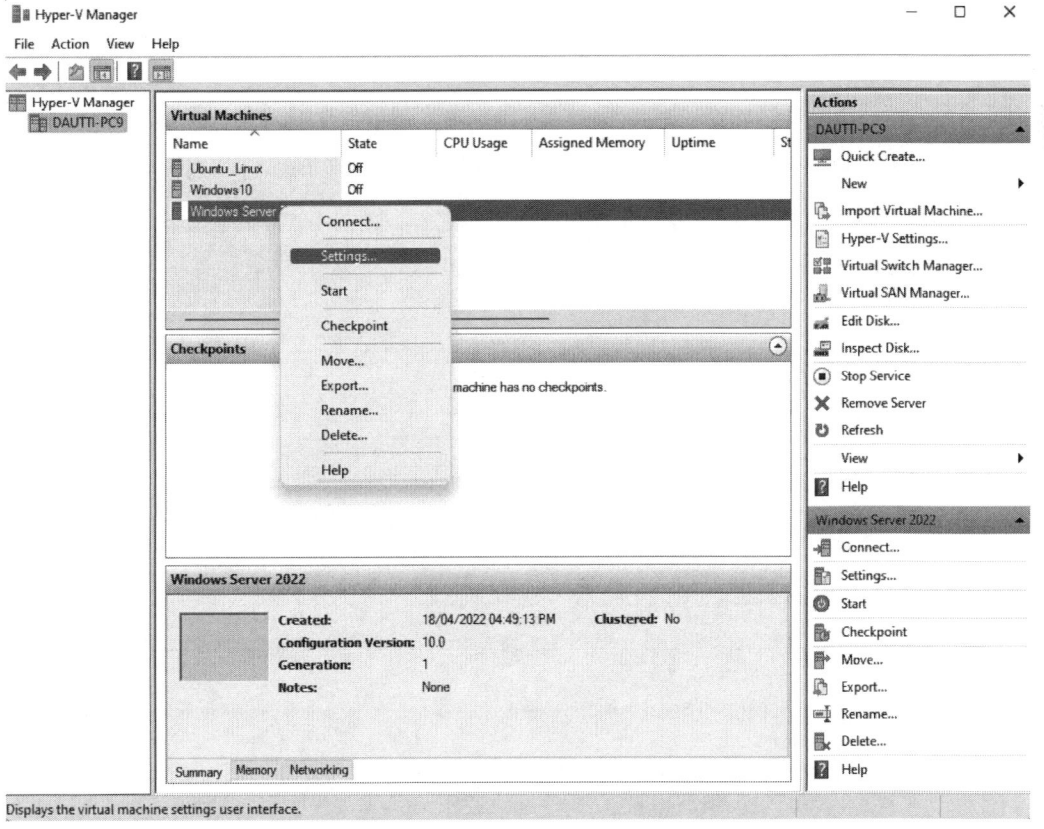

Figure 8.6 – VM settings in Hyper-V Manager

4. In the left-hand pane, under **Add Hardware**, click **Memory**.

5. You can set a fixed or dynamic amount of memory, as shown in the following screenshot:

   - Enter the amount in MB within the **RAM** text box to set a fixed amount of memory.
   - To set a dynamic amount of memory, tick the **Enable Dynamic Memory** checkbox and set the amount of memory for **Minimum RAM** and **Maximum RAM**:

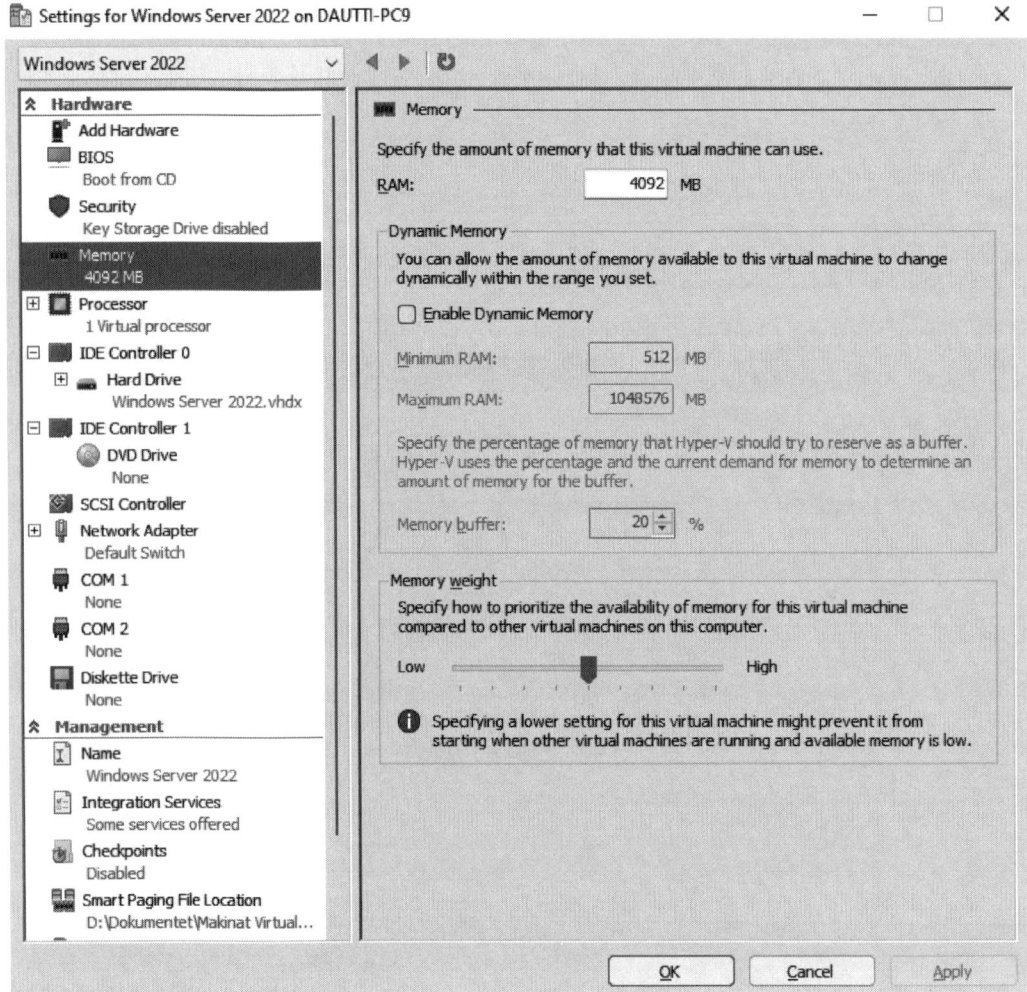

Figure 8.7 – Managing virtual memory in Hyper-V Manager

6. Click **OK** to close the **VM Settings** window.

Now, let's learn how to set up a virtual network to connect VMs.

Setting up a virtual network

Like in the physical network, where a switch is required to connect computers, similarly, in a virtual network, a virtual switch is required to connect VMs. There are three types of virtual switches available in Hyper-V:

- The **external switch** binds the physical network adapter so that the VMs can access the physical network.
- The **internal switch** can only be used by the VMs on that physical server and between VMs and the physical server.
- The **private switch** can only be used by the VMs on that physical server.

To set up a virtual switch in Windows Server 2022 using Hyper-V Manager, follow these steps:

1. Click the **Start** button. Then, choose **Windows Administrative Tools** in the **Start** menu.
2. In the **Windows Administrative Tools** window, click **Hyper-V Manager**.
3. In the **Hyper-V Manager** window, click **Virtual Switch Manager...** in the **Actions** pane, as shown in the following screenshot:

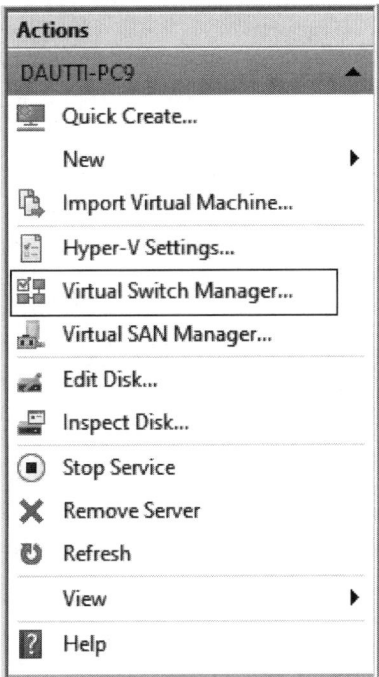

Figure 8.8 – Creating a virtual switch

4. Select the type of virtual switch that you want to create, and then click **Create Virtual Switch** (see *Figure 8.9*).

5. Enter a name for the new virtual switch.

6. Enter notes for the new virtual switch.

7. Select the connection type that you want the virtual switch to connect to.

8. Enable virtual LAN identification:

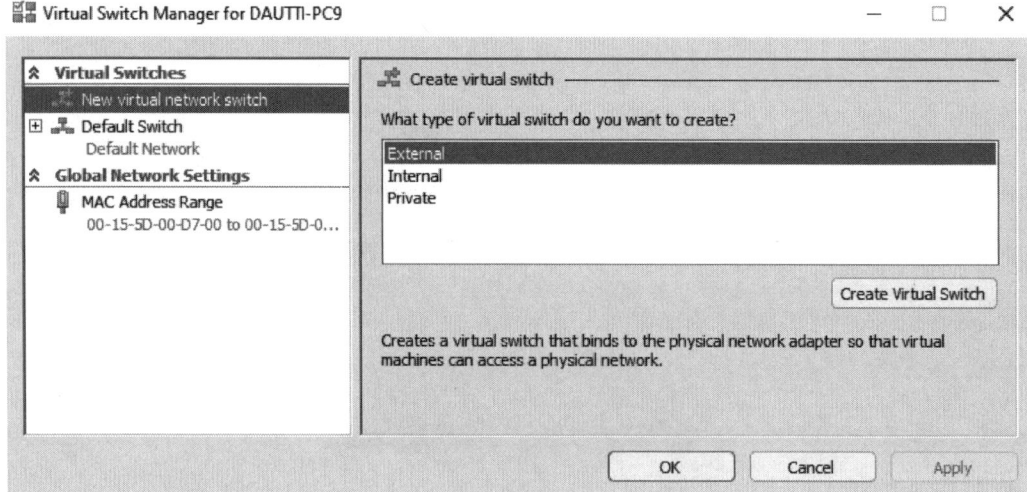

Figure 8.9 – Virtual switch properties

9. Click **OK** to close the **Virtual Switch Manager** window.

In the next section, we'll briefly learn about checkpoints, which help VMs recover if something goes wrong during updates, installations, or configurations.

Understanding checkpoints

Hyper-V offers various features and capabilities to facilitate the administration of the virtual environment. These include checkpoints (formerly known as snapshots). In Hyper-V, checkpoints are called **restore points**. A checkpoint is a point in time created when a new application or an update is installed on a VM. A checkpoint helps revert a VM to an early state whenever something goes wrong with its execution.

To set up a checkpoint for a specific VM, follow these steps:

1. Click the **Start** button. Then, choose **Windows Administrative Tools** in the **Start** menu.
2. In the **Windows Administrative Tools** window, click **Hyper-V Manager**.
3. In the **Hyper-V Manager** window, right-click the VM and select the **Checkpoint** option from the context menu, as shown in the following screenshot:

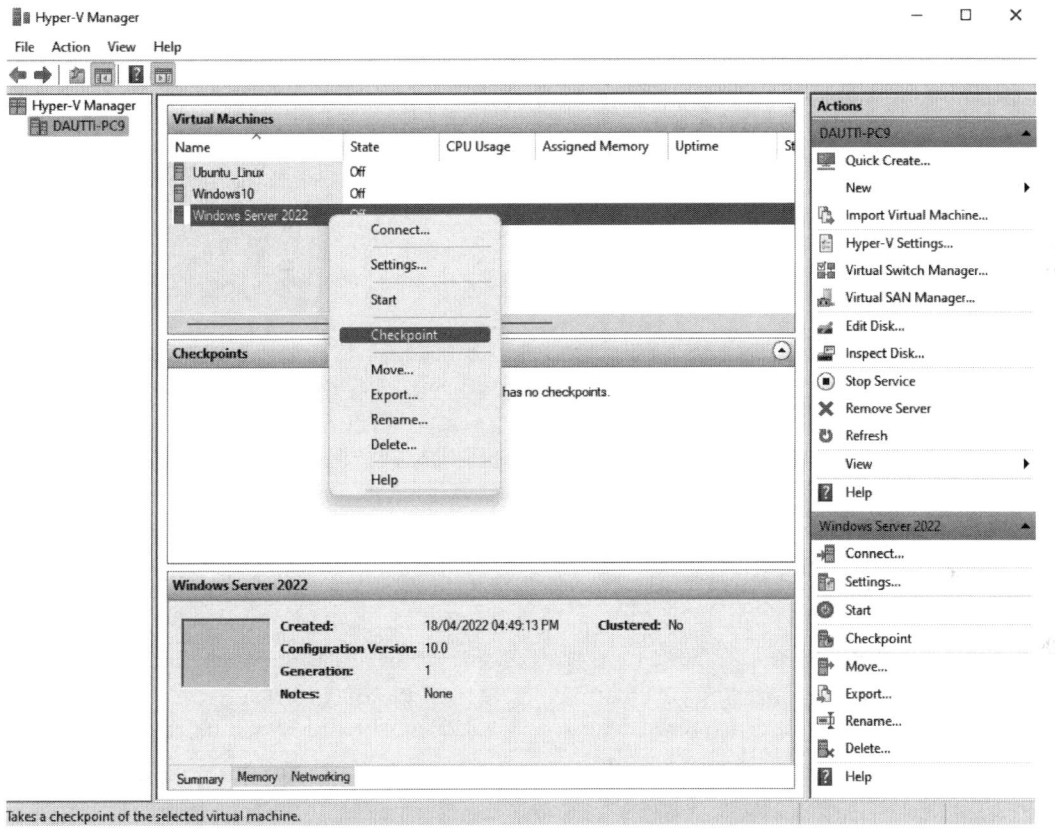

Figure 8.10 – Creating a checkpoint

4. Shortly, you will notice that the checkpoint has been created within the **Checkpoints** section.

When creating a checkpoint for a running VM, you will receive confirmation that the checkpoint has been created, as shown in the following screenshot:

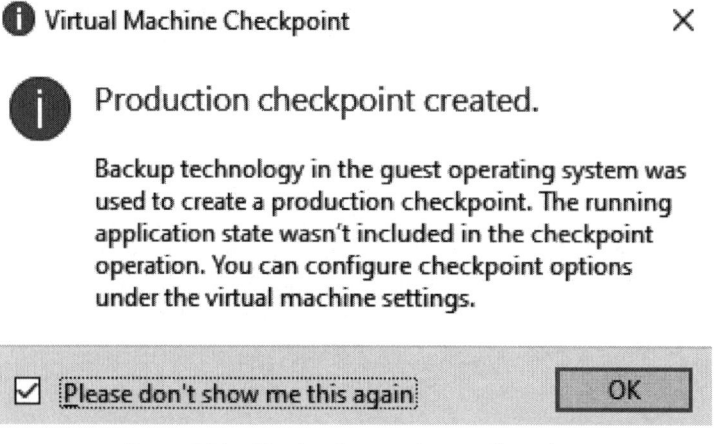

Figure 8.11 – Checkpoint creation confirmation

Windows Server 2022 offers two types of checkpoints, as shown in *Figure 8.12*:

- **Production checkpoint**, which does not include information about running applications
- **Standard checkpoint**, which captures the current state of applications

This checkpoint discussion has highlighted concepts such as data and application consistency. In the case of data consistency, each user sees a consistent view of the data – for example, visible changes that have been made by the users and other users' transactions in an organization's database. In the case of application consistency, however, this depends on the way the application's state is captured. Such a practice allows you to coordinate backups across its constituents. An example of this is checkpoints in Hyper-V, which serve to back up VMs, containers, and cloud services:

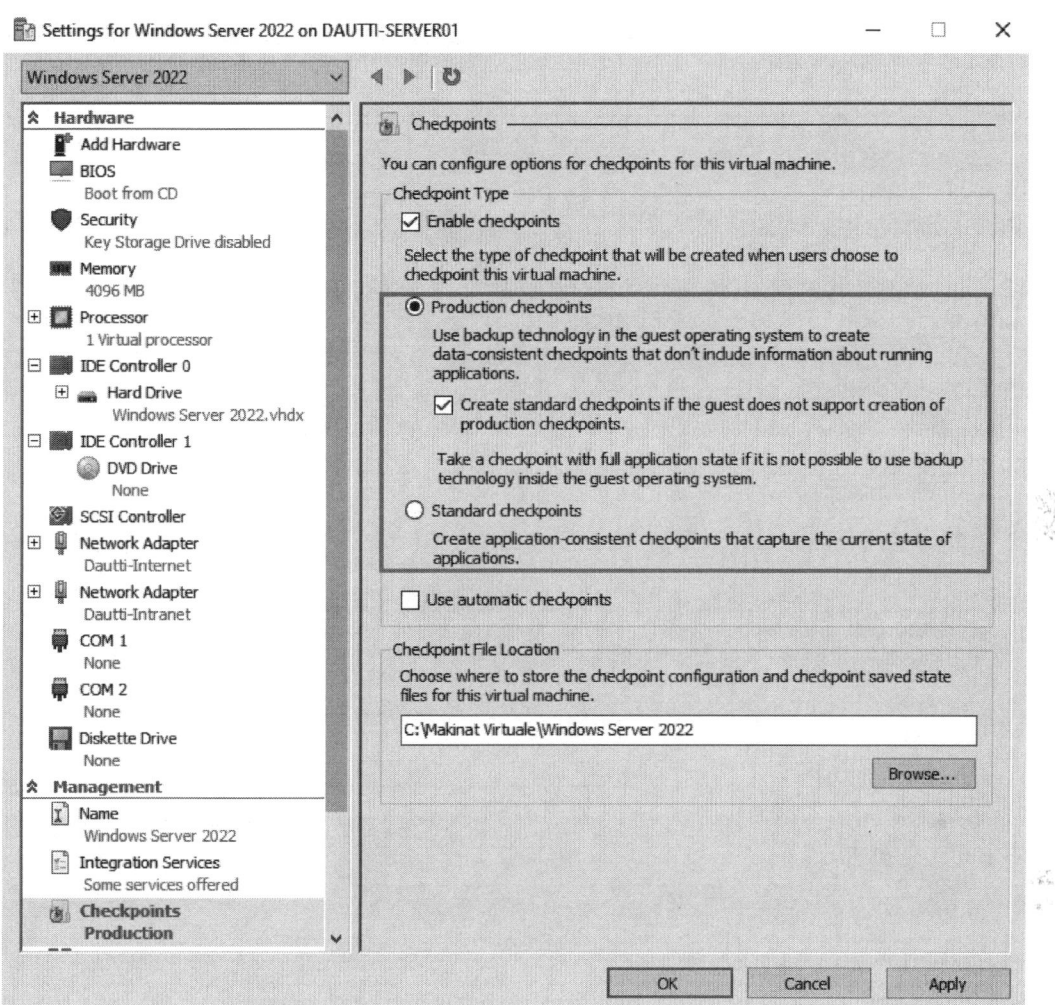

Figure 8.12 – Checkpoint types

Next, let's understand VHD and VHDX files. These represent the VM's disk, which is where an OS, apps, and data can be installed and stored.

VHD and VHDX formats

When Hyper-V was introduced in Windows Server 2008, it supported the VHD format with a disk storage capacity of up to 2 TB. Due to that, VHD became Hyper-V's *native disk storage*. However, with Windows Server 2012, Microsoft introduced a new Hyper-V feature, that of a VHDX with a disk storage capacity of up to 64 TB. Like that, VHDX replaced VHD by making the latter a *legacy disk storage* format that Hyper-V still supports on Windows Server 2022.

Next, let's understand **physical-to-virtual (P2V)** conversion, which can be used to virtualize on-premises workloads.

P2V conversion

Nowadays, virtualization has become the norm. Because of that, organizations are migrating their physical servers to virtual servers for reduced cost, ease of management, and future expansion. Thus, knowing that VMs use VHDs, Microsoft engineers have developed the **Disk2vhd** app, as shown in the following screenshot, to convert a physical disk drive into a VHD to complete the P2V conversion. Then, using Hyper-V Manager, you can set up a VM with a converted VHD:

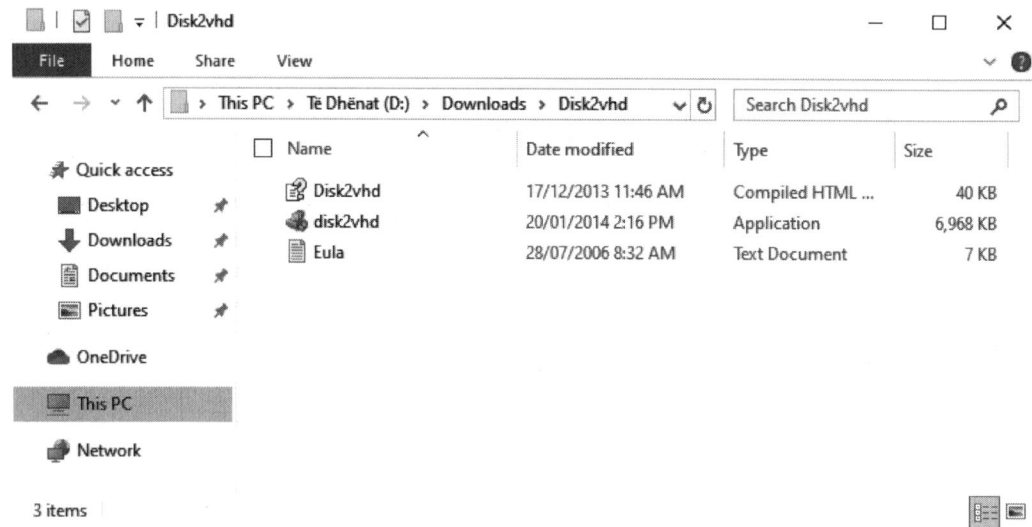

Figure 8.13 – The Disk2vhd app allows you to convert a physical disk drive into a VHD

> **Important note**
> You can download the Disk2vhd app from `https://docs.microsoft.com/en-us/sysinternals/downloads/disk2vhd` to run the conversion of P2V.

Next, let's understand **virtual-to-physical (V2P)** conversions.

V2P conversion

Despite our reasons for V2P conversion, it is good to remind ourselves that the trend is P2V conversion because of the technological era. Hypervisor manufacturers, including Microsoft, will not encourage you to conduct V2P conversion. This might be why hypervisor manufacturers now offer P2V conversion tools. However, you may find adequate tools for V2P conversion from various sources on the internet.

Another common enterprise option for V2P conversion would be migration. In that instance, you would want to install Windows Server 2022 on a physical server and then migrate settings and applications from a virtual server to a physical server.

> **Important note**
> You can download the EZ Gig IV cloning software from `https://www.apricorn.com/upgrades/ezgig` to perform V2P conversion. It works in three simple steps: select your source drive, select your destination drive, and press the **Start Clone** button.

Now, let's configure the VM settings that have been applied to that VM alone. This will help you understand the difference between Hyper-V and VM settings.

Configuring VM settings

Right-click the desired VM and select **Settings** from the context menu to set up VM settings. Among the VM settings that are available, the following ones can be set up:

- **Add Hardware** allows devices to be added to your VM.
- **BIOS** allows the boot order to be placed.
- **Security** allows state and VM migration traffic to be encrypted.
- **Memory** helps the VM's memory to be set.
- **Processor** supports the number of virtual processors to be set up.
- **IDE Controller 0** allows hard drives and CD/DVD drives to be added to the first IDE controller.
- **IDE Controller 1** allows hard drives and CD/DVD drives to be added to the second IDE controller.
- **SCSI Controller** allows hard drives to be added to/removed from an SCSI controller.
- **Network Adapter** allows you to configure the network adapter to be specified.
- **COM 1** allows the first virtual COM port to be configured.

- **COM 2** allows the second virtual COM port to be configured.
- **Diskette Drive** allows the virtual floppy disk file to be specified.

These settings can be seen in the following screenshot:

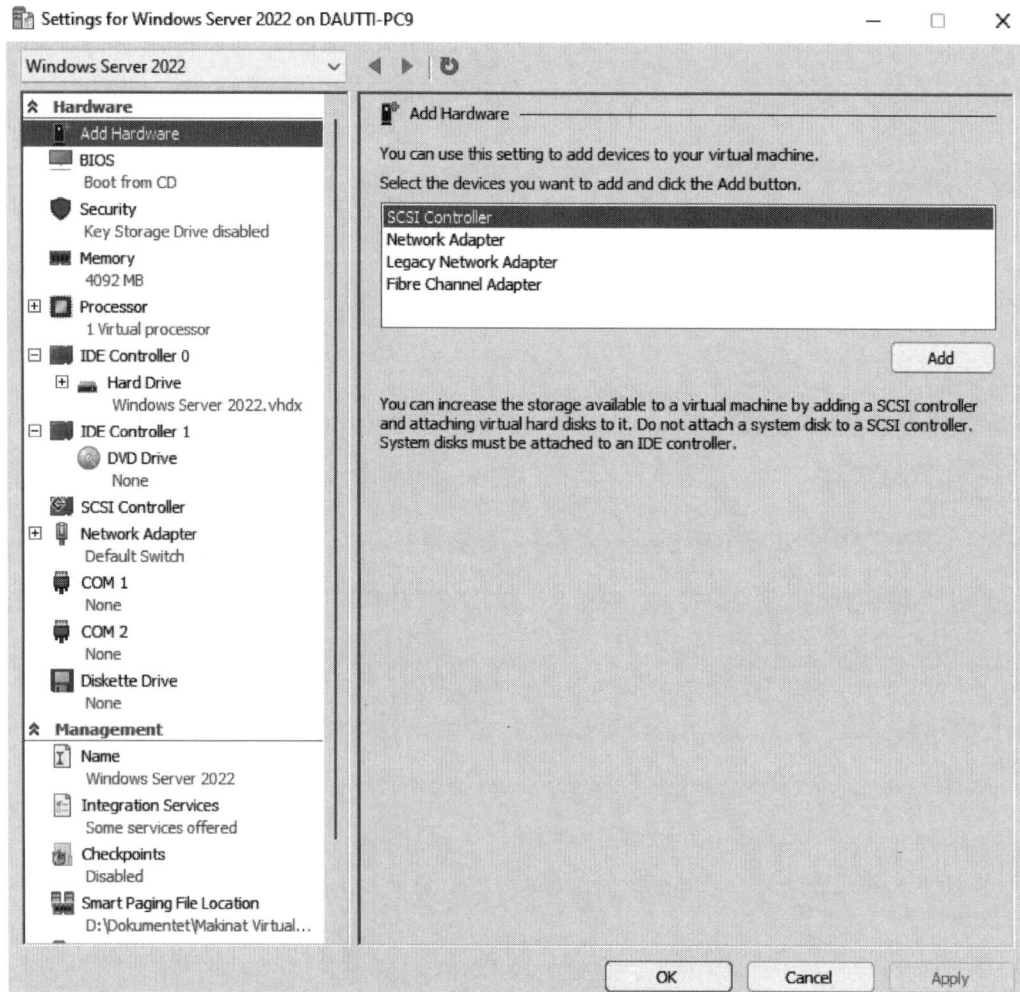

Figure 8.14 – Establishing VM settings

Now, let's learn how to manage VMs since there will usually be more than one VM in Hyper-V.

Managing VMs

Regarding VM management, the **Actions** pane and the VM's context menu offer many options for you to consider. From **New** down to **Help**, the **Actions** pane, which acts as a one-stop resource, helps you create VMs, establish Hyper-V settings, create virtual switches and virtual SANs, edit and inspect disks, stop services, remove servers, and refresh, as shown in the following screenshot:

Figure 8.15 – The Actions pane in Hyper-V Manager

Unlike the **Actions** pane, the VM's context menu options only focus on VMs. So, for example, from **Connect...** to **Rename...**, you can manage the selected VM, as shown in the following screenshot:

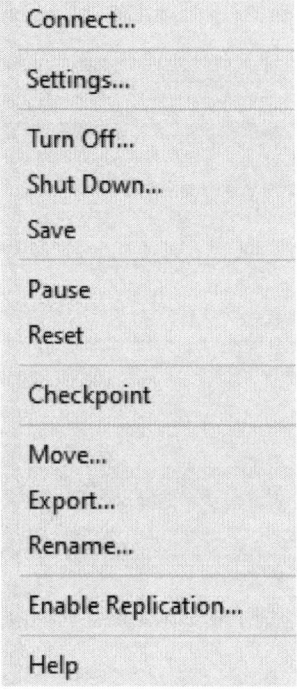

Figure 8.16 – Context menu in Hyper-V Manager

In this section, you became familiar with Hyper-V Manager and learned about its many options and capabilities. Now, let's learn how to install the Hyper-V role in Windows Server 2022 by completing this chapter's exercise.

Chapter exercise – installing Hyper-V on Windows Server 2022

To install the Hyper-V role on Windows Server 2022 using Server Manager, follow these steps:

1. Click the **Start** button. Then, choose **Server Manager** in the **Start** menu.
2. In the **Server Manager** window, click **Add roles and features**.
3. On the **Before You Begin** page, click **Next**.
4. On the **Installation Type** page, click **Next**.

5. On the **Server Selection** page, click **Next**.
6. Select the **Hyper-V** role, as shown in the following screenshot:

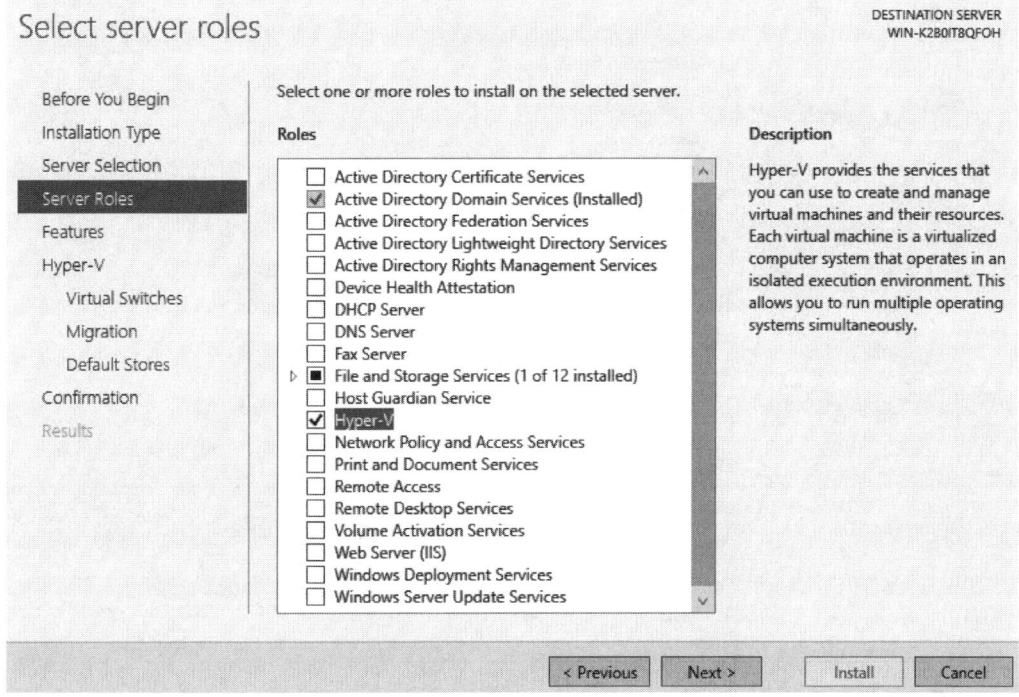

Figure 8.17 – Selecting the Hyper-V role

7. Click the **Add Features** button to add features that are required for Hyper-V.
8. There is no feature to add, so click **Next**.
9. On the **Hyper-V** definition page, click **Next**.
10. Select the available network adapter and click **Next**.
11. Select **Allow this server to send and receive live migrations of virtual machines** and click **Next**.
12. Set up the path where you will store the VMs and click **Next**.
13. Confirm the installation selections for the Hyper-V role by clicking **Install**.

14. When the installation process completes, click **Close**, as shown in the following screenshot. The server will restart automatically:

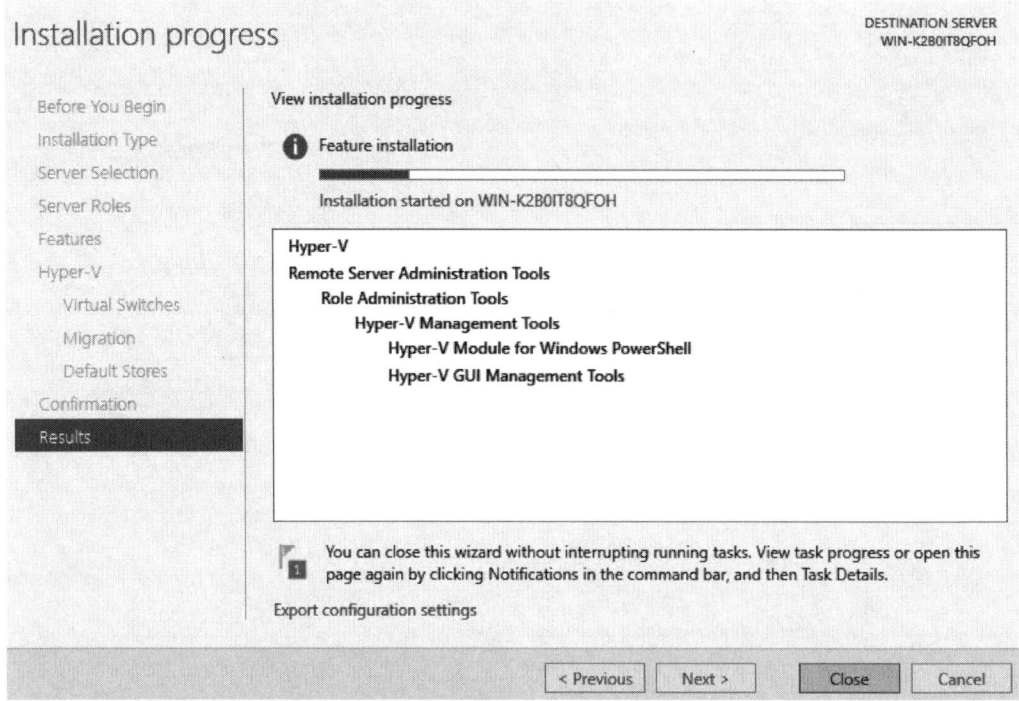

Figure 8.18 – Installing the Hyper-V role on Windows Server 2022

The Hyper-V role will be installed and ready to be used.

Summary

In this chapter, you learned about Hyper-V, through which you can enable virtualization on a server and set up and run VMs.

Then, you learned about virtualization concepts and the components and capabilities of Hyper-V, which can help you virtualize physical servers and thus be able to run the client/server networking services from the organization's virtualized IT infrastructure.

Finally, this chapter concluded with an exercise that provided instructions on installing the Hyper-V role.

In the next chapter, we will look at storing data in Windows Server 2022 while exploring the various storage technologies available.

Questions

Answer the following questions to test your knowledge of this chapter:

1. Hyper-V provides services you can use to create and manage VMs and their resources. (True | False)
2. _____ is based on a hierarchical format where the first level represents the hypervisor as the main element of the Hyper-V virtual platform.
3. Which of the following are virtualization modes in Hyper-V? (Choose 2)

    A. Fully virtualized mode

    B. Paravirtualized mode

    C. Production checkpoints

    D. Standard checkpoints

4. Checkpoints allow you to make backups of the disk image at a specific time so that when unexpected situations arise, you can revert your VM to a previous state. (True | False)
5. Components such as _____ and _____, through logical channels for communication through the VMBus, enable communication between the root portion and the branch OSs.
6. Which of the following are checkpoint types in Hyper-V? (Choose 2)

    A. Production checkpoints

    B. Standard checkpoints

    C. Inspect disk

    D. Edit disk

7. Organizations migrate their physical servers to virtual servers (P2V) for cost, ease of management, and future expansion. (True | False)
8. _____ is an administration tool that you can use to manage VMs.
9. Which of the following are elements of the Hyper-V architecture? (Choose 2)

    A. Hypervisor

    B. Root

    C. Branch

    D. Snapshot

10. Discuss nested virtualization.
11. Discuss P2V conversion.
12. Discuss V2P conversion.

Further reading

To learn more about the topics that were covered in this chapter, take a look at the following resources:

- *Virtualization documentation*: https://docs.microsoft.com/en-us/windows-server/virtualization/virtualization

- *Hyper-V Virtual Hard Disk Format Overview*: https://docs.microsoft.com/en-us/previous-versions/windows/it-pro/windows-server-2012-r2-and-2012/hh831446(v%3Dws.11)

- *Disk2vhd v2.01*: https://docs.microsoft.com/en-us/sysinternals/downloads/disk2vhd

9
Storing Data in Windows Server 2022

This chapter is designed to give you a brief understanding of storage technologies and related topics. From a technical point of view, a disk is considered one of the server's core hardware components. For example, it stores, accesses, provides, and manages digital data, files, and services. That said, you can tell how vital these technologies are in the computer world.

This chapter will teach you about physical interfaces and disk controllers, storing data in a storage medium, and the system types that are used in network environments. Various storage concepts and protocols such as **Data Deduplication (Dedup)**, **Storage Spaces Direct (S2D)**, **Software-Defined Storage (SDS)**, **Small Computer System Interface (SCSI)**, **Internet Small Computer System Interface (iSCSI)**, **Fibre Channel (FC)**, **Fibre Channel over Ethernet (FCoE)**, and others will be covered in this chapter. Hence, you will learn how to manage a server's storage using Server Manager and Windows PowerShell.

Furthermore, you will be introduced to the concepts and types of **Redundant Array of Independent Disks (RAID)**. After that, you will learn about volatile and non-volatile storage technologies such as RAM, ROM, **hard disk drives (HDDs)**, **solid-state drives (SSDs)**, optical drives, and flash memory drives and cards. Finally, this chapter will conclude with an exercise on enabling Dedup in Windows Server 2022.

The following topics will be covered in this chapter:

- Understanding storage technologies
- Understanding RAID
- Understanding disk types
- Chapter exercise – enabling Dedup on Windows Server 2022

Technical requirements

To complete the exercise in this chapter, you will need the following equipment:

- A PC with Windows 11 Pro, at least 16 GB of RAM, 1 TB of HDD, and access to the internet
- A virtual machine with Windows Server 2022 Standard (Desktop Experience), at least 4 GB of RAM, 100 GB of HDD, and access to the internet

Understanding storage technologies

Can you imagine an application server without RAM or a file server without an HDD? These questions may sound odd because RAM and HDD are inseparable from today's computer systems. This section aims to show you how vital storage technologies are to computers and that they come in various types, shapes, and sizes for different purposes. Besides the fact that storage technologies are an objective of the certification exams, their importance is absolute in **information and communications technology** (**ICT**). Therefore, this chapter is dedicated to storage technologies.

In most cases, the server will likely require ample storage space besides high processing power, sufficient RAM, and several network connections. Whether it's a single server or a cluster of servers, IDE, SAS, SCSI, DAS, NAS, SAN, and RAID represent various storage technology options.

Let's begin by understanding what the different storage types are.

Exploring different storage types

As noted earlier, storage technologies are numerous, and there are multiple ways and methods to use them. Because of that, different types of storage technologies exist. Thus, the following list will help you become familiar with the main categories of storage technologies:

- **Optical disks** offer large capacities and read-and-write accepted speeds. However, they continue to play the role of backup media for data.
- **HDDs** offer large capacities and high read-and-write speeds. However, after a long period of being the first choice for OSs, they seem to have lost. Thus, HDDs are mainly used today to store client/server app data.
- **SSDs** are becoming a popular storage technology with growing capacities and extraordinary read-and-write speeds. As a result, SSDs have become the secondary storage choice where an OS is installed in today's computers.

Next let's get acquainted with the **Advanced Technology Attachment** (**ATA**), **Parallel ATA** (**PATA**), **Serial ATA** (**SATA**), and **Small Computer System Interface** (**SCSI**) interfaces, which are used to connect storage technologies.

The ATA, PATA, SATA, and SCSI interfaces

Like many technologies for storing data, various interfaces are available to connect these storage devices to computers. Therefore, when acronyms such as ATA, PATA, SATA, and SCSI are mentioned, you can immediately determine the interface type and its usage. For example, ATA, also known as **Integrated Drive Electronics** (**IDE**), is a legacy interface for connecting HDDs, optical disk drives, floppy disk drives, and related storage technologies to computers. The two most popular types of ATA interfaces are as follows:

- **PATA**: This uses a 40-pin connector and cable to transfer data to connect the storage device to the computer's motherboard. It uses Molex as a power connector to connect the storage device to the computer's power supply. The disk controller resides on a drive itself.

- **SATA**: This represents a replacement for the PATA interface and is widely used in PCs. It uses a 7-pin connector to transfer data and connect the storage device to the computer's motherboard. It also uses a 15-pin connector to connect the storage device to the computer's power supply. Like PATA, the disk controller is also located on the SATA drive.

- These can be seen in the following figure:

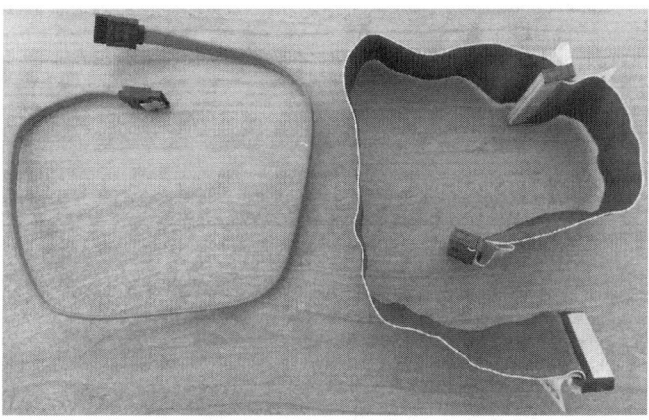

Figure 9.1 – PATA and SATA data cables

Another type of interface is SCSI. Pronounced *scuzzy*, SCSI is another interface that connects storage devices and peripheral devices to computers. The two most popular types of SCSI are **SCSI Parallel Interface** (**SPI**) and **Serial-Attached SCSI** (**SAS**). While SPI is considered the legacy version of SCSI, SAS is the modern version and provides high data transfer rates, and is widely used in servers.

Next, we will learn about the **Peripheral Component Interconnect** (**PCI**) and **PCI Express** (**PCIe**) expansion cards.

PCI and PCIe

In the mid-90s, PCI replaced IBM's **Industry Standard Architecture (ISA)**, a 16-bit internal bus specification for PCs. Unlike ISA, Intel's PCI offered 32-bit and 64-bit internal bus specifications, enabling more data to be communicated to RAM. Later, with the increased demand for faster speeds, PCI was replaced by **PCI Express (PCIe)**, as shown in the following figure. PCIe is an internal serial bus standard with four connections: PCIe x1, PCIe x4, PCIe x8, and PCIe x16. It transmits data in full-duplex mode, sending and receiving data simultaneously over communication paths known as **lanes**:

Figure 9.2 – The PCIe slot

Next, let's become familiar with **local storage**, which usually refers to a computer's internal disk.

Understanding local storage

Local storage refers to the HDD or SSD directly attached to the server. As the name suggests, **Direct-Attached Storage (DAS)** refers to a single disk or a group of disks known as a disk subsystem (see *Figure 9.3*) directly attached to a server. Therefore, you will not be mistaken if you think of your computer's HDD as DAS. In addition to internal storage devices, external storage devices connected to computers with any of the interfaces mentioned earlier are also considered DAS:

Figure 9.3 – DAS system

Now that you are familiar with local storage, let's learn about **network storage**.

Understanding network storage

In contrast to local storage, network storage, as the name implies, refers to the storage device that's connected to a computer network to provide data storage capabilities and access to users. Therefore, it can be understood that it is the network device, and thus interfaces that are common in networking are used. **Network Attached Storage (NAS)** and **Storage Area Network (SAN)** are the best examples of network storage. We will discuss these in the following sections.

Network-attached storage

As the name implies, NAS is a network appliance that connects to computers and servers through a switch and acts as dedicated storage in an organization's network, as shown in the following diagram:

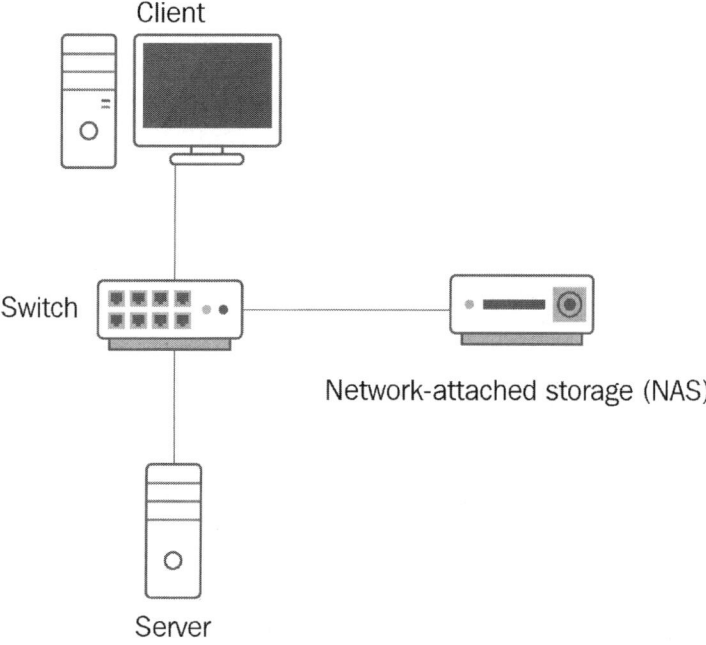

Figure 9.4 – NAS system

NAS is mainly used as the file server and connects to the network through a standard Ethernet connection. It brings flexibility so that organizations can rely entirely on NAS for file-sharing services without the need to use other servers for such tasks.

Storage area network

Like DAS and NAS, SAN is another in-store storage technology. However, unlike DAS and NAS, SAN is considered a dedicated network of storage devices because, as its name suggests, SAN is the type of LAN made of disks instead of computers. SAN is a shared pool of storage space that can be accessed by servers and clients in an organization. It mainly uses an Ethernet connection or **Fiber Channel** (**FC**) connected to **Host Bus Adapter** (**HBA**) interfaces. Hence, the HBA is an expansion card that's added to a server to connect to a SAN. In addition, proprietary protocols or **Simple Network Management Protocols** (**SNMPs**) provide management for SANs, as shown in the following diagram:

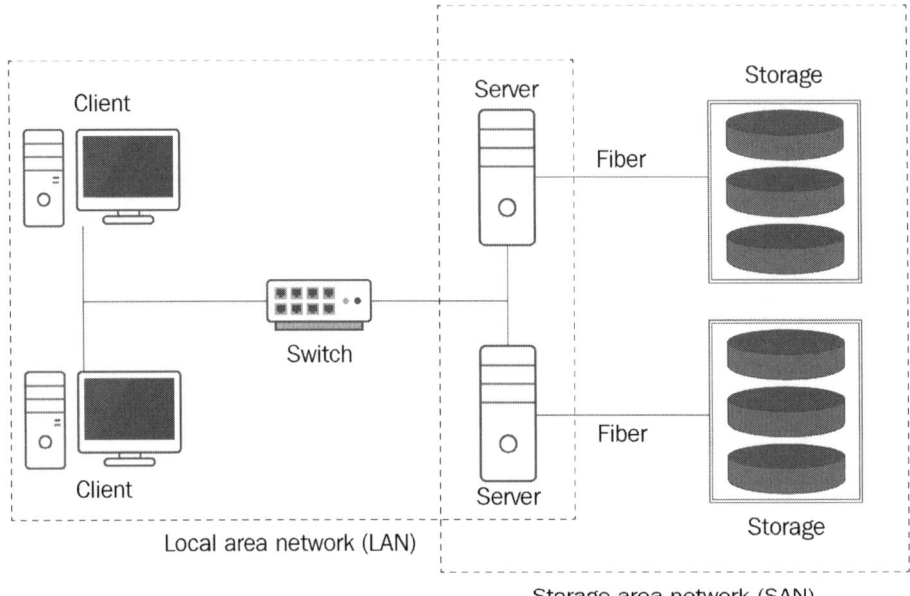

Figure 9.5 – SAN system

Now that you have been introduced to local and network storage, let's get acquainted with block- and file-level storage. Both approaches represent how data is stored in storage technologies.

Differentiating between block-level storage and file-level storage

The following table will make it easier to understand and, at the same time, compare the file-level storage and block-level storage technologies:

File-Level Storage	Block-Level Storage
Data is stored and accessed in the form of files and folders	Data is stored in blocks representing volumes, which the OS then manages
Used by NAS	Used by SAN

Table 9.1 – Block-level versus file-level storage

Next, let's learn about adapters and controllers so that you know how reading and writing take place in storage technologies.

Adapter and controller types

The disk controller is an electronic circuit located below the disk sealing part, as shown in the following figure. It performs operations such as spinning disks, moving heads for reading and writing, and communicating data to and from RAM:

Figure 9.6 – Disk controller in HDD

Next, let's understand the serial bus and data transmission technologies that transfer data to and from storage technologies.

Serial bus technologies

For data transmission, both parallel and serial communications are used. Thus, a string of 8 bits in the parallel transmission is usually transmitted at a time equal to 1 byte. In contrast, only 1 bit is generally sent at a given time in serial transmissions. Nevertheless, even though more bits are transmitted through parallel communication, serial transmission is the most widely used storage technology. This is because it is far more convenient for the disk controllers to handle bits arriving one by one (that is, serial transmission) rather than all together (that is, parallel transmission). Hence, the disk's read-and-write head reads and writes 1 bit within the given time. Because of that, the serial method of transmission has proved to be more pragmatic by eliminating overhead processing, signal skewing, and crosstalk. As a result, the most widely used serial interfaces in storage technologies are SATA, SAS, FC, and USB technologies.

Next, let's look at which storage protocols you will find easier to understand if you draw a parallel with networking communication protocols.

Storage protocols

Storage protocols allow you to store and retrieve data in/from storage systems. Mainly, storage technologies such as NAS and SAN utilize storage protocols. The most used storage protocols are as follows:

- **Small Computer System Interface (SCSI)** is a storage protocol that's utilized in block-level storage systems. For example, the OS uses the SCSI protocol to read and write data on a SCSI controller that manages storage devices.
- **Internet Small Computer System Interface (iSCSI)** places the standard SCSI protocol in an IP packet, thus extending its functionalities throughout the organization's network.
- **Fibre Channel (FC)** is another way of extending the functionalities of the standard SCSI protocol by enabling storage consolidations and high-speed transmission over longer distances.
- **Fibre Channel over Ethernet (FCoE)** does the same for the FC protocol as iSCSI does for the SCSI protocol. That being said, FCoE extends the functionalities of the FC protocol across Ethernet networks.

> **Important note**
> You can learn more about SCSI at `https://www.lifewire.com/small-computer-system-interface-scsi-2626002`.

Now, let's learn about file-sharing protocols, which represent examples of networking communication protocols themselves.

File-sharing protocols

File-sharing protocols enable data sharing over LANs, WANs, and the internet. Simply put, file-sharing protocols provide the standard for file requests between clients and servers. The most widely used file-sharing protocols are as follows:

- **Server Message Block (SMB)** represents a file-sharing protocol that's heavily used by Windows OSs. For example, Microsoft's **Common Internet File System (CIFS)** is an SMB dialect, which specifies the SMB protocol's particular implementation.
- **Network File System (NFS)** is a file-sharing protocol mainly used by Unix and Linux distros.
- **File Transfer Protocol (FTP)** enables file sharing by transferring files from site to site over the internet.

- **Hypertext Transfer Protocol (HTTP)** allows file sharing over a **World Wide Web (WWW)** service.
- **Secure Shell (SSH)** enables remote file sharing over a secure connection.

> **Important note**
> You can learn more about SSH at `https://www.ssh.com/ssh/`.

Next, we will learn about the HBA and FC switches for connecting network storage technologies such as SAN.

The HBA and FC switches

As explained earlier in this chapter, the HBA is an interface that enables fiber connectivity to the server. In contrast, the FC switch is a Layer 3 network switch. Therefore, both network components are compatible with the FC protocol. Moreover, a high-speed FC network medium connects the two, thus creating the FC fabric. The FC fabric consists of one or more FC switches and, as such, it constitutes the SAN topology.

Next, we will learn about the iSCSI hardware, another technology for connecting network storage technologies.

The iSCSI hardware

iSCSI, a form of block-level storage, uses an IP to send SCSI commands over TCP/IP networks. iSCSI works so that the clients, known as **initiators**, use the IP protocol to send SCSI commands called **command descriptor blocks (CDBs)** to storage devices known as **targets**. In SANs, the **logical unit number** (LUN) represents a logical disk. In iSCSI, TCP port 860 is reserved for the iSCSI system port, whereas TCP port 3260 is iSCSI's default port.

Next, let's learn about **Storage Spaces Direct (S2D)** to understand storage pools.

S2D

S2D, introduced in Windows Server 2016, is a supported feature in Windows Server 2022. In addition, it was updated to offer storage tiering when using fast media such as SSDs or NVMe for caching. Furthermore, S2D allows you to group disks into storage pools, thus creating software-defined **storage spaces**:

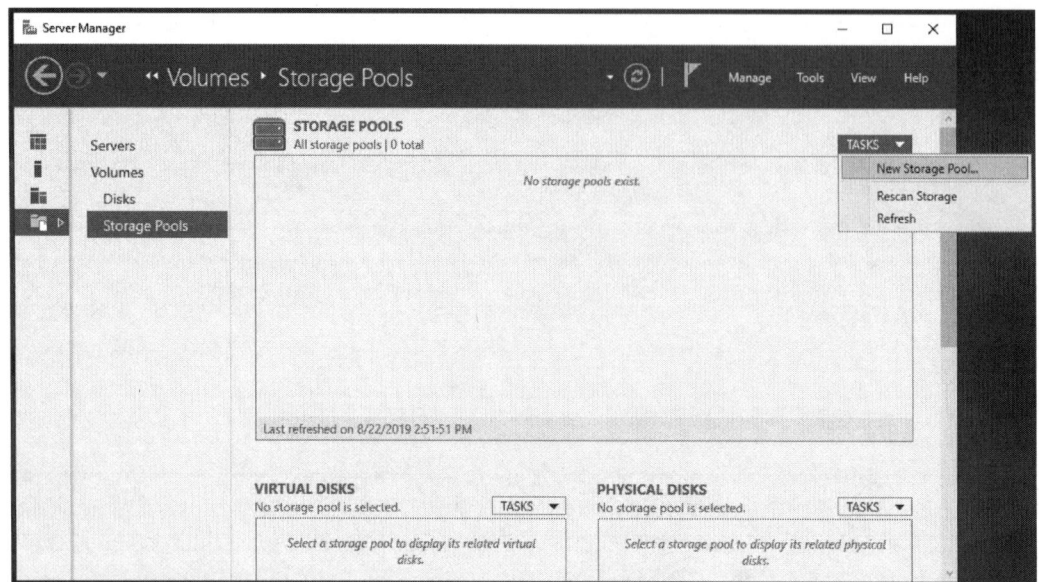

Figure 9.7 – Creating a new storage pool in Windows Server 2022

Now, let's learn how the Dedup feature removes duplicated data.

Dedup

The idea behind the **Data Deduplication** server role, known as **Dedup**, is to provide disk space savings. Dedup is a technique that removes duplicated data from a dataset, thus storing a single copy of identical data on a disk. First, it analyzes the data to identify duplicated data in the dataset. Then, the original file is stored in storage media while the duplicated files are replaced with a reference that points to the original file:

Understanding storage technologies   255

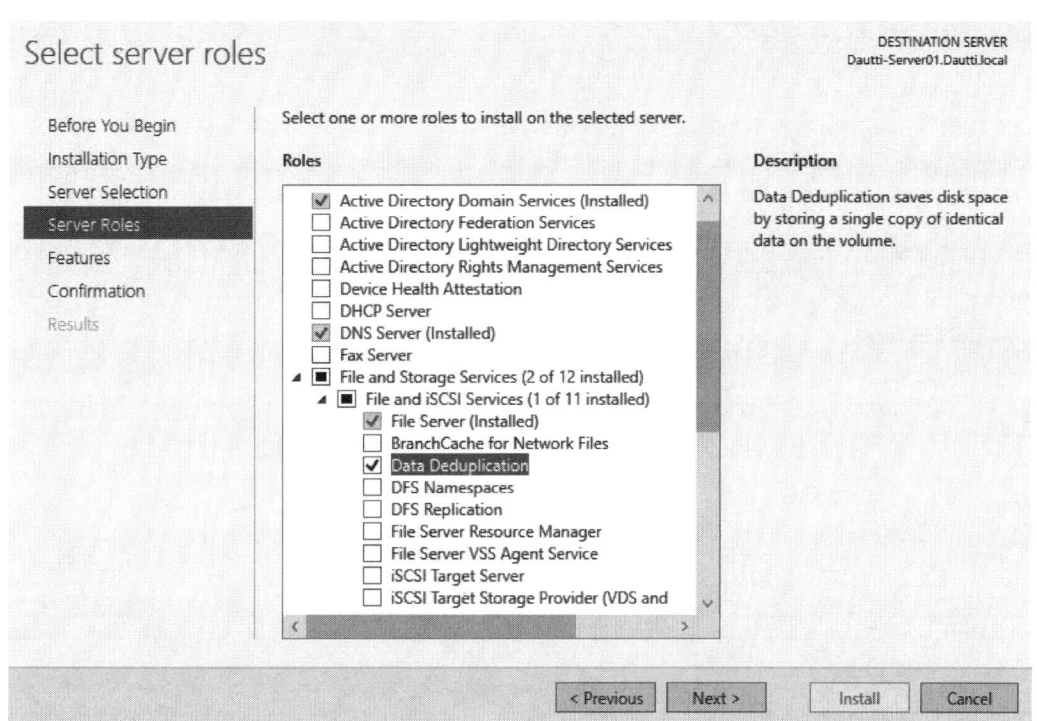

Figure 9.8 – Installing Dedup in Windows Server 2022

> **Important note**
> Cluster rolling upgrades let you upgrade the server OS in a cluster without the need to stop Hyper-V.

Next, let's learn how storage tiering helps store data in high-performing storage.

Storage tiering

Another interesting built-in feature in Microsoft's Windows Server is **storage tiering**, which allows you to automatically transfer the most frequently accessed files to faster storage. In simpler terms, this lets you combine high-performance storage with low-performance storage (for example, HDDs with SSDs) to reduce storage costs. Thus, the storage tiering agents will place the most accessed files on the faster storage, while rarely accessed files will be placed in slower storage.

Now, let's learn how to manage storage with Server Manager and Windows PowerShell.

Managing storage

Before taking on storage management with **Server Manager** (`servermanager.exe`), ensure that the **File and Storage Services** role has been added to the server. As you know, to add **File and Storage Services**, use the **Add roles and features** option within **Server Manager**. The following screenshot shows how to manage storage using **Server Manager** on the local server:

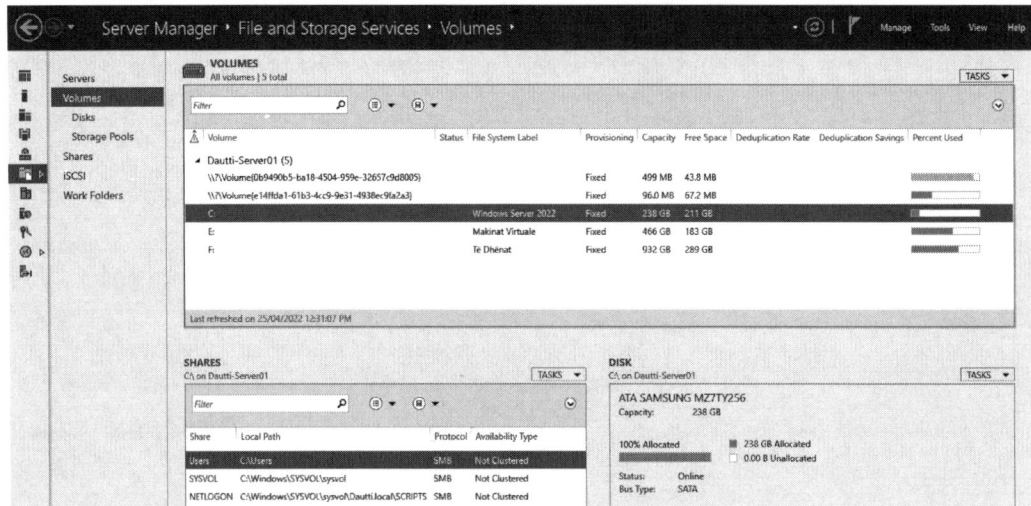

Figure 9.9 – Managing storage with Server Manager

Besides Server Manager, the Windows Server storage can also be managed through Windows PowerShell. Among the cmdlets that can be used to manage the storage in Windows Server using the Windows PowerShell, there's `Get-Disk`, `Get-Partition`, and `Get-Volume`, as shown in the following screenshot:

Figure 9.10 – Managing storage with Windows PowerShell

This section taught you about storage technologies, adapter and controller types, storage protocols, file-sharing protocols, and managing storage with Server Manager and Windows PowerShell. In the next section, you will learn about RAID.

Understanding RAID

Whether you have come across the term **Redundant Array of Independent Disks** or **Redundant Array of Inexpensive Disks**, know that you are dealing with the concept of fault tolerance in a Windows Server. That said, RAID is a technology that combines a considerable number of physical disks into a single logical unit so that it can protect data in the case of disk failure. At the same time, note that RAID is not a backup solution and should never be considered.

Let's look at the various types of RAIDs to understand storage redundancy.

Types of RAID

There are a considerable number of RAID types. The most widely used RAID types are as follows:

- **RAID 0** is known as disk striping. It offers higher read and writing performance but is not fault-tolerant. For example, on Windows Server 2022, you can create a striped volume if you convert the disk from basic into dynamic.

- **RAID 1** is known as disk mirroring and requires at least two disks for its implementation while offering excellent read and write performance. It works so that all the data on disk A is mirrored on disk B. In the case of disk failure, the RAID controller uses any available disks.

- **RAID 5** is disk striping with parity and requires three disks for implementation at a minimum. As a result, it represents the most fault-tolerant RAID that's available. In addition, the parity data is spread across all disks, meaning that RAID 5 can withstand the failure of a single disk.

- **RAID 01 or 10** (*beware, it is not ten; instead, it is one zero*) is the combination of striping and mirroring because it combines RAID 0 with RAID 1. Because of that, it requires a minimum of four disks for its implementation. In the case of disk failure, the rebuild time is very fast since the striping is spread across all drives.

Next, let's compare hardware RAID and software RAID.

Hardware versus software RAID

Concerning the RAID deployment, there are two types of RAID:

- A **hardware RAID** deployment (see *Figure 9.11*) is an expensive solution and requires configuration before installing the OS. It is an electronic board that either the manufacturer of the server or the server's technician will plug into the appropriate slot on the server's motherboard:

Figure 9.11 – The RAID controller

- A **software RAID** deployment is a cheaper solution that's configured after installing an OS. Specifically, it is an application that you buy from a particular vendor.

Next, let's understand **Software-Defined Storage** (**SDS**), where the software is used to provide a storage solution.

SDS

Due to budget constraints, an organization cannot afford to own NAS or SAN storage systems, so the alternative solution is SDS. Organizations can create SDS with local storage using S2D in Windows Server 2022. Therefore, SDS helps separate the software that manages the storage from the storage hardware itself. In addition, such a solution offers excellent flexibility in using various storage technologies.

Now, let's learn how redundancy while using S2D enables fault tolerance and storage efficiency.

Resiliency using S2D

The fault-tolerance approach in S2D is resiliency, which offers a mirror with parity. This is similar to RAID software in deployment but with notable differences in Windows Server 2022. This is because S2D provides fault tolerance and storage efficiency.

Now, let's learn about the **high availability** (**HA**) standard.

High availability

HA is a characteristic of a system that's available anytime and anywhere. However, specific standards were established to achieve a HA requirement that relies upon numerous parameters. Hence, from backup to fault tolerance and resilience to reliability, all storage media and devices must be operational to make the system highly available.

The following table shows the HA standard:

Availability (%)	Downtime Per Month	Downtime Per Year
99%	7.20 hours	3.65 days
99.9%	43.2 minutes	8.76 hours
99.99%	4.32 minutes	52.6 minutes
99.999%	25.9 seconds	5.26 minutes
99.9999%	2.59 seconds	31.5 seconds

Table 9.2 – High availability standard

In this section, you learned about the different types of RAID, the difference between hardware and software RAID, SDS, resiliency using S2D, and HA. In the next section, you will learn about disk types to understand the various storage technologies available for storing data.

Understanding disks

As you may know, a disk is a physical component or an object in which data is stored. However, *storing* data involves writing and reading data on a disk. Therefore, to become familiar with disk types, it would help if you get acquainted with their classifications and technical specifications. That way, you will understand any disk type's storage features and characteristics. We will learn about the HDDs first.

HDDs

The HDD, as shown in *Figure 9.12*, is considered a secondary storage computer component that uses an electromotor to spin the disk. It also contains a magnetic read-and-write head and metal platters that permanently store data. Each platter includes tracks and sectors. The starting point for storing data in HDDs is the outer track. The read-and-write head is located above the disk platter at a distance of microns, thus never touching the disk. If it does, then physical damage occurs.

The data storage capacity is measured in bytes (nowadays, GB and TB), whereas the disk spinning speed is measured in **Rotations Per Minute** (**RPM**). The most common RPM rates for PCs and laptops are from 5,400 (5.4 K) RPM to 7,200 (7.2 K) RPM, while for servers, the most common RPM rates are from 10,000 (10 K) RPM to 15,000 (15 K) RPM.

Usually, the HDD is located inside the computer's case and is mounted in a drive bay. However, there are also external HDDs, which are mainly used for storing and backing up data and not running the OS. In the event of HDD disposal, it is recommended to perform disk shredding, primarily if such disks were used to store data in financial organizations or banks:

Figure 9.12 – An HDD

Understanding disks

> **Important note**
> You can learn more about HDDs at `https://www.computerhope.com/jargon/h/harddriv.htm`.

Next, let's learn about SDDs, a storage technology similar to HDDs but with different technical characteristics.

SSDs

Like HDD, an SDD, as shown in *Figure 9.13*, is another secondary storage technology for PCs and servers. However, unlike HDDs, SSDs are considered to be memory modules with no moving parts. As a result, they use less voltage – usually 5V – than HDDs, which commonly use 12V to spin the disk platters.

SSDs are noiseless, more physically reliable, and provide faster data access. Recently, SSDs have been matching HDDs as far as capacity is concerned. Because of that, nowadays, SSDs have become the first choice for secondary storage in PCs and servers where an OS is installed and running. Therefore, many manufacturers have turned HDDs into the first choice for storing the application's data on PCs and servers. Moreover, SSD drives are also part of NAS and SAN solutions:

Figure 9.13 – An SSD

> **Important note**
> You can learn more about SSDs at `https://www.lifewire.com/solid-state-drive-833448`.

Next, let's learn about the **Optical Disk Drive** (**ODD**) to become familiar with lands and pits.

ODDs

Unlike HDDs, which use electromagnetic fields to read and write data from/to disk platters, an ODD, as shown in *Figure 9.14*, utilizes a laser beam under a specific wavelength to read and write data from/to **compact disks (CDs)**. However, always differentiate between ODDs and **optical disks (ODs)**, such as CDs or DVDs. The former is where a medium such as a CD or DVD is inserted. CDs contain tracks in the form of a spiral. In contrast to HDDs, which use the outer track, the inner track is the starting point for storing data on CDs and DVDs. As with HDDs, optical disks are measured by capacity in bytes. Usually, a CD's storage capacity is between 650 MB and 700 MB, while a DVD's storage capacity ranges from 4.7 GB to 8.5 GB. The speed of ODs is measured in KB/s and is determined by an X symbol equal to 150 KB/s. Therefore, if your optical drive has a speed of 24x, its speed is *24 x 150 KB/s = 3600 KB/s = 3.6 MB/s*. There are three recording types of optical disks, as follows:

- **CD-ROM and DVD-RAM** are read-only optical disks.
- **CD-R and DVD-R/DVD+R** are write-once optical disks.
- **CD-RW and DVD-RW/DVD+RW** are rewritable optical disks.

The following figure shows an ODD:

Figure 9.14 – An ODD

DVDs and Blu-ray disks are the most widely used optical disks for writing purposes. The latter has been designed to supersede DVD technologies, thus achieving tremendous storage capacities. For example, a single-layer Blu-ray disk holds 25 GB, a dual-layer Blu-ray disk holds 50 GB, a triple-layer Blu-ray disk has 100 GB, and a quadruple-layer Blu-ray disk holds 128 GB.

> **Important note**
>
> You can learn more about OD types at `https://www.ifixit.com/Wiki/Optical_Disc_Types`.

Next, let's learn about basic disks based on the **Master Boot Record** (**MBR**) and **GUID Partition Table** (**GPT**) partition schemes.

Basic disk

Once you have installed the OS on the server's **hard disk** (**HD**), the HD structure is in its *basic configuration*. This means that the primary disk configuration is organized into partitions. As you have learned so far, a basic disk is based on the MBR and GPT partition schemes, and, as such, one partition cannot be extended on one or more physical disks. Instead, a partition can be extended by adding unallocated space from the same physical disk.

Next, we will look at dynamic disks to understand how read-write performance can be increased.

Dynamic disk

Dynamic disk configuration was introduced to overcome the limitations of the basic disk by providing increased read and write performance with disk striping and by operating with volumes instead of partitions. This means that the volumes in a dynamic disk configuration can be extended to more than one physical disk, thus creating five volumes, as follows:

- A **simple volume** is a part of a physical disk that functions as if it were a separate unit.
- A **mirrored volume** is a fault-tolerant volume that copies data to two or more physical disks.
- A **striped volume** combines areas of free space from many hard disks into a logical volume.
- A **spanned volume** combines areas of unallocated space from multiple disks into a logical volume.
- **RAID-5 volumes** stores data with parity information across multiple physical disks.

Now that you know what basic and dynamic disks are, let us learn how to convert a basic disk to a dynamic one in order to have dynamic volumes in infrastructure.

Converting a basic disk into a dynamic disk

In Windows Server 2022, to convert a basic disk into a dynamic disk, you must complete the following steps:

1. Right-click the **Start** button.
2. Select **Disk Management**.

3. Right-click the preferred disk, and, from the context menu, select **Convert to Dynamic Disk...**, as shown in the following screenshot:

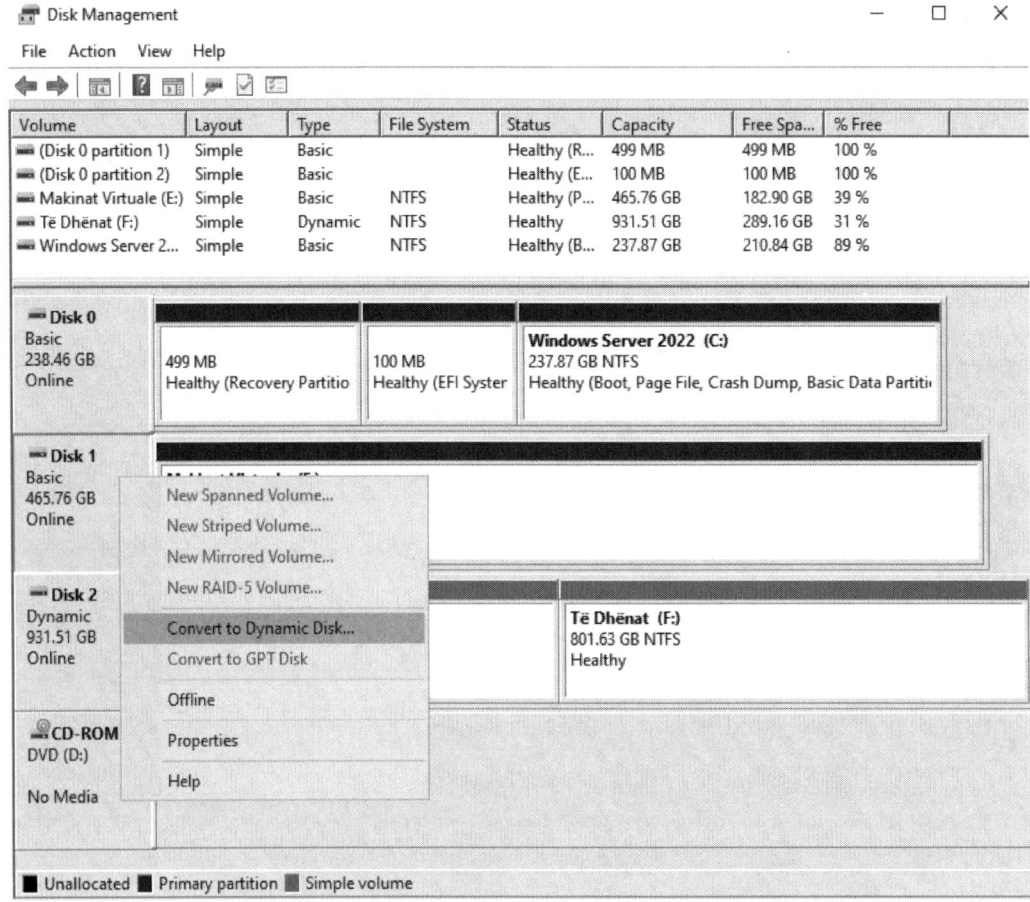

Figure 9.15 – Converting a basic disk into a dynamic disk

4. If you have more than one disk, then, from the **Convert to Dynamic Disk** window, select the relevant disks and click **OK**.
5. Click **Convert** in the **Disks to Convert** window.
6. After reading the information in the **Disk Management** dialog box, click **Yes**.
7. Shortly after, the conversion will be completed.

Next, let's learn about mount points to increase the folder size.

Mount points

A mount point is established when attaching an unallocated partition to an empty folder. Such an operation helps increase the folder size if the partition runs out of space. Therefore, in Windows Server 2022, Disk Management (`diskmgmt.msc`) can be used to set up a mount point, as shown in the following screenshot:

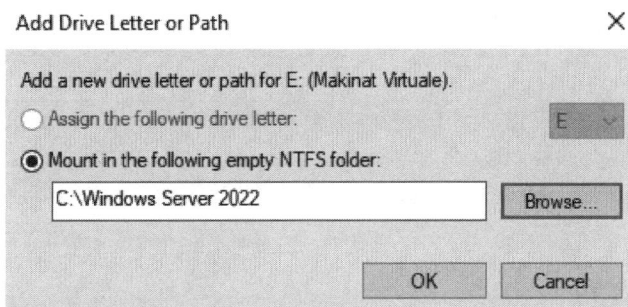

Figure 9.16 – Creating a mount point with Disk Management in Windows Server 2022

Next, let's learn about filesystems, which we can use to store and organize data in storage devices.

Filesystems

A filesystem organizes files on physical media, such as HDDs, CDs, and flash drives. Furthermore, a filesystem helps run operations on files and folders, such as naming, renaming, copying, moving, and deleting. In addition, a filesystem allows you to specify the place on a disk where the file is logically placed for storage and retrieval. Hence, the stored information would have been impossible to identify and retrieve without a filesystem. By taking a look at the following list, you will become familiar with the well-known filesystems that the Windows OS uses:

- **File Allocation Table (FAT)**: This is the earliest filesystem that was used by both MS-DOS and Windows. As its name implies, it is based on a table that contains a map of clusters. A cluster is a unit of logical storage on the hard disk. FAT32 is the latest version of FAT.

- **New Technology File System (NTFS)**: This filesystem was introduced in the 1990s with Windows NT 3.1 and is still used on Windows 11 and Windows Server 2022. Among the features that NTFS offers, there's disk quotas, **Encrypting File System (EFS)**, journaling, the **Volume Shadow Copy Service (VSS)**, and security. NTFS is a native filesystem on Windows Server 2022.

- **Resilient File System (ReFS)**: This filesystem was introduced in Windows Server 2012 and was meant to be the successor of NTFS. However, nowadays, it is mainly used on disks, partitions, or volumes where client/server application data is stored. ReFS is available as a disk format option in Windows Server 2022. The new features that ReFS offers include resiliency, performance, and scalability.

- **Extended File Allocation Table (exFAT)**: This new version of FAT was developed primarily with portable storage devices such as USB flash drives and SD cards. Interestingly, exFAT is platform-independent, thus enabling drives that have been formatted with an exFAT filesystem to be supported by Mac computers.

> **Important note**
> A journaling filesystem is an important feature that maintains data integrity by keeping track of changes being made to data in a separate log, thus making it possible to restore data whenever a power outage or disk crashes occur. Microsoft has removed `Journal.dll` from Windows Server 2016. Therefore, it is not supported in Windows Server 2022 either.

Next, let's learn how to mount a **virtual hard disk (VHD)** so that it looks and operates like a physical disk.

Mounting a VHD

Based on the concept of the mount point, which was explained earlier in this chapter, a mounted VHD drive is a mount point that represents a drive mapped to an empty folder on a volume that uses NTFS. Primarily, mounted VHD drives function like any other drive, except they use the drive path instead of drive letters.

In Windows Server 2022, you can attach a VHD to your server using **Disk Management**. To do so, follow these steps:

1. Press the Windows key + R.
2. In the **Run** window, enter `diskmgmt.msc` and hit *Enter*.
3. In the **Disk Management** window, click the **Action** menu and select **Attach VHD**, as shown in the following screenshot:

Figure 9.17 – Attaching a VHD using Disk Management in Windows Server 2022

4. The attached VHD will be displayed in the **Disk Management** window shortly after.

Next, let's learn about **Distributed File Systems** (**DFS**), a filesystem that's distributed in multiple locations.

DFS

If you have wondered how to synchronize files between servers in your organization automatically, then DFS will help. Moreover, it helps organize many distributed file shares into a DFS. Such sharing is done in an authorized and controlled way. With DFS, data stored in shared folders on different servers can be grouped into logically structured namespaces. This makes it possible for users to access the data as if it were stored on local computers. In Windows Server 2022, DFS is part of the **File and Storage Services** role. Thus, to install the DFS role in Windows Server 2022, you should expand the **File and Storage Services** role, expand **File and iSCSI Services**, and select **DFS Namespace**, **DFS Replication**, and **File Server Resource Manager**, as shown in the following screenshot:

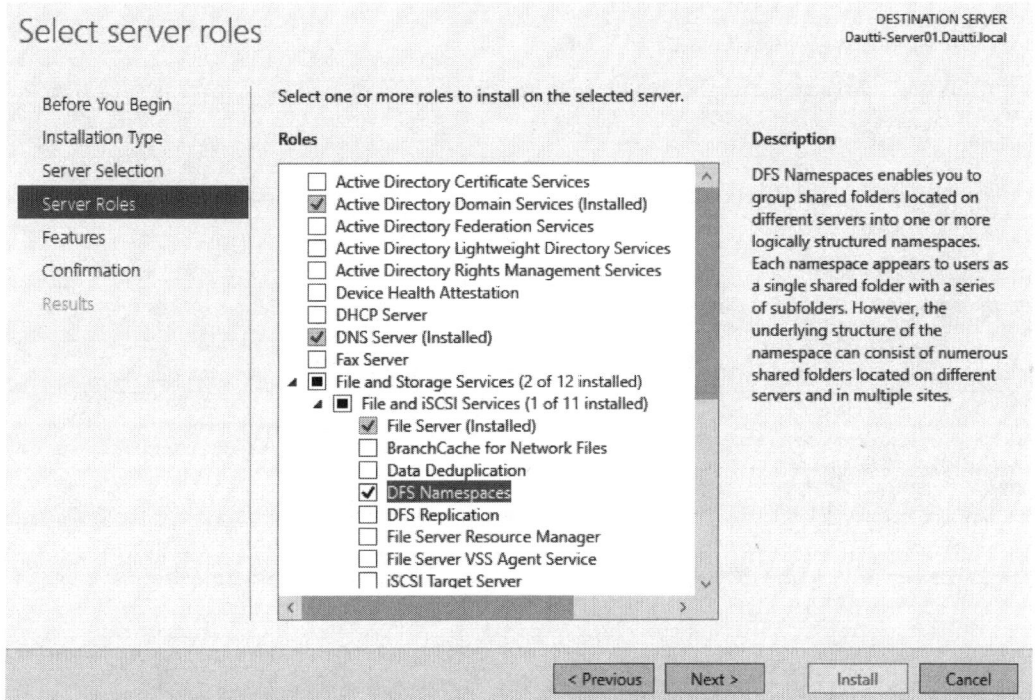

Figure 9.18 – Installing DFS in Windows Server 2022

In this section, you learned about disks, it's various types and components. The next section will teach you how to add the Data Deduplication role or Dedup feature on Windows Server 2022.

Chapter exercise – enabling Dedup on Windows Server 2022

This chapter's exercise will teach you how to install the Data Deduplication role or Dedup feature on Windows Server 2022. To add the Dedup feature to Windows Server 2022, follow these steps:

1. Click **Add roles and features** within the **Server Manager | WELCOME TO SERVER MANAGER** section.
2. In the **Before You Begin** window, click **Next**.
3. Click **Next** in the **Installation Type** window.
4. In the **Server Selection** window, click **Next**.
5. In the **Server Roles** window, expand **File and Storage Services**.
6. Then, expand **File and iSCSI Services**.
7. Select **Data Deduplication**, as shown in *Figure 9.8*.
8. In the **Features** window, click **Next**.
9. In the **Confirmation** window, click **Install**.
10. When you notice that the installation has been completed, click **Close**, as shown in the following screenshot:

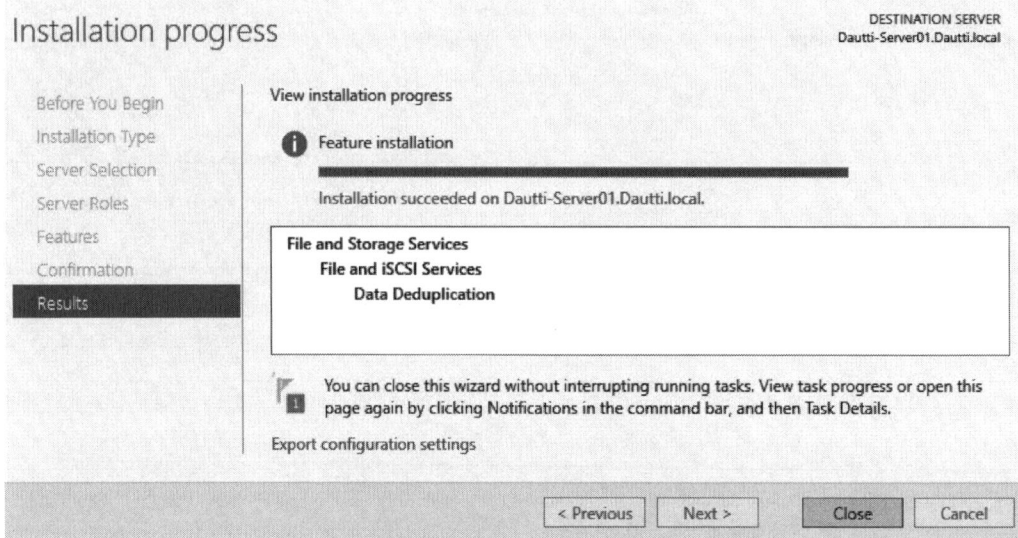

Figure 9.19 – Closing the Add Roles and Features Wizard

With that, you have successfully installed the Dedup feature on Windows Server 2022. However, to use Dedup at the volume level, you need to enable it, which requires some additional configurations.

Summary

This chapter taught you about the storage technologies available on servers and enhanced Windows Server 2022-related storage technology management features.

After exploring the various storage technologies, you now understand the multiple ways and methods of working with local storage, network storage, block-level storage, and file-level storage; adapter and controller types; serial bus technologies, storage protocols; file-sharing protocols; and more. You also briefly looked at managing storage with Server Manager and Windows PowerShell. Then, you learned about RAID and the difference between hardware and software RAIDs, which help enable HA for the services in your organization. Finally, this chapter concluded with an exercise that provided instructions on enabling the Dedup feature in Windows Server 2022.

The next chapter will teach you about tuning and maintaining the performance of Windows Server 2022.

Questions

Answer the following questions to test your knowledge of this chapter:

1. DFS allows you to data from your server in an authorized and controlled way. (True | False)
2. _____ is a network appliance that connects to computers and servers through a switch and acts as dedicated storage in an organization's network.
3. Which of the following are network storage technologies? (Choose two)

    A. DAS

    B. NAS

    C. RAM

    D. ROM

4. Block-level storage stores data in files and folders representing volumes managed by the server operating system. (True | False)
5. _____ is an electronic circuit that resides on a hard disk and performs operations such as spinning disks, moving heads for reading and writing, and transferring data to and from RAM.
6. Which of the following are storage protocols? (Choose two)

    A. SCSI

    B. FC

    C. PATA

    D. SATA

7. The HDD is a computer component that uses a motor to spin the disk, has a magnetic read-and-write head, and has metal platters that permanently store data. (True | False)

8. _____ is a characteristic of a system that never fails, thus being available at all times.

9. Which of the following are RAID types? (Choose two)

    A. RAID 1

    B. RAID 5

    C. RAID 15

    D. RAID 20

10. SCSI, pronounced *scuzzy*, is another interface that connects storage and peripheral devices to computers. (True | False)

11. _____ is a legacy interface that connects HDDs, ODDs, floppy disk drives, and related storage technologies to computers.

12. Which of the following are optical disks? (Choose two)

    A. CD-ROM

    B. DVD-RAM

    C. EPROM

    D. POST

13. Discuss Dedup in Windows Server 2022.

14. Discuss S2D in Windows Server 2022.

15. Discuss DFS in Windows Server 2022.

Further reading

To learn more about the topics that were covered in this chapter, take a look at the following resources:

- *Hard drive*: `https://www.computerhope.com/jargon/h/harddriv.htm`
- *Overview of Disk Management*: `https://docs.microsoft.com/en-us/windows-server/storage/disk-management/overview-of-disk-management`
- *Optical Disc Types*: `https://www.ifixit.com/Wiki/Optical_Disc_Types`
- *Difference Between Basic Disk and Dynamic Disk*: `http://www.differencebetween.net/technology/difference-between-basic-disk-and-dynamic-disk`
- *What is RAID (Redundant Array of Inexpensive Disks)?*: `https://stonefly.com/resources/what-is-raid`

Part 4: Keeping Windows Server 2022 Up and Running

Part 4 covers tuning, maintaining, updating, and troubleshooting in Windows Server 2022. Upon completing this part, you will be able to use task manager, performance, and resource monitor to tune and maintain Windows Server 2022. In addition, you will learn correct troubleshooting methodology, and how to use Windows Update and Event Viewer to update and troubleshoot Windows Server 2022.

This part of the book comprises the following chapters:

- *Chapter 10, Tuning and Maintaining Windows Server 2022*
- *Chapter 11, Updating and Troubleshooting Windows Server 2022*

10
Tuning and Maintaining Windows Server 2022

This chapter is designed to teach you considerations for server hardware and best practices regarding performance monitoring methodologies. Understanding the importance of a server's role in a computer network and possessing knowledge of a server's hardware components has two benefits: it helps with server hardware selection and troubleshooting server hardware-related issues.

Furthermore, this chapter will teach you server performance monitoring methodologies and procedures. Performance monitoring will help you identify the cause of server performance issues early. In that way, you will be able to react promptly to avoid further degradation of server performance. As for the server, the baseline represents a snapshot of the server's performance under a typical workload, enabling the compilation of a detailed report about the server's overall performance.

This chapter then concludes with an exercise on performance logs and alerts. In a nutshell, the following topics will be covered in this chapter:

- Understanding server hardware components
- Understanding performance monitoring
- Understanding logs and alerts
- Chapter exercise—working with the Performance Logs & Alerts service

Technical requirements

To complete the lab for this chapter, you will need the following equipment:

- A PC with Windows 11 Pro, at least 16 **gigabytes** (**GB**) of **random-access memory** (**RAM**), 1 **terabyte** (**TB**) of **hard disk drive** (**HDD**), and access to the internet
- A **virtual machine** (**VM**) with Windows Server 2022 Standard, at least 4 GB of RAM, 100 GB of HDD, and access to the internet

Understanding server hardware components

From a technical point of view, a server is nothing more than a computer. As you open up the computer's case, you will notice various components that make up the computer's internal hardware. In computer jargon, these parts are called **hardware components**—in our case, the server's hardware components.

Now, the question arises: why should we pay that much attention to these hardware components? Does this sound like the right question? First, let me remind you of something you may already know! A server's primary task is not data processing; the server's role (see *Chapter 6, Adding Roles to Windows Server 2022*) includes providing network services and handling user requests to access services. That is why you should know the server hardware and pay attention when selecting the server's hardware.

With that in mind, let's examine server hardware components, their roles, and their impacts on a server's performance to strengthen further the assertion that considerations should be made when dealing with server hardware.

Processor

Among the many internal components of the processor's architecture (shown in *Figure 10.1*), there is no doubt that speed is one of the determining factors in choosing a suitable processor. The processor's speed is measured in **hertz** (**Hz**). Today's processor speeds are measured in **gigahertz** (**GHz**). However, as noted at the beginning of this section, speed is just one crucial component among others when choosing a processor. Therefore, we would be mistaken if we relied only on the speed factor. Thus, the following factors should also be considered:

- **Cache** refers to the processor's memory. The larger the cache capacity, the more data can be retrieved from the memory to be processed. As such, cache memory is essential to the performance of the **central processing unit** (**CPU**). Modern processors have three types of cache: L1, L2, and L3, where **L** stands for **level**. While L1 and L2 are inside the processor's architecture, L3 is outside. Therefore, their numbers determine the speed of each cache memory.

- **Core** refers to the processor's processing unit. In the past, processors had just one core. Today, processors have two, four, eight, or more cores. Thus, to better understand the cores in a processor, consider mounting two processors in a single package where each processor would be presented through its core. Does that make sense? Of course, in this case, we would have a dual-core processor. So, the more cores a processor has, the more multiprocessing is done.

- **Word size** has to do with the processor's internal architecture that defines the data bus size, the instruction size, and the address size. Today, there are 32-bit and 64-bit processors. These numbers determine the processor's word size and the amount of data the processor receives from the memory to be processed and sent back to the memory.

- **Registers** refer to the processor's high-speed memory. They usually have a small capacity but are very high-speed and store small amounts of data needed during processing, such as the address of the next instruction to be executed. Since speed was mentioned, registers are considered the fastest component, not just inside the processor alone but in the computer system.
- **Virtualization technology** refers to the processor's ability concerning the virtualization concept, if the processor supports virtualization, where many operating systems simultaneously share processor resources efficiently.
- You can see an example of quad-core processors in the following screenshot:

Figure 10.1 – Intel's Xeon quad-core processors

No doubt, a processor plays a significant role in a server; however, it will not be able to function without memory. So, let's learn more about memory.

Memory

The primary storage is a hardware component that temporarily stores data, allowing the processor to access data quickly and acting as a communication bridge between applications and peripheral devices. Both RAM and **read-only memory** (**ROM**) are the computer's primary storage. While the former is volatile, the latter is considered to be non-volatile.

We can get more on the characteristics of these two types of primary storage from the following table:

RAM	ROM
Volatile	Non-volatile
Data is lost when the power goes off	Information is kept when the power goes off
Known as working memory as it loads the operating system and apps	Known as hardware initialization memory as it runs a **power-on self-test** (**POST**)

Table 10.1 – RAM versus ROM

Every computer is equipped with RAM. However, when it comes to servers, while the physical size of the server's RAM module may seem to be almost the same as that for PCs (see *Figure 10.2*), the chips in a memory module are not of the same number. Instead, the server's RAM modules have one chip more. That is because usually, servers use a type of RAM called **error-correcting code** (**ECC**) memory. ECC RAM enables the detection and correction of memory errors. Other advanced types of RAM for servers include **single-device data correction** (**SDDC**) and **double-device data correction** (**DDDC**), which enable multiple memory errors to be detected and corrected simultaneously. However, with all of these advantages of RAM modules for servers, one disadvantage is their price. RAM for servers is generally expensive due to the characteristics mentioned in this section.

In the following screenshot, you can see an example of ECC RAM modules placed in memory banks:

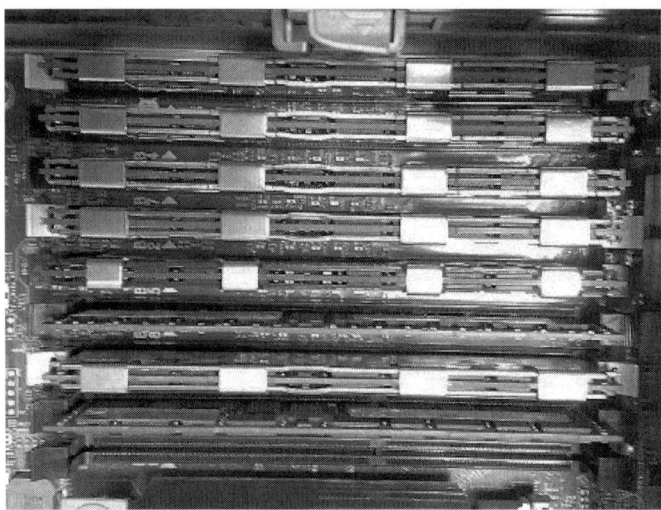

Figure 10.2 – ECC RAM modules placed in memory banks

While RAM is considered primary storage and is volatile, the disk is regarded as a secondary memory and is non-volatile. Therefore, let's learn more about the disk next.

Disk

As you may already know, servers must be up and running most of the time. Additionally, data stored on a server's disks (see *Figure 10.3*) and services should be available to users. Therefore, for servers to be operational, they must have hardware that enables such a requirement. That said, servers are equipped with so-called **direct-attached storage** (**DAS**). As outlined in *Chapter 9, Storing Data in Windows Server 2022*, DAS is a group of disks connected to the server. Moreover, *hot-swappable* technology replaces the damaged disk with a new disk while the server is running. These and other features make the server a machine that provides **high availability** (**HA**) of data and services.

In the following screenshot, you can see an example of SAS HDDs:

Figure 10.3 – SAS HDDs

Moving forward, let's get to know the network interface.

Network interface

As the name implies, a network interface is the connection point between the computer and the network. From that, a network interface is not meant to be just a physical representation but can also be logical because, as such, it is in VMs represented by software. Servers usually have more than one network interface (see *Figure 10.4*). You must consider adding **network interface cards** (**NICs**) to the server if they do not. Being equipped with multiple NICs presents tremendous benefits. Some of these server benefits are outlined here:

- **NIC teaming** enables you to increase bandwidth from/to the server.
- **Network Load Balancing** (**NLB**) allows you to distribute the network load across servers.
- **Network separation** enables you to separate intranet traffic from internet traffic.

You can see an example of server network interfaces here:

Figure 10.4 – Server network interfaces

Usually, a network interface represents a physical port, while network architecture represents a computer network design. Let's get acquainted with 32-bit and 64-bit architectures next.

32-bit and 64-bit architectures

At first glance, it may seem like there is only a difference in numbers, but in reality, there is more than that in the background. Thus, in computers, the difference between 32-bit and 64-bit depends on processing power. As a first finding, it can be said that computers with 32-bit processors are mainly considered legacy computers, while computers with 64-bit processors are considered modern because they are fast and secure. Further, since computers are digital and use a binary number system, a 32-bit processor will access 2^{32} data from RAM, while a 64-bit processor will access 2^{64} data from RAM. And that for sure is a significant difference in processing power!

On a computer with a 64-bit processor, 32-bit and 64-bit applications can be installed, but not on computers with a 32-bit processor. As for security, driver signing is mandatory on computers with 64-bit processors (remember that driver signing is covered in *Chapter 4, Post-Installation Tasks in Windows Server 2022*), ensuring that its installation will not cause any reliability or security issues. From all that was said and other essential considerations, the recommendation is to consider 64-bit hardware for the server without compromising 64-bit software such as the operating system, applications, device drivers, and other utilities. All this enables an increase in the overall performance of the server.

In general, the secondary memory in computers is fixed (that is, non-removable); however, there are also removable storage technologies. So, let's take a closer look at removable drives.

Removable drives

First, the removable drive is considered both external and portable storage technology that can be plugged into and unplugged from the computer while running. Furthering the concept of hot-swappable technology (explained earlier in the *Disk* section of this chapter) enables removable drives to be plugged into servers using **Universal Serial Bus (USB)** and **Institute of Electrical and**

Electronics Engineers (**IEEE**) 1394 ports. Thus, CDs, DVDs, HDDs, floppy drives (an obsolete technology), USB flash drives, and backup drives are some of the removable storage technologies used with servers.

The server might sometimes require a graphics card to process video data. Let's learn more about that in the next section.

Graphics cards

Generally speaking, servers are backend computing machines and, hence, do not necessarily have advanced **graphics cards**. That is because we are dealing with the client/server architecture, where the server is usually considered the backend. Regardless, everything depends on the purpose that the server serves. So, if a server is involved in graphics and video processing, it might require an advanced graphics card, as shown in *Figure 10.5*. From that, a graphics card—often called a video card or graphic adapter—is an internal hardware component that makes images appear on the screen. Thus, advanced graphics cards are widely used for games, 3D animation, AutoCAD, Archicad, and others types of computer work. It is worth mentioning that modern graphic cards contain processors called **graphics processing units** (**GPUs**) and memory called video memory. That is also evident in the graphic card presented in the following screenshot:

Figure 10.5 – AMD's Radeon video graphics adapter

Sometimes, servers may heat up while processing heavy loads, thus coolers are required to keep the system optimally cooled. Let's learn more about that in the following section.

Cooling

Without excluding other server components, processors and HDDs are the hardware components that generate the most thermal heat. Therefore, in addition to processor coolers, servers are also equipped with multiple additional coolers known as **case coolers**, often called **case fans**, as shown in the next screenshot. As the name suggests, a case cooler is an electric fan that draws the thermal heat out of the computer case by blowing it. As mentioned earlier, servers have internal components that generate extreme thermal heat. Therefore, in addition to electric fans in the power supply, case coolers—often in racks—are installed as additional coolers to maintain the optimum temperature of servers mounted in the rack. Additionally, air conditioners maintain the optimum temperature in the server room and data centers:

Figure 10.6 – A cooling system in a server's case

Moving forward, let's understand how a server uses a power supply.

Power supply

As with any server hardware component we have mentioned so far, the **power supply** is considered a crucial component that takes **alternating current** (**AC**) from the wall outlet, converts it to **direct current** (**DC**), and supplies the motherboard, adapters, and peripheral devices. In addition to providing power to the server, another role of the power supply is to reduce the voltage by using an input power transformer, usually by lowering it to the voltage required by the load. The load here is expressed by internal components such as multiple processors and disks, network interfaces, large motherboards, graphic cards, **redundant array of independent disk** (**RAID**) cards, optical drives, backup tapes, and so on. These are considered regular power consumers. Then, another power supply role is to provide cooling and facilitate airflow through the case. Because power supplies are considered a **single point of failure** (**SPOF**), servers are equipped with redundant power supplies. Therefore, depending on a server's form factor, most servers are equipped with two or more **power supply units** (**PSUs**), as shown in the following screenshot. Even more important is that these power supplies have hot plugs that enable replacing the one that failed while the server is working:

Figure 10.7 – A server's PSUs

Finally, let's learn about physical ports and their relevance to a server.

Physical ports

Due to the server's different components, having multiple and diverse physical ports is natural. Thus, each physical port has a specific role. But the question is, what is a physical port? A **physical port** is a communication point that allows us to connect cables to the computer. Therefore, as noted so far, it is a characteristic of servers to have multiple physical ports. Some of the ports (see the following screenshot) located at the rear of servers include AC power connectors, Ethernet ports, **Peripheral Component Interconnect Express (PCIe)** ports, USB ports, **high-density-15 (HD-15)** video connectors, management ports, and (although rarely seen nowadays) legacy ports such as serial ports, parallel ports, and **Personal System/2 (PS/2)** ports:

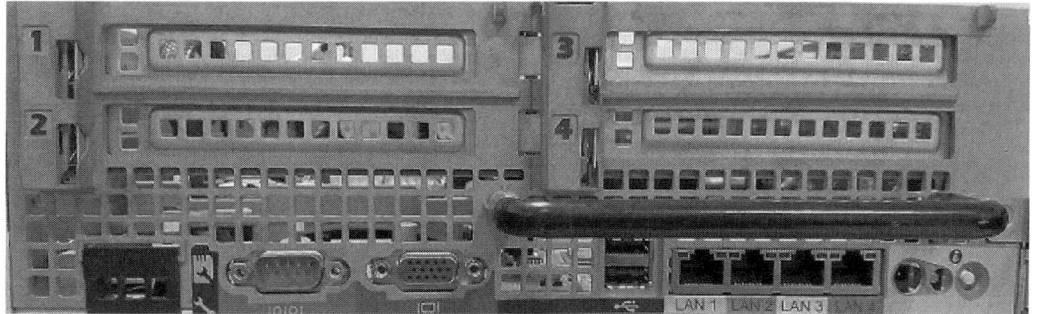

Figure 10.8 – Various ports to connect a variety of devices

This section has enabled you to get acquainted with various server hardware components and understand the difference between 32-bit and 64-bit architectures. In addition, the following section will help you to learn about monitoring the server's performance.

Understanding performance monitoring

If we consider the saying *"prevention is better than cure"* and apply this to server maintenance, we will understand how vital performance monitoring is. Monitoring the server's performance helps identify server problems at an early stage of occurrence. Then, the appropriate measures can be taken to prevent issues from becoming costly in both time and business.

A clear plan with the right tools is required for effective performance monitoring. That also means setting up a metric by which performance monitoring will be measured. Such a metric needs baseline information that will help evaluate the actual performance of servers, determine when hardware and software upgrades are needed, and assess whether an upgraded system is working better than the previous one.

> **Important note**
> Microsoft's *TechNet* website, https://technet.microsoft.com/en-us/, provides valuable information about Microsoft products, including monitoring.

Let's understand the performance monitoring methodology in a bit more detail.

Performance monitoring methodology

So far, we have learned that performance monitoring keeps the server healthy. However, if there is no clear plan for its implementation, the results obtained from performance monitoring will be based on multiple assumptions. From a business perspective, the wrong data results in bad business decisions. Simply put, performance monitoring must be based on facts and not assumptions. An approach known as a performance monitoring methodology is used to prevent that from happening.

In that regard, the performance monitoring methodology helps in conducting research. A questionnaire is modeled to understand the process and achieve a goal. For that reason, you need to make sure that your questionnaire contains questions including the following:

- What is the purpose of the server?
- What are the services that the server is providing?
- Which components do you want to monitor?
- What is the metric of component performance?
- Which tool will you use for system analysis and data collection?

> **Important Note**
> You can read more about monitoring and tuning servers at https://msdn.microsoft.com/en-us/library/bb742410.aspx.

Now, let's take a look at performance monitoring procedures.

Performance monitoring procedures

From what has been said so far, it can be understood that the performance monitoring methodology helps develop procedures. In that instance, it has to do with server performance monitoring procedures that are well-structured activities. Thus, to monitor the server's performance, you may want to consider the following guidelines:

- Document the server's hardware, software, and configuration.
- Establish the server's baseline.
- Upgrade the server's hardware and software.
- Perform the server's baseline and compare that with previous baselines .
- Identify server bottlenecks.
- Take concrete steps to fine-tune the server's performance.

Before implementing these procedures, we need to implement a baseline first. So, let's learn how to do that.

Server baselines

In general, a server's performance must be monitored. Therefore, you should ask a few questions about server performance monitoring as a system administrator, which may include the following questions:

- How do you know when servers are working under their load?
- Do you have a sample to compare their performance against?

Questions such as these may encourage you to approach such activity with more dedication. However, before explaining how to establish a baseline and considering when to create a baseline, it is good to know: what is a baseline? In short, a **server baseline** represents a snapshot of a server's performance under a normal workload. It enables you to compile a detailed report on the performance of various server components under normal workload conditions. Without neglecting other server components, the main reason for a baseline is to collect the following performance information:

- Processor utilization
- Memory utilization
- Disk read-and-write operations
- Network connection utilization

That implies that performance monitoring is not solely confined to collecting information from the hardware, as mentioned earlier. Moreover, the fact that many computer networks serve different purposes strengthens the opinion that the parameters we monitor need to be different too.

> **Important note**
>
> As mentioned in the information box of the *Processing GPOs* section of *Chapter 7, Group Policy in Windows Server 2022*, be aware that Microsoft publishes new baselines. Hence, in April 2022, Microsoft released the security configuration baseline settings for Windows 11, Windows 10 version 21H2, and Windows Server 2022. The new security baseline can be downloaded from `https://www.microsoft.com/en-us/download/details.aspx?id=55319`.

Performance Monitor

As the name indicates, Performance Monitor is a Windows **Microsoft Management Console** (**MMC**) program that monitors the server's performance. It enables visualizing the performance information in real time or from a log file. In addition, the examined performance information is displayed in a line graph, histogram bar, or report format.

To run Performance Monitor in Windows Server 2022, complete the following steps:

1. Press the Windows key + R.
2. Enter `perfmon.exe` and press *Enter*.
3. Shortly, a **Performance Monitor** window will appear, as shown in the following screenshot:

Understanding performance monitoring 285

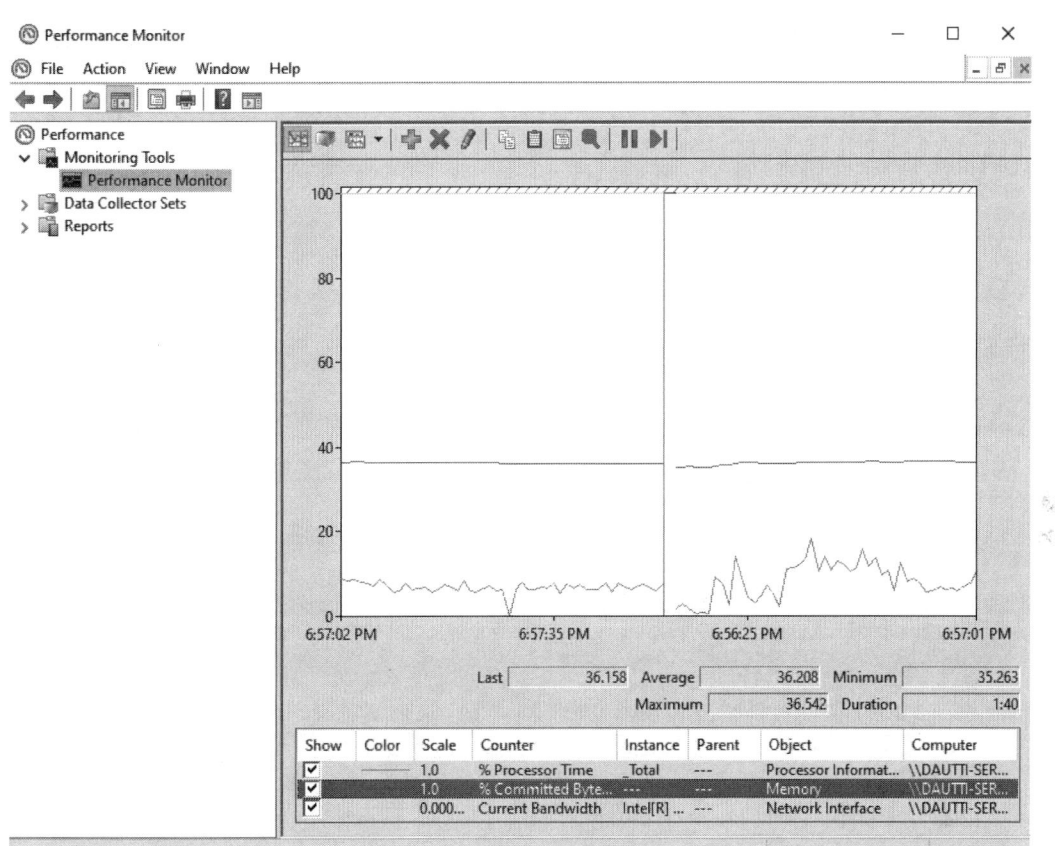

Figure 10.9 – Performance Monitor in Windows Server 2022

> **Tip**
> You can open Resource Monitor from within Performance Monitor.

Next, let's learn about Resource Monitor, an alternative to Performance Monitor.

Resource Monitor

If you have owned a computer with a Windows operating system for a while now, it may have happened that just recently, you started to feel that it is working slower than when you first bought it. Regardless, you do not need to worry because you can use Resource Monitor to determine the cause of your computer's slow performance. The same thing happens with servers too. So, whenever it is proven that server performance has decreased considerably, Resource Monitor is at your disposal to view real-time usage of both hardware and software resources.

To run Resource Monitor in Windows Server 2022, complete the following steps:

1. Press the Windows key + R.
2. Enter `resmon.exe` and press *Enter*.
3. Shortly, a **Resource Monitor** window will appear, as shown in the following screenshot:

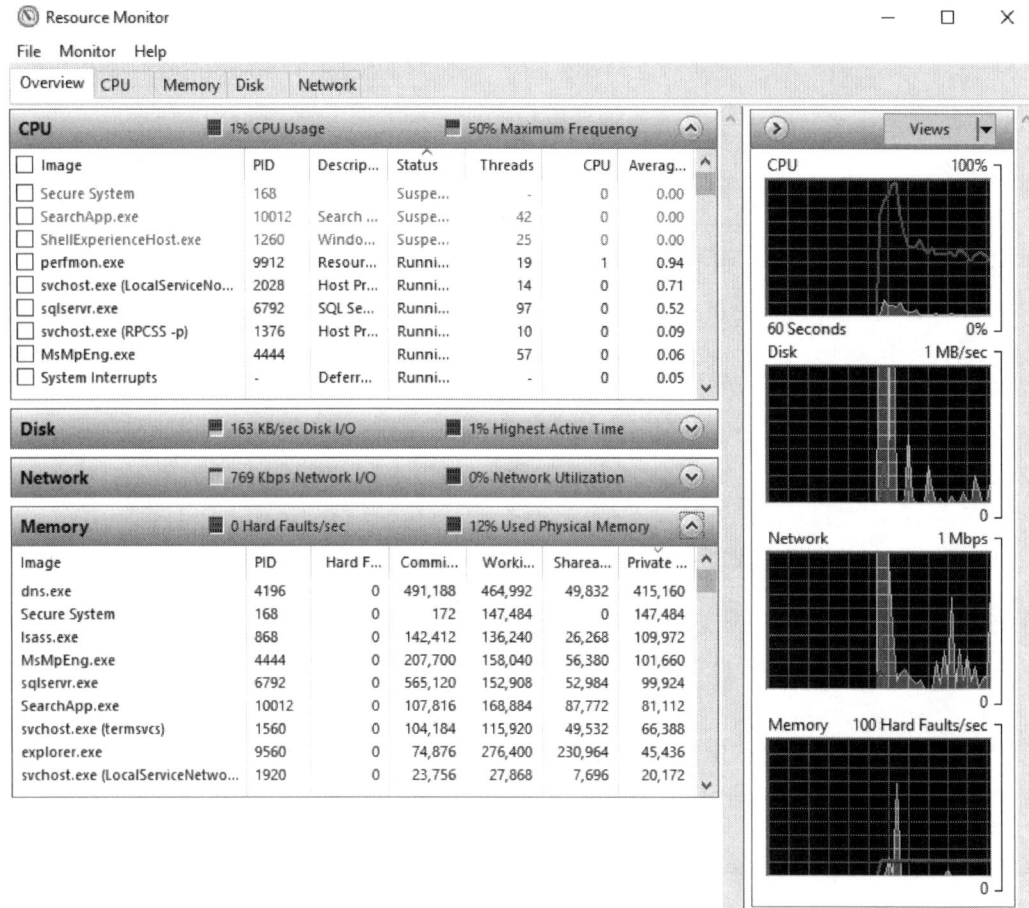

Figure 10.10 – Resource Monitor in Windows Server 2022

Besides Performance Monitor and Resource Monitor, Task Manager is an essential monitoring tool. So, let's quickly learn about it.

Task Manager

In addition to Performance Monitor and Resource Monitor, Task Manager enables monitoring of processes, performance, and services currently running on a server. It allows users to start/stop applications and background processes. Moreover, it offers a visual representation in a summary and graph views regarding the server's performance.

To run Task Manager in Windows Server 2022, complete the following steps:

1. Right-click the taskbar and, from the context menu, select **Task Manager**.
2. Shortly, a **Task Manager** window will appear, as shown in the following screenshot:

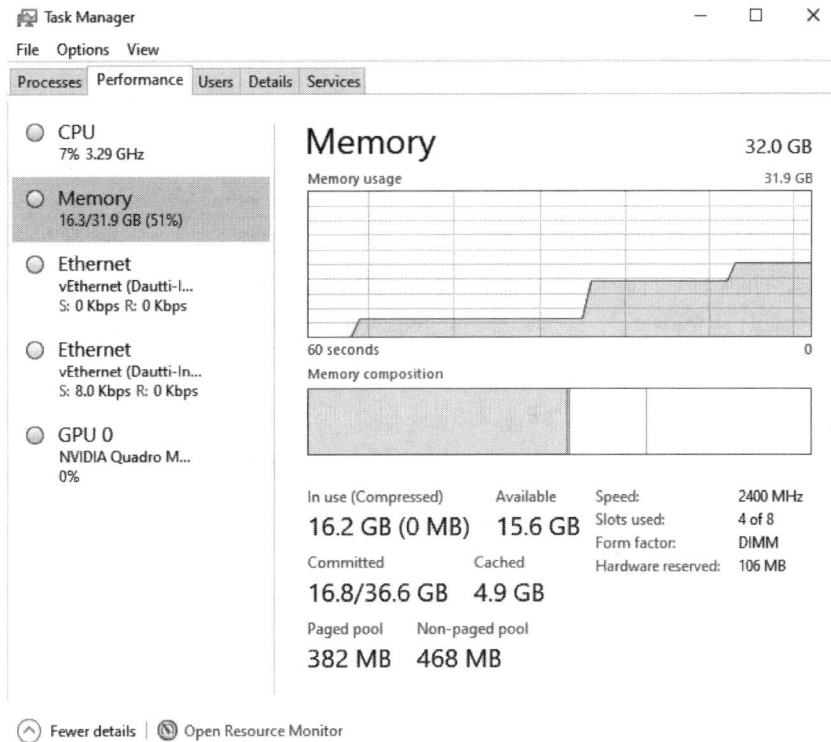

Figure 10.11 – Task Manager in Windows Server 2022

> **Important note**
> You can open Resource Monitor from within Task Manager as well.

Now, let's understand performance counters and how to use them.

Performance counters

In Performance Monitor, you can use counters and instances of selected objects to collect data for the server hardware you are keeping your eye on. Counters provide performance information on how well an operating system, application, service, or driver is working. Objects have counters to measure different aspects of performance, where each object has at least one instance representing an individual copy of a particular type of object. That way, you can determine server bottlenecks and react promptly to avoid further degradation of server performance. Then, you can fine-tune the server's performance by undertaking the appropriate measures. The following section will walk you through setting up a Data Collector Set.

Setting up Data Collector Sets

To set up a Data Collector Set in Windows Server 2022, complete the following steps:

1. With the **Performance Monitor** window open, expand **Monitoring Tools** and select **Performance Monitor**.
2. Right-click **Performance Monitor** and select **New | Data Collector Set**.
3. Enter the name for your Data Collector Set and click **Next**.
4. Specify where you want to save it by clicking **Browse** and **Next**.
5. Set the **User in Run** option and select either **Start this data collector set now** or **Save and close**.
6. Click **Finish**.
7. Right-click **Graph** and select **Add Counters…**.
8. Select counters from the **Available counters** section and click the **Add** button to add them to the list in the **Added counters** section.
9. Repeat *step 8* to add more counters, as shown in the next screenshot.
10. Click **OK** to close the window:

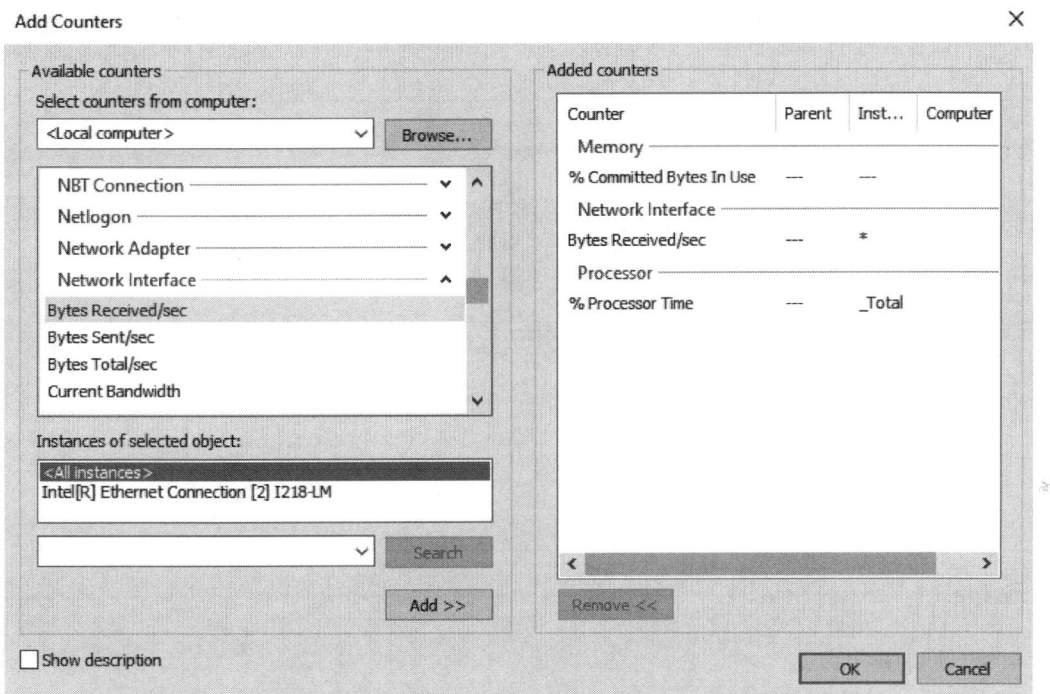

Figure 10.12 – Performance Monitor counters in Windows Server 2022

> **Important note**
> You can find information about performance monitoring thresholds and other helpful information for various technologies at https://www.manageengine.com/network-monitoring/network-performance-monitoring.html.

This section has taught you how to monitor server performance. In the next section, we will learn about logs and alerts.

Understanding logs and alerts

It is worth mentioning that performance monitoring activities are ongoing. As such, this requires dedication and patience from system administrators. However, it's necessary to recognize the importance of the performance monitoring methodology to maintain a server's ongoing operations, and the correct tools must be used to keep the server up and running. For that reason, tools such as logs and alerts play an essential role in that process. While logs are helpful for detailed analysis and record archiving, alerts enable system administrators to be informed on time about occurrences, thus taking the right actions to maintain the server's up-and-running status.

With Performance Monitor, a system administrator can collect performance information, log that information automatically, and set up alerts. The logged performance information can then be used for analysis or exported to a spreadsheet or a more advanced analytical program for subsequent analysis and reporting. To configure performance logs and alerts in Windows Server 2022, use the approach outlined in the following section.

In this short section, we have learned about performance logs and alerts. In the next section, you will carry out this chapter's performance logs and alerts exercise.

Chapter exercise – working with the Performance Logs & Alerts service

In this exercise, you will learn how to start the Performance Logs & Alerts service, access the Performance Monitor logs folder, create performance data logs, and set up performance counter alerts.

Starting the Performance Logs & Alerts service

To start the Performance Logs & Alerts service in Windows Server 2022, complete the following steps:

1. Press the Winkey + *R*.
2. Enter `services.msc` and press *Enter*.
3. From the list of services, locate the **Performance Logs & Alerts** service (see *Figure 10.13*) to check its status.
4. If it is stopped, then right-click and select **Start**.
5. Close the **Services** window.

You can see an overview of the **Services** window in the following screenshot:

Chapter exercise – working with the Performance Logs & Alerts service

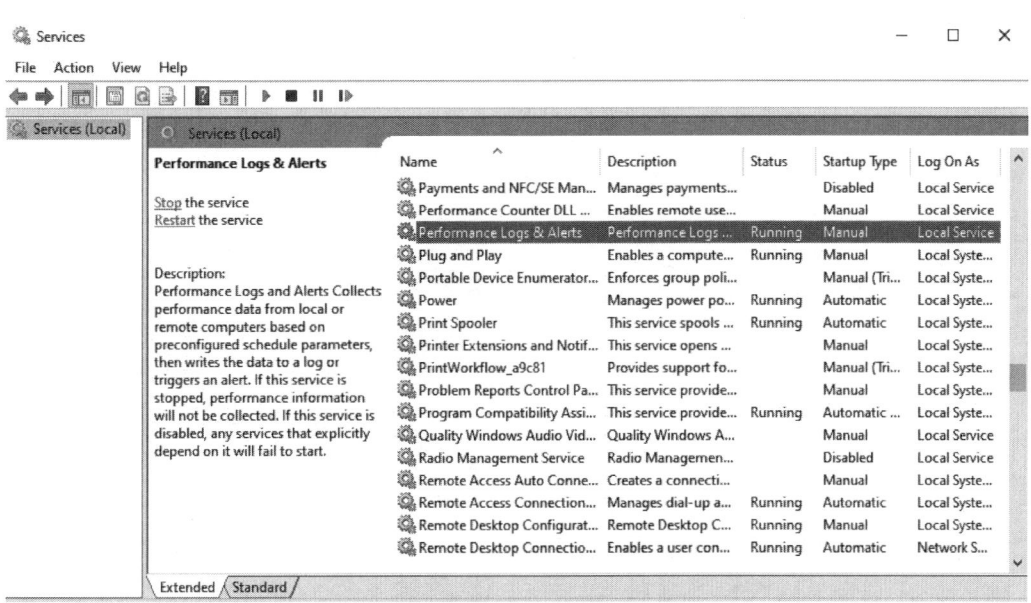

Figure 10.13 – Performance Logs & Alerts service in Windows Server 2022

Now that we have started the Performance Logs & Alerts service, let's move on to our next task of accessing the Performance Monitor logs folder.

Accessing the Performance Monitor logs folder

To access the Performance Monitor logs folder, `PerfLogs`, in Windows Server 2022, complete the following steps:

1. Press the Windows key + *R*.
2. Enter `C:` and press *Enter*.

3. The `PerfLogs` folder appears, as shown in the following screenshot:

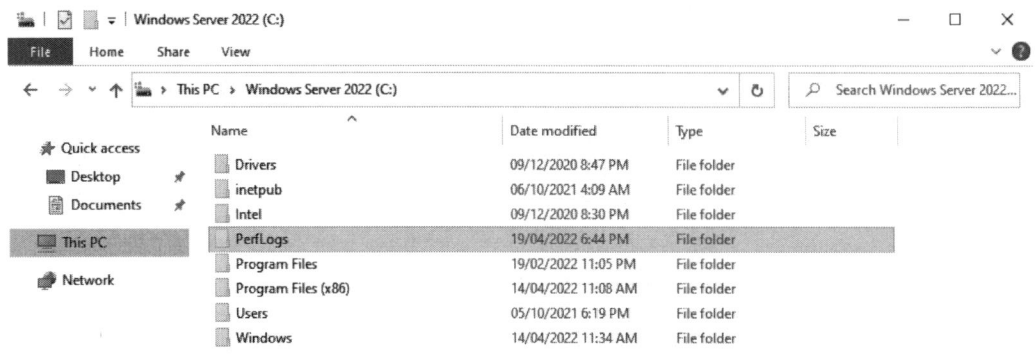

Figure 10.14 – PerfLogs folder in Windows Server 2022

Let's now create some performance data logs using Performance Monitor.

Creating performance data logs

To create performance data logs in Windows Server 2022, complete the following steps:

1. With Performance Monitor open, expand **Data Collector Sets** and select **User Defined**.
2. Right-click **User Defined** and select **New | Data Collector Set**.
3. Enter the name for your Data Collector Set.
4. Choose the **Create manually (Advanced)** option and then click **Next**.
5. Choose the **Create data logs** option and the **Performance counter** sub-option, then click **Next**.
6. Click the **Add…** button to add counters, as shown in the following screenshot, specify a time interval, and then click **Next**:

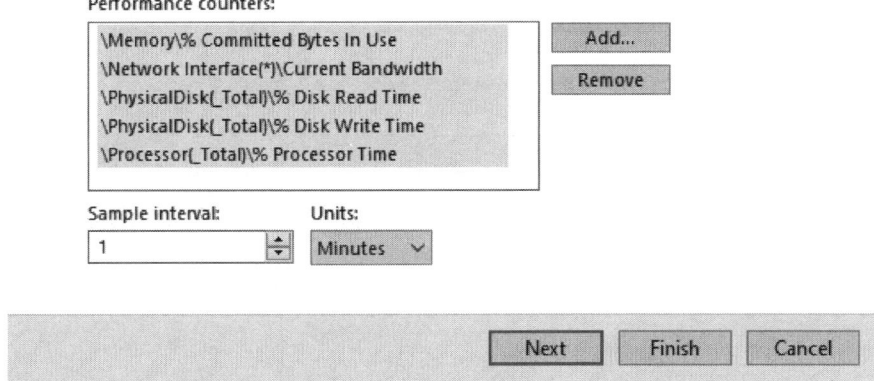

Figure 10.15 – Adding performance counters

7. Check that the default folder for saving data logs is the `PerfLogs` folder, and then click **Next**.
8. Set the **User in Run** option and select the **Start this data collector set now** option.
9. Click **Finish**.

Let's now set up performance counter alerts and complete this exercise.

Setting up performance counter alerts

To set up a performance counter alert in Windows Server 2022, complete the following steps:

1. Repeat *steps 1 to 4* from the *Creating performance data logs* section.
2. Choose **Performance Counter Alert**, and then click **Next**.
3. Click the **Add…** button to add counters, as shown in the following screenshot, specify an alert limit, and then click **Next**:

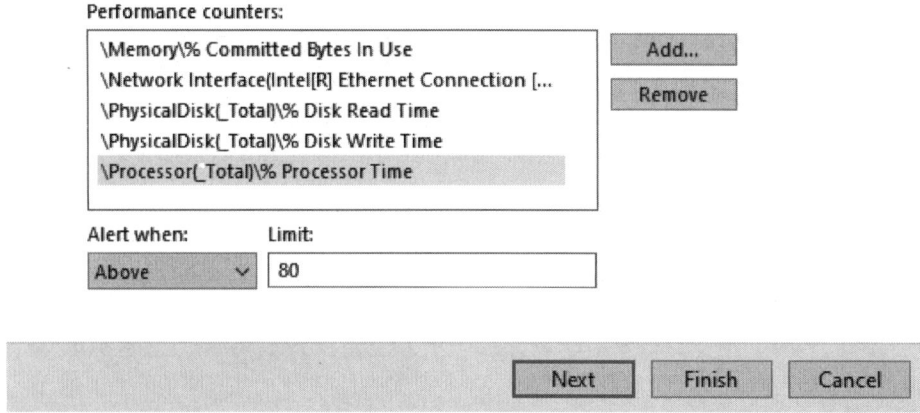

Figure 10.16 – Setting up a performance counter alert

4. Set the **User in Run** option and select the **Start this data collector set now** option.
5. Click **Finish**.

That was an excellent exercise as it covered various examples of performance logs and alerts in Windows Server 2022.

Summary

In this chapter, you have learned about a server's hardware components and how to maintain and monitor a server's performance through Windows Server 2022 native tools and utilities.

This module provided valuable information on server hardware, including key components such as processor, memory, disk, and network. In addition, using Performance Monitor, Resource Monitor, Task Manager, and performance counters will help monitor the server's performance, specifically server hardware performance.

Finally, the chapter concluded with an exercise that provided instructions to run the Performance Logs & Alerts service.

The next chapter will teach you about updating and troubleshooting Windows Server 2022.

Questions

Provide answers to the following questions:

1. Servers provide network services and handle user requests to access services. (True | False)
2. _____ represents a snapshot of the server's performance under a normal workload.
3. Which of the following is related to processors? (Choose two)

    A. Cache

    B. Cores

    C. NIC teaming

    D. Hot-swap

4. Task Manager enables you to monitor the server's processes, performance, and services currently running. (True | False)
5. _____ is the hardware that generates the most thermal heat.
6. Which of the following benefits from having multiple NICs on the server? (Choose two)

    A. NLB

    B. Network separation

    C. Word size

    D. Virtualization technology

7. Because power supplies are not considered a SPOF, servers are not equipped with a redundant power supply. (True | False)
8. _____ has to do with the processor's internal architecture and defines the data bus size, the size of the instructions, and the address size.
9. Which of the following Windows MMCs programs are used for performance and resource monitoring? (Choose two)

    A. Performance Monitor

    B. Resource Monitor

    C. Server Manager

    D. Device Manager

10. Counters provide performance information on how well an operating system, application, service, or driver works. (True | False)

11. _____ helps identify server problems at an early stage of development and take the necessary steps to prevent them from turning into costly issues in terms of time and business.

12. Which of the following are considered to be a server's primary storage? (Choose two.)

    A. RAM

    B. ROM

    C. HDD

    D. USB flash drive

13. Discuss Performance Monitor and Resource Monitor.

14. Discuss performance logs and alerts.

Further reading

For further information, have a look at the following resources:

- *Hardware requirements for Windows Server*: `https://docs.microsoft.com/en-us/windows-server/get-started-19/sys-reqs-19`

- *Server Hardware Performance Considerations*: `https://docs.microsoft.com/en-us/windows-server/administration/performance-tuning/hardware/`

- *How to: Use Performance Monitor to Collect Event Trace Data*: `https://docs.microsoft.com/en-us/dynamics365/business-central/dev-itpro/administration/monitor-use-performance-monitor-collect-event-trace-data`

11
Updating and Troubleshooting Windows Server 2022

This chapter will teach you how to update and troubleshoot Windows Server 2022. Such things are considered among the most challenging tasks of working with servers. However, as you progress through this chapter, you will notice that even the most difficult tasks have been simplified and are easy to run with a plan and strategy. Thus, in the business world, understanding the importance of troubleshooting, updating, monitoring, and maintaining servers will give you a significant chance of establishing a high business continuity standard that will significantly increase the business's competitive advantage in the market.

This chapter will introduce you to the Windows Server startup process, advanced boot options and Safe Mode, backup and restore, disaster recovery plans, updating Windows Server 2022, server hardware, and third-party software. Event Viewer will also be covered in this chapter, which allows you to review different logs on Windows Server 2022, thus helping you troubleshoot and solve any problems you may experience. In that way, you will minimize downtime, which is expressed in money lost from a business perspective.

The chapter will conclude with an exercise on monitoring and managing Windows Server 2022 logs using Event Viewer.

The following topics will be covered in this chapter:

- Understanding updates
- Understanding the troubleshooting methodology
- Understanding the startup process
- Understanding business continuity
- Chapter exercise – using Event Viewer to monitor and manage logs

Technical requirements

To complete the exercise in this chapter, you will need the following equipment:

- A PC with Windows 11 Pro, at least 16 GB of RAM, 1 TB of HDD, and access to the internet
- A virtual machine with Windows Server 2022 Standard, at least 4 GB of RAM, 100 GB of HDD, and access to the internet

Understanding updates

As is the case, after every Windows OS installation, it is recommended to check the Windows Update service for any new updates. The aim is to install security updates to protect the Windows OS from malicious attacks, download driver updates for specific hardware, add new features, and enhance the current ones. Moreover, updating the Windows OS helps resolve Windows issues and bugs. Therefore, for the reasons mentioned here and others, updating Windows Server 2022 remains one of the first and foremost activities to be undertaken after installing it on a new or used server.

Understanding Windows Update

Every second Tuesday of each month, known as **Patch Tuesday**, Microsoft releases new updates, including the latest features, security updates, and fixes for Microsoft OSs and programs, including Windows Server 2022. Everything is distributed through Microsoft's Windows Update server and received via a Windows Update feature. These can also be found on their official website: https://update.microsoft.com. In addition, a notification is displayed periodically in both the system tray and **Notification Center** stating that *you need some updates*, as shown in the following screenshot:

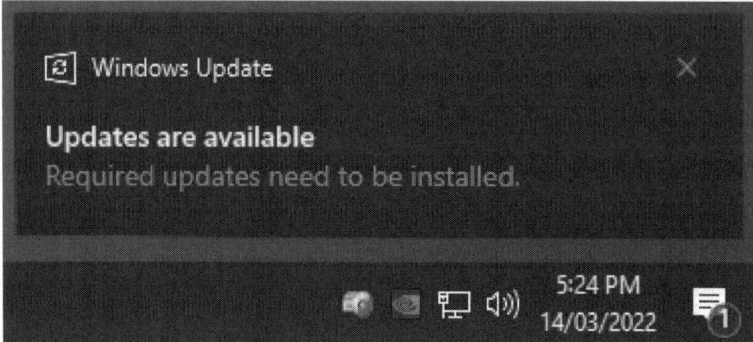

Figure 11.1 – Windows Update notification

When it comes to accessing the **Update & Security** settings, it is no different from Windows Server 2019, although there is a slight change in terms of theme and the options on the **Windows Update** page of Windows Server 2022. As a result, the following options are available (see *Figure 11.2*):

- **Pause updates for 7 days**: This option lets you pause the Windows Server 2022 update for a week. Once you click on that option, the **Windows Update** page in Windows **Settings** will let you know that the Windows Server 2022 update has paused. However, if you want to continue this pause for another week, click on **Pause updates for 7 more days**.

- **Change active hours**: This option lets you set up active hours so that **Windows Update** will not restart the server during operational hours, even if it requires a restart to install the updates.

- **View update history**: This option lets you display the list of updates and their statuses. It also allows you to uninstall updates and access recovery options.

- **Advanced options**: This option lets you choose how updates are installed. You have two choices: **Give me updates for other Microsoft products when I update Windows** and **Defer feature updates**.

- These options can be seen in the following screenshot:

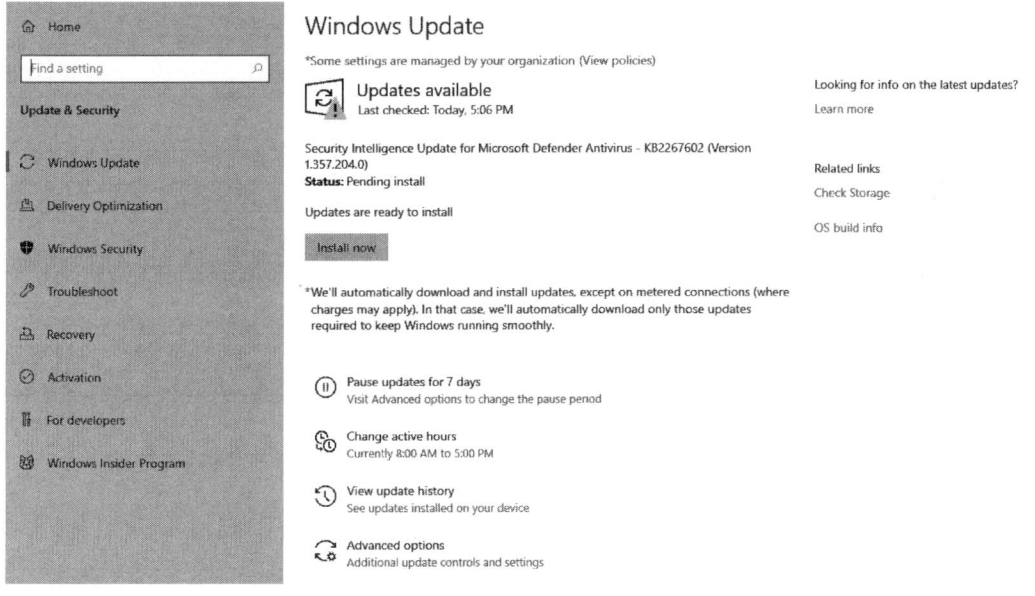

Figure 11.2 – Windows Update user interface

Now that we have provided a general overview of **Windows Update**, let's learn how to update Windows Server 2022. To do this using Windows Update, follow these steps:

1. Press the Windows key + *I* to open the Windows **Settings** section.
2. In the Windows **Settings** window, click on **Update & Security**.
3. Under **Update status**, click on the **Check for updates** button. It will begin checking for updates.
4. If **Windows Update** finds new updates, it will prompt you to install them, as shown in the following screenshot:

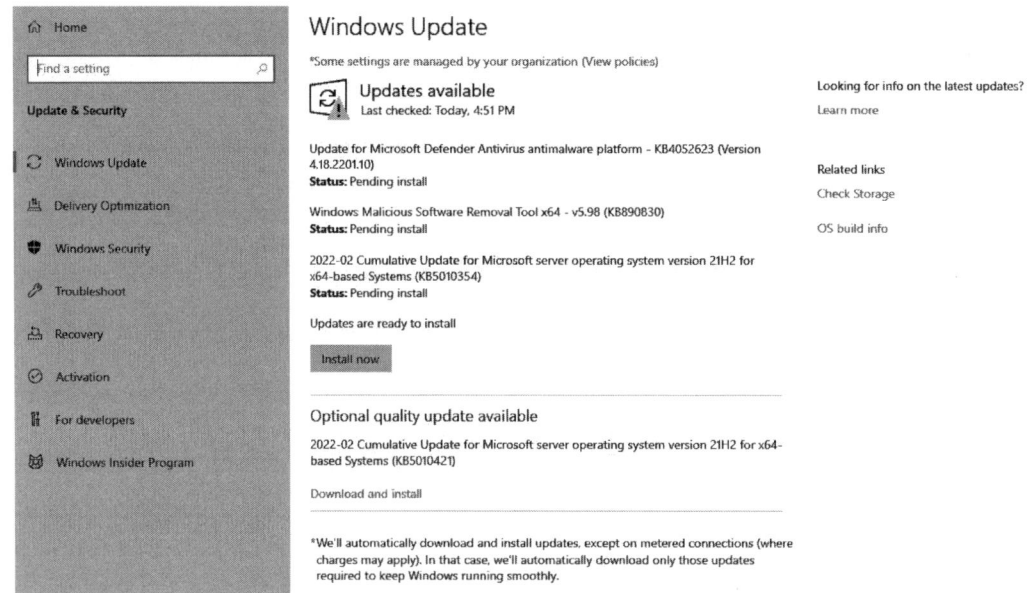

Figure 11.3 – Updating Windows Server 2022

5. In most cases, you will need to restart the server for the updates to take effect.

> **Important note**
>
> As you might know, Windows 10 has introduced a new way of providing updates, known as **Windows as a service**. This means that Windows 10 will constantly evolve, and thus new releases will be delivered through **Windows Update**. But what if you do not want to receive these new releases? Well, the option that Microsoft offers is called **Defer feature updates**. You can learn more about this exciting feature at https://www.onmsft.com/news/mean-defer-feature-updates-windows-10.

With that, we've learned how to update Windows Server 2022. Now, let's learn how to update Microsoft programs in Windows Server 2022.

Updating Microsoft programs

It is common to have Microsoft programs running on a server powered by Windows Server 2002. Microsoft Exchange, SQL Server, and SharePoint are just some client/server programs that can run on Windows Server 2022 for specific purposes, such as email, database, and collaboration servers. For performance and security reasons, it is good to update Microsoft programs regularly. To do so, follow these steps:

1. Press the Windows key + *I* to open the Windows **Settings** section.
2. In the **Settings** window, click on **Update & Security**.
3. Under the **Update settings** section, click on **Advanced options**.
4. Enable the **Receive updates for other Microsoft products when you update Windows** option, as shown in the following screenshot:

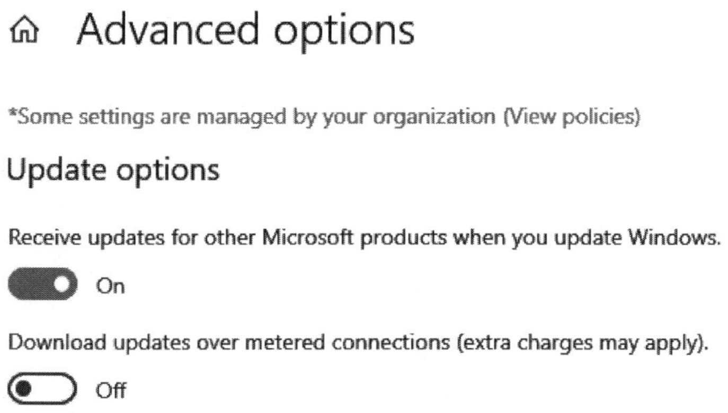

⌂ Advanced options

*Some settings are managed by your organization (View policies)

Update options

Receive updates for other Microsoft products when you update Windows.

 On

Download updates over metered connections (extra charges may apply).

 Off

Figure 11.4 – Updating Microsoft programs

5. Close the **Settings** window.

As we have learned, Microsoft programs are usually updated via Microsoft's **Windows Update** service. However, that may not be the case with third-party programs. Let's demonstrate how to update such programs.

Updating third-party programs

In addition to Microsoft programs, a server powered by Windows Server 2022 can also run third-party programs, such as an Oracle database, an Apache web server, a VMware ESXi, and so on. Because of that, for a system administrator, it is essential to understand the differences between updating Microsoft programs and third-party programs. As we have learned so far, for the most part, on Windows-based servers, Windows Update is responsible for updating both Microsoft OSs and programs. However, that is not the case with third-party programs. This is because if each software company is unique in how it develops software, then the procedure for updating that type of software shall be exceptional. Follow these steps to update a third-party program such as Adobe Acrobat Reader DC on Windows Server 2022:

1. From the **Start** menu, open the **Adobe Acrobat Reader DC** program.
2. In the **Help** menu, select **Check for Updates...**
3. Shortly afterward, **Adobe Acrobat Reader DC Updater** will check for new updates. When one is found, it will indicate that there is an update that can be downloaded.
4. Click on the **Download** button.
5. The icon in the system tray will silently download the update.
6. Depending on your internet connection speed, **Adobe Acrobat Reader DC Updater** will let you know when the update is ready to be installed.
7. Click on the **Install** button, as shown in the following screenshot:

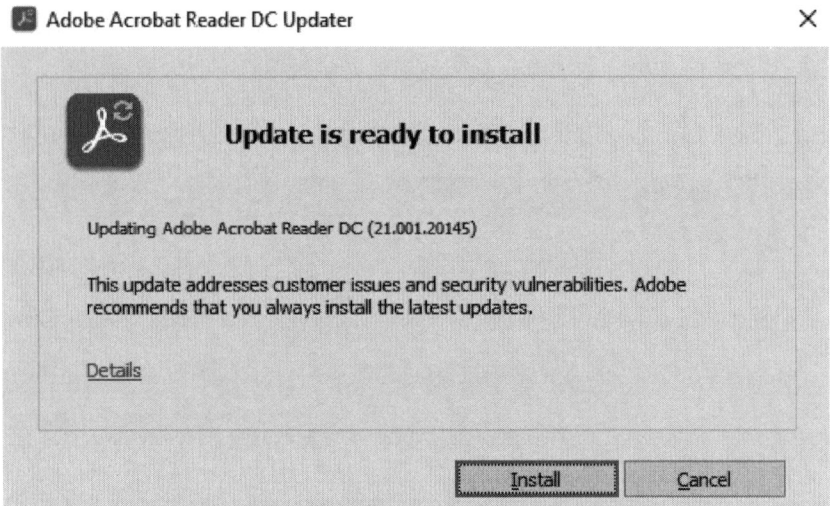

Figure 11.5 - Updating the third-party Acrobat program

8. When prompted, click the **Yes** button on the UAC dialog box.
9. Once installation completes, click the **Close** button to close **Adobe Acrobat Reader DC**.

Now that we have learned how to update Microsoft and third-party programs, it's time to learn how to update device drivers.

Updating the device drivers

In the *Updating the device drivers* section of *Chapter 4, Post-Installation Tasks in Windows Server 2022*, you learned how to update device drivers using Device Manager. However, besides using Device Manager, in this section, you will learn how to configure the Windows Update feature to automatically check for the latest drivers and updates. Follow these steps:

1. Press the Windows key + *R*, enter `Control Panel`, and then press *Enter*.
2. Click on **Hardware** | **Devices and Printers**.
3. From within the **Devices and Printers** window, right-click on the server's name, and then click on **Device installation settings**, as shown here:

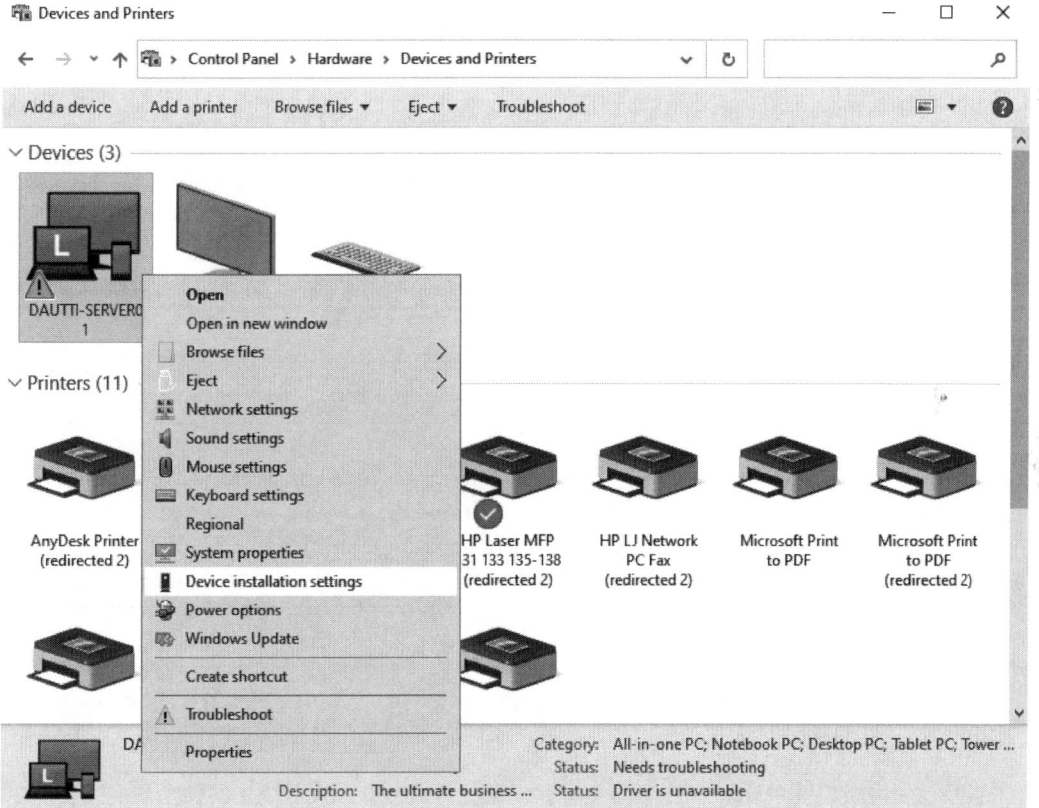

Figure 11.6 – Device installation settings

4. Select the **Yes (recommended)** option, as shown here:

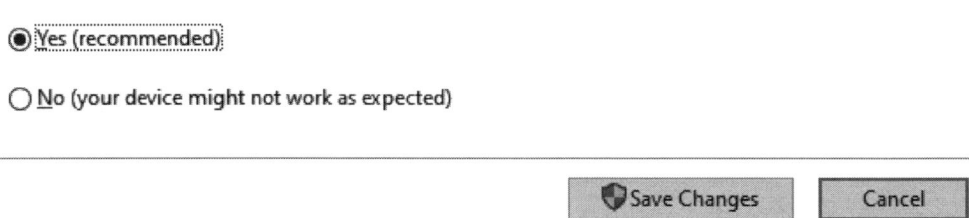

Figure 11.7 – Setting up a device driver update

5. Click on **Save Changes** to close the **Device installation settings** dialog box.

Microsoft's product updates need to be managed and organized. This can be achieved with the help of **Windows Server Update Services** (**WSUS**). We'll learn more about this in the next section.

Getting to know WSUS

As the successor to **Software Update Services** (**SUS**), WSUS allows system administrators to manage the distribution of Microsoft's product updates to their organization's computers. WSUS works so that its infrastructure can download updates, patches, and fixes to an organization's server. Then, the server approves and distributes the updates to other organizations' computers. Using WSUS, system administrators can approve or cancel updates, set the installation of updates on a given date, and generate reports to determine what updates are required for each computer. In addition, the organization's computers do not need to refer to Microsoft Update anymore since WSUS provides the updates.

In Windows Server 2022, WSUS is a role that can be added using **Server Manager**. Thus, to add the WSUS role, follow these steps:

1. Press the Windows key + R, enter `servermanager.exe`, and press *Enter*.
2. In the **Server Manager** console, select **Add Roles and Features**.
3. In the **Before You Begin** step, click **Next**.
4. In the **Installation Type** step, ensure that role-based or feature-based installation is selected, and then click **Next**.

5. Select a server from the server pool in the **Server Selection** step and click **Next**.
6. In the **Server Roles** step, select **Windows Server Update Services**, as shown in the following screenshot, and click **Next**:

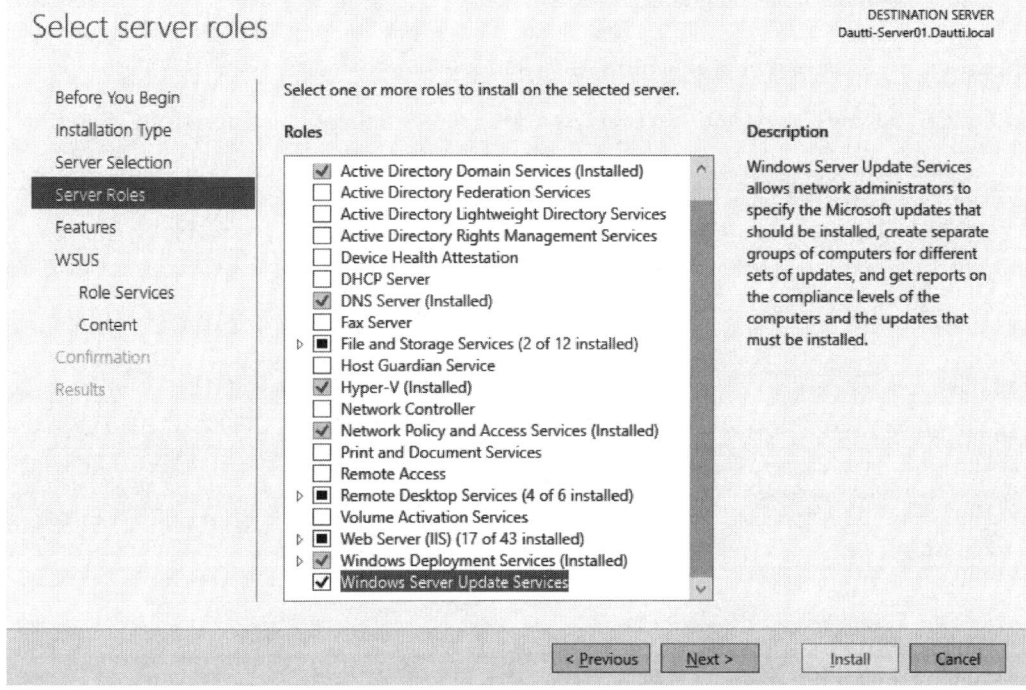

Figure 11.8 – Installing Windows Server Update Services

7. In the **Features** step, there is no need to add features; therefore, click **Next**.
8. The **WSUS** step presents the description and things to note regarding the WSUS installation. Click **Next**.
9. In the **Role Services** step, **WID Connectivity** and **WSUS Services** are selected by default. Click **Next**.
10. In the **Content** step, enter the name of the local or network location where the updates will be stored. Click **Next**.
11. In the **Confirmation** step, click **Install**.
12. When the installation is complete, click **Close** to close **Add Roles and Features Wizard**.

In this section, you learned how to update Windows Server 2022, Microsoft programs, non-Microsoft programs, and device drivers using the Windows Update feature. In the next section, we will understand the troubleshooting methodology.

Understanding the troubleshooting methodology

Troubleshooting in IT is a skill that you will master with time. Every time you solve a problem, you gain more confidence, become more experienced, and establish a more extensive knowledge base. That is why learning and practicing means a lot in IT – you are practicing the troubleshooting itself while learning how to troubleshoot. With that in mind, the more you refine your mastery, the greater your chances of overcoming issues and solving problems.

Let's begin by understanding the troubleshooting methodology's best practices, guidelines, and procedures.

Best practices, guidelines, and procedures

In IT, *best practices* are well-defined methods that are applied wherever problems occur. These best practices ensure that an organization's policy and procedure management are handled effectively and efficiently. Among the best practices that are well-known, there are accredited management standards such as ISO 9000 and ISO 14001. Therefore, by following best practices, servers will run more efficiently, network services will be more reliable, client/server applications will be more secure, and network infrastructures will be more scalable. In addition, *guidelines* represent suggestions, recommendations, or best practices for satisfying the policy standard. Finally, *procedures* are the step-by-step instructions that detail how to implement the components of the policy.

> **Important note**
> You can learn more about the ISO 9000 standard at `http://asq.org/learn-about-quality/iso-9000/overview/overview.html` and the ISO 14001 standard at `http://asq.org/learn-about-quality/learn-about-standards/iso-14001`.

Next, let's understand the troubleshooting process so that we can overcome any issue.

Troubleshooting process

As explained in the previous section, troubleshooting is an activity in which the server technician deals with solving a specific server problem. CompTIA's six-step troubleshooting model is the troubleshooting methodology used by Microsoft product support services engineers among dozens of available methods. The steps are as follows:

1. Discover the problem by gathering as much technical information as possible.
2. Evaluate the system configuration by asking questions to determine whether hardware, software, or network changes have been made recently.

3. List and track the possible solutions by isolating the problem by removing or disabling hardware or software components.
4. Execute a plan through testing solutions and ensure that you have a plan B.
5. Check the results. If the problem has not been solved, go back to *step 3*.
6. Take a proactive approach by documenting any changes you have made while troubleshooting the problem.

> **Important note**
> You can learn more about troubleshooting, particularly the detection method, at `https://technet.microsoft.com/en-ca`.

Besides these processes, troubleshooting is based on two approaches, which we will learn about next.

The systematic versus the specific approach

It is good to clarify from the outset that, when troubleshooting a particular server problem, the progress and success of the troubleshooting process itself depends very much on the approach we take toward solving that problem. Based on that, in general, the troubleshooting process and problem-solving techniques recognize two main methods:

- A **systematic approach** is an effective troubleshooting methodology because it is based on structured steps toward solving the problem, regardless of the type of problem.
- A **specific approach** is based primarily on knowledge and prior experience solving the same/similar problems. In this approach, guesswork comes into play.

Now, let's examine various troubleshooting procedures in Windows Server 2022.

Troubleshooting procedures

No matter how skillful you might be, troubleshooting is a skill that depends a lot on experience, expertise, and expressiveness. In addition, it requires specific guidelines that demand that a server technician is organized and takes on a logical approach to problem-solving with computers and servers. The procedures that need to be considered when dealing with the troubleshooting process include the following:

- Consider checking the documentation to see whether the problem has occurred in the past.
- Check any available logs, including Event Viewer.
- Consider searching the Microsoft **Knowledge Base** (**KB**) articles.

- Consider using utility programs
- Consider running a backup before trying out any solutions

The utilities that a server technician considers when troubleshooting server problems are as follows:

- The **Advanced Boot Options** menu, including **Safe Mode**
- Windows Repair
- Memory Diagnostics
- System Information
- Device Manager
- Task Manager
- Performance Monitor
- Resource Monitor
- Event Viewer

Now, let's examine the **Information Technology Infrastructure Library** (**ITIL**), which allows you to tailor your IT services to your business needs.

ITIL

As you may know, due to the application of computer technology and networking increasing in businesses, the departments of information and communication technology in the 80s started becoming the driving force in increasing efficiency and productivity. Such a reality required recognizing IT as a service that would apply consistent practices across organizations' entire IT service life cycles. Thus, ITIL was born and is considered the foundation for IT service management. ITIL, a well-structured framework, consists of best practices that guide IT organizations to design, implement, operate, and manage IT services. All of these ITIL practices are presented in the form of publications. The latest version of ITIL is v4 and was released throughout 2019 and 2020. At the same time, these publications constitute the ITIL core books. The basic concepts described in these books are as follows:

- **Service strategy**: Describes the goals and objectives for both businesses and customers
- **Service design**: Includes IT practices such as policies, architectures, and documentation
- **Service transition**: Emphasizes change management
- **Service operation**: Describes IT management
- **Continual service improvement**: Includes policy improvements and updates

ITIL enables organizations to tailor IT services to business needs, thus making IT an essential driver in today's economy.

> **Important note**
> You can learn more about ITIL at `https://www.axelos.com/best-practice-solutions/ITIL`.

Next, let's learn about Event Viewer and its benefits as a source of troubleshooting information.

Event Viewer

Event Viewer (see *Figure 11.9*), as the name suggests, is considered a tool for solving various problems in Windows OSs. In technical terms, Event Viewer is an MMC snap-in that enables system administrators to monitor server events. This makes Event Viewer a viable source of troubleshooting information whenever software, hardware, and network-related issues impact the server. However, it is worth mentioning that experienced system administrators are already accustomed to even a properly functioning system showing various warnings and errors in Event Viewer. This does not necessarily mean that a system administrator should ignore warnings and errors in Event Viewer. On the contrary, the system administrator needs to have a basic working knowledge of the tool and know when it can be helpful. Thus, from applications to forwarded events, five types of logs can be monitored with Event Viewer:

- The **Application** log includes applications or program events.
- The **Security** log includes events triggered by security-related activities, such as an invalid login attempt or accessing a folder with denied permissions. In addition, it requires auditing to be enabled.
- The **Setup** log includes application setup events.
- The **System** log includes events triggered by Windows system components.
- The **Forwarded Events** log includes events triggered by remote computers. It requires you to create an event subscription.

- These can be seen in the following screenshot:

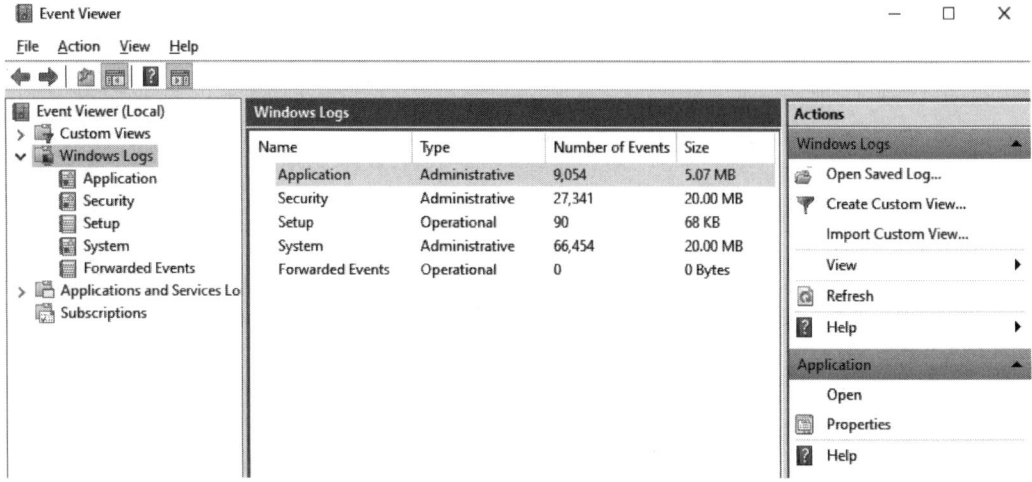

Figure 11.9 – Event Viewer

In this section, you learned about the methodologies, procedures, best practices, and approaches that will help you keep your server up and running. In the next section, we will introduce the startup process.

Understanding the startup process

Although it is entirely technical, identifying and understanding the hardware components and the startup process has tremendous benefits. It helps troubleshoot hardware-related problems, thus keeping downtime to a minimum. To diagnose and troubleshoot a server startup problem, the server technician must understand what happens during startup. To understand what occurs during server startup, let's look at the **Basic Input/Output System** (**BIOS**).

BIOS

Imagine that you are in front of a server and have just pressed the button to turn it on. Although you may move back a little because of the noise generated by the cooling fans, I am sure that at this stage, it will be interesting to know what is happening inside the server. As soon as the DC flows through the server's internal architecture, the so-called ROM chip on the server motherboard will activate the BIOS to enable you to access and set up the server's hardware. That being said, the BIOS (see *Figure 11.10*) is a program that controls the functionality of the server's hardware. Other than identifying and configuring the hardware in a server, the other essential task of the BIOS includes identifying the boot devices:

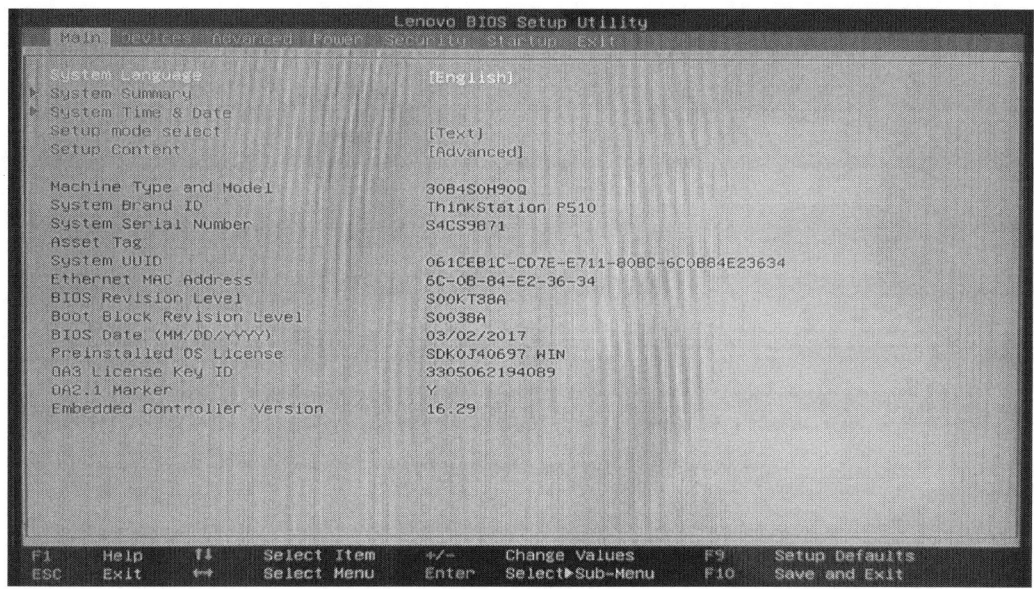

Figure 11.10 – Lenovo BIOS Setup Utility

> **Important Note**
> You can find the BIOS boot options information in the *Understanding boot options* section of *Chapter 3*, *Installing Windows Server 2022*.

Over time, BIOS was eventually replaced by the **Unified Extensible Firmware Interface** (**UEFI**). We will learn more about this next.

UEFI

Unlike computers in the past, modern computers do not have a legacy BIOS; instead, they are equipped with UEFI, as shown in the following screenshot. To overcome the BIOS limitations regarding current hardware support during the booting process, UEFI Consortium has developed UEFI. Unlike BIOS, which is limited to a 16-bit processor mode and 1 MB of addressable memory, UEFI supports 32-bit or 64-bit processor modes and can access the entire computer's memory. Also, in contrast to BIOS, which uses **Master Boot Record** (**MBR**) and supports disks up to 2 TB in size, UEFI uses **GUID Partition Table** (**GPT**), which enables support for disks more prominent than 2 TB in size. Moreover, UEFI can be quickly updated by downloading the firmware updates directly from the manufacturer's website:

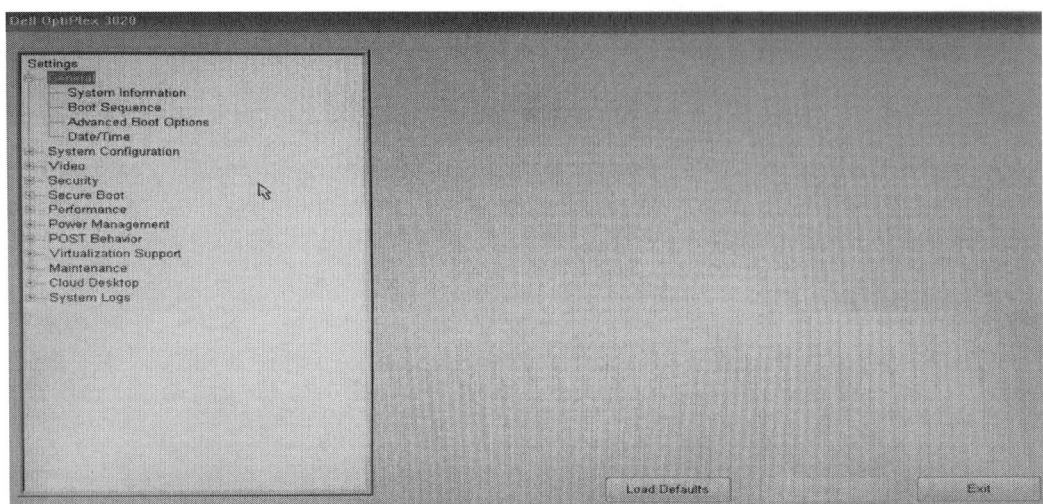

Figure 11.11 – The UEFI setup utility

Now, let's look at the **Trusted Platform Module** (**TPM**), which helps secure the hardware.

TPM

Since Windows Vista, when Microsoft introduced the BitLocker feature for disk encryption, TPM has been present on computers with Windows OSs. This is because TPM supports encrypting disks by BitLocker by providing hardware security for the latter. From a technical standpoint, TPM is a chip on the computer's motherboard, which Windows uses to store the encryption key whenever BitLocker encrypts the drives. Thus, BitLocker uses TPM to help ensure the integrity of early startup components by providing that no changes were made to the BIOS, boot sector, and boot manager. Once TPM has verified no changes, it releases the decryption key to the Windows OS bootloader. If TPM detects changes, it blocks any volume protected by BitLocker, and the disk will remain protected. This reveals the idea of using the TPM chip to provide security at the lowest level of an OS, where the Windows kernel is located. An example of such a practice is Windows 11, which has a mandatory Secure Boot and TPM requirement.

The following screenshot shows the TPM Management console in Windows Server 2022, which can be initiated by entering `tpm.msc` in the *Run* dialog box:

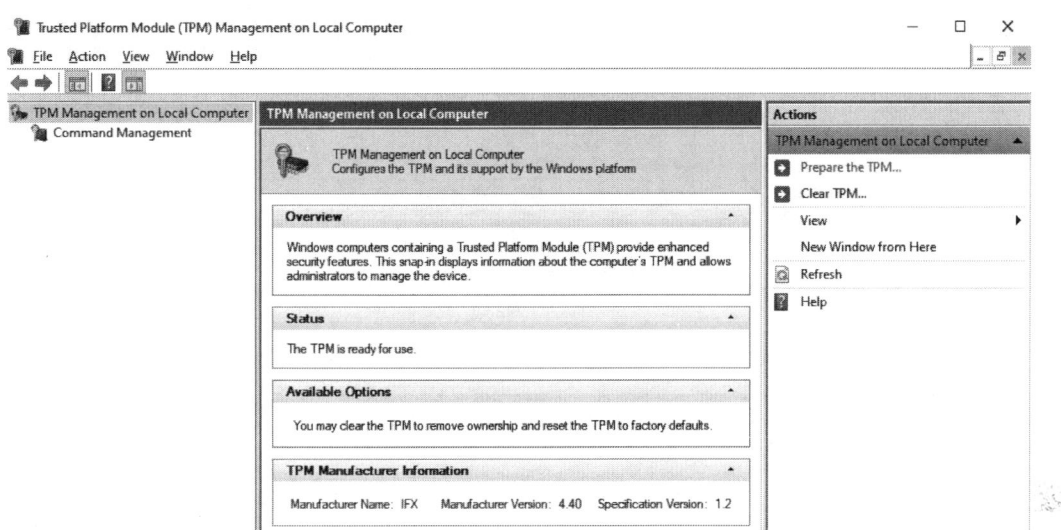

Figure 11.12 – The UEFI setup utility

Now, let's explore **Power-On Self-Test** (**POST**), a diagnostic test that verifies whether or not the server hardware is working correctly.

POST

The BIOS performs a hardware test known as POST when booting a server. POST is a diagnostic test that verifies that the server hardware is working correctly. Therefore, learning the POST's beeps during server hardware initialization is beneficial, regardless of the BIOS manufacturer. In addition, it is recommended that you keep an eye on the components, such as the processors, memory, and graphics cards, as they are the first three components to be examined by POST. If any of these components are faulty, then the server boot fails.

> **Important note**
> You can learn about the various beep codes of different BIOS manufacturers at https://www.computerhope.com/beep.htm.

Now, let's look at the MBR, a legacy partition scheme.

MBR

Once POST finishes verifying that the server hardware is working correctly, the BIOS hands control over the first boot device. Next, the BIOS looks after the boot device that contains the MBR. The MBR is created when disk partitions are made; however, the MBR resides outside the disk partitions. The MBR is located on the first disk sector and contains the information to identify and boot the OS. Here, the MBR has either **NT Loader** (**NTLDR**), **Boot Manager** (**BOOTMGR**), or both, depending on the Windows OS installed on the server's disk. This determines the progress of loading the OS into RAM, as shown in the following table:

NTLDR (Windows NT to Windows Server 2003)	BOOTMGR (Windows Vista to Windows Server 2022)
BOOT.INI	Boot Configuration Data (BCD)
NTDETECT.COM	WinLoad.exe
NTOSKRNL.EXE	NTOSKRNL.EXE
HAL.DLL	Boot-class device drivers

Table 11.1 – NTLDR versus BootMgr

> **Important note**
> You can find additional information about the MBR, including details about GPT, in the *Understanding partition schemes* section of *Chapter 3, Installing Windows Server 2022*.

Now that we understand the MBR, let's look at **Boot Configuration Data** (**BCD**).

BCD

BCD represents a store consisting of specific files that control an OS boot. BCD provides a standard boot option interface for the newest Windows OSs, including Windows Vista to Windows Server 2022, independent of the firmware. As a result, it is more secure than the previous boot option (`Boot.ini`) and enables administrators to assign permissions to manage boot options. Moreover, BCD is available at runtime and during all stages of system configuration. For example, `Bcdedit.exe`, as shown in the following screenshot, is a file that's used for the BCD data store. Similar to `boot.ini`, `bcdedit.exe` is located inside the disk partitions:

```
Administrator: Windows PowerShell

Windows PowerShell
Copyright (C) Microsoft Corporation. All rights reserved.

PS C:\Users\Administrator> Bcdedit.exe

Windows Boot Manager
--------------------
identifier              {bootmgr}
device                  partition=\Device\HarddiskVolume1
description             Windows Boot Manager
locale                  en-US
inherit                 {globalsettings}
bootshutdowndisabled    Yes
default                 {current}
resumeobject            {660a0452-bcad-11e9-b710-00155d007501}
displayorder            {current}
toolsdisplayorder       {memdiag}
timeout                 30

Windows Boot Loader
-------------------
identifier              {current}
device                  partition=C:
path                    \Windows\system32\winload.exe
description             Windows Server
locale                  en-US
inherit                 {bootloadersettings}
recoverysequence        {660a0454-bcad-11e9-b710-00155d007501}
displaymessageoverride  Recovery
recoveryenabled         Yes
allowedinmemorysettings 0x15000075
osdevice                partition=C:
systemroot              \Windows
resumeobject            {660a0452-bcad-11e9-b710-00155d007501}
nx                      OptOut
PS C:\Users\Administrator>
```

Figure 11.13 – Running bcdedit.exe

In a multiple boot scenario, the MBR contains both NTLDR and BOOTMGR, which means that both `boot.ini` and `bcdedit.exe` are present to display the respective OS's list. In that case, `bootsect.exe` (refer to the *Boot sector* section later in this chapter) can be used to update the MBR for hard disk partitions, which require the switch between NTLDR and BOOTMGR.

Naturally, the sections for MBR and BCD precede the bootloader. Now, let's explore the bootloader.

Bootloader

In its simplicity, a bootloader, often called a bootstrap loader or boot manager, is a program that boots a computer. The bootloader appears after POST verifies that the computer hardware is functional. Located in the MBR, the bootloader loads the Windows OS kernel into memory or disk. For example, in Windows OSs, there are two types of bootloaders:

- **NTLDR** is the legacy Windows bootloader from Windows NT to Windows Server 2003.
- **BOOTMGR** is the newest Windows bootloader from Windows Vista to Windows Server 2022.

Let's explore the boot sector, which contains the information needed to boot the server (the bootloader).

Boot sector

In the *HDD* section of *Chapter 9*, *Storing Data in Windows Server 2022*, tracks and sectors were mentioned. The tracks look like concentric circles, and there are many on a disk, whereas the sectors are the track's divisions whose size depends on the filesystem the server's OS is running. So far, you have most likely established an initial understanding of the boot sector. Of course, this is a sector on a server's disk that contains the information required to boot that server. Technically, the boot sector is located in the first sector of the first disk track, and usually, it contains the MBR, which then contains the bootloader.

Now, let's learn about the boot menu, which is used when more than one OS runs on a computer.

Boot menu

If multiple Windows OSs are running on a computer, this is known as **multi-booting**. Usually, such a machine displays a boot list that lists all the OSs running on it every time that computer is turned on. Boot.ini (see *Figure 11.14*), a text file, is responsible for displaying the boot menu. It is mainly utilized by OSs, including Windows NT and Windows Server 2003. Unlike the MBR, boot.ini is located inside the disk partitions, precisely at the root partition (that is, the C partition). The path to boot.ini is C:\boot.ini, which contains the boot options such as bootloader and the OS. As we in *Table 11.1*, the equivalent of boot.ini in the post-Vista OSs is BCD:

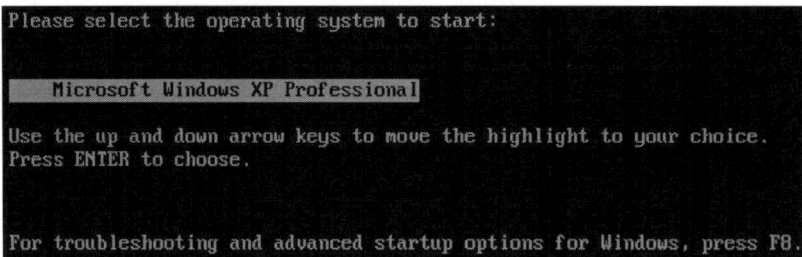

Figure 11.14 – Boot.ini displays the list of OSs

Now, let's look at Safe Mode, which represents a diagnostic mode and uses a minimal set of drivers and services.

Safe Mode

Often, users experience malfunctions when attempting to boot Windows OSs. For example, the OS does not boot when attempting to turn on the computer. However, by pressing the *F8* key without overthinking anything, **Windows Advanced Options Menu** can be accessed from where the **Safe Mode** option can be selected. This is done because **Safe Mode** represents a diagnostic mode in Windows OSs that uses a minimal set of drivers and services. However, note that the *F8* key option can only be used in pre-Vista Windows OSs such as Windows NT to Windows Server 2003. **Advanced Startup Options** enables Windows OS recovery in post-Vista Windows OS such as Windows Vista to Windows Server 2022, including access to **Safe Mode**:

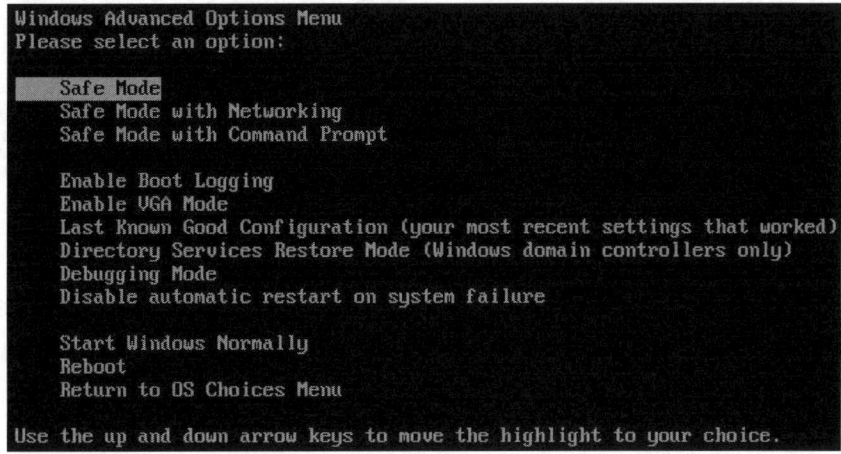

Figure 11.15 – Windows Advanced Options Menu in Windows XP Professional

In Windows Server 2022, follow these steps to access the **Safe Mode** option from the **Advanced Options** menu:

1. While holding down the *Shift* key, restart Windows Server 2022 by clicking on **Restart** from the **Power** option.
2. On the **Choose an option** screen, select **troubleshooting**.
3. On the **Advanced options** screen, choose **Startup Settings**.
4. Click on the **Restart** button on the **Startup Settings** screen.

5. The **Advanced Boot Options** screen will be displayed shortly afterward, as shown here:

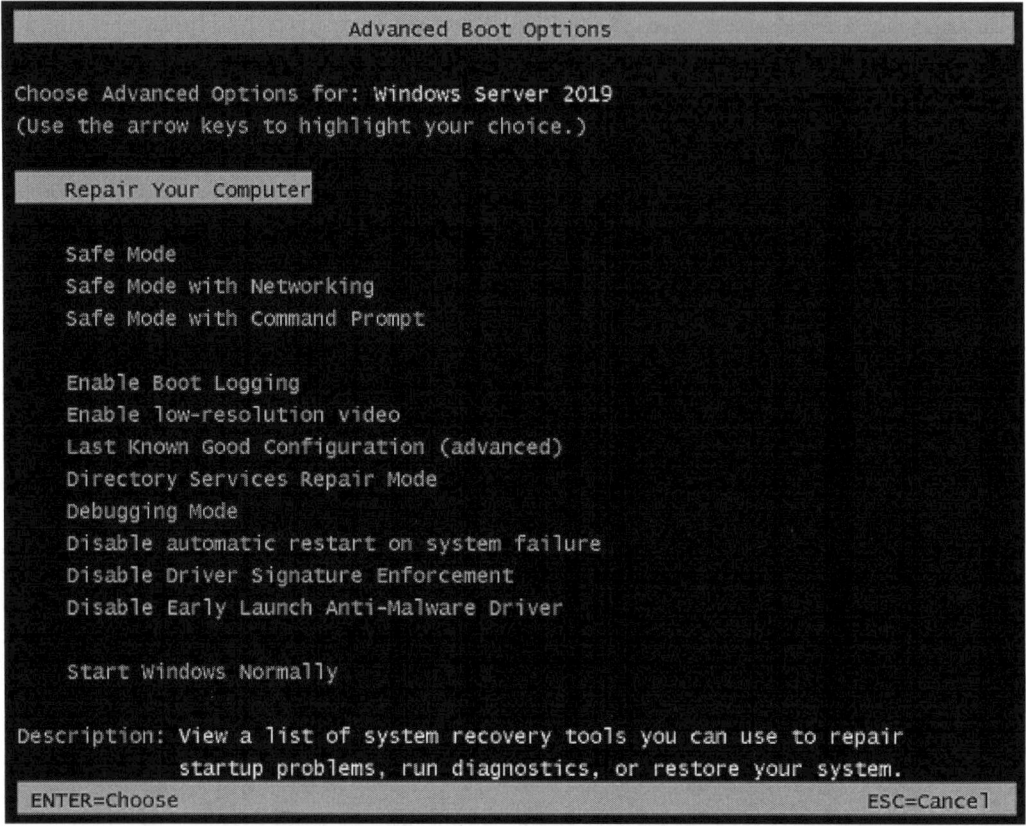

Figure 11.16 – The Advanced Boot Options menu

In this section, you learned about partition schemes, the bootloader, the boot sector, the boot menu, and Safe Mode. In the next section, we will explore business continuity.

Understanding business continuity

As a system administrator in this digital age, you must understand that any period of downtime will mean a loss of profit for the company. Therefore, your primary responsibility is to minimize downtime as much as possible. This can be achieved by adequately assessing the components that can fail and taking the appropriate measures to avoid such failure.

Let's start by learning about **disaster recovery plans** (**DRPs**).

DRP

A DRP is a well-structured plan that ensures an organization will continue to provide services or recover from a disastrous situation as soon as possible. If a business cannot prevent unexpected events, it can minimize the losses if that business is prepared. Therefore, DRP is a proactive method for maintaining business continuity in such situations. The following is a list of things that organizations should consider when compiling DRP:

- Make an inventory of all hardware and software.
- Analyze all potential threats and vulnerabilities.
- Establish the organization's priorities.
- Define the organization's tolerance in case of a disaster.
- Review how the disaster was handled in the past.
- Acknowledge that the staff matter more than data recovery and services.
- Execute DRP DRY tests regularly.
- Have management approve the DRP.
- Never forget to update the DRP.

Now, let's understand data redundancy, which helps restore services in a natural disaster.

Data redundancy

Data redundancy is a process that allows you to store the same set of data in multiple locations and update it automatically. But what if the data updates are not successfully implemented? Data inconsistency problems occur, leading to more issues, such as data integrity. Such matters can further worsen the situation and potentially harm the organization's extensive data and multiple data storage locations.

Now, let's explore clustering, which merges the processing power of several servers.

Clustering

Clustering refers to a group of servers that combine processor power, RAM, storage capacity, and network interfaces to achieve high availability of services. Clustering recognizes the following two most common practices:

- **Failover clustering** requires a minimum of two servers and works on the active-passive principle, where one server is active, and the other is passive. Usually, it is applied to databases, mail servers, and, in general, backend processing environments.

- **Load-balancing clustering**: This requires a minimum of two servers; however, servers are merged into one virtual server, exchanging heartbeats. As far as users are concerned, they access a single server; as far as backend processing is concerned, the loads are distributed between the servers. Usually, it is applied to web servers and, in general, frontend processing environments.

Now, let's examine redirection, which facilitates accessing documents in a network environment.

Folder redirection

System administrators can use folder redirection to redirect the folder on a local computer, or a shared folder on a network, to a new location. With folder redirection, the data stored on the server can be accessed by users similar to how it would be if kept on a local computer.

In Windows Server 2022, you can create a **Group Policy Object** (**GPO**) to redirect a folder, as shown in *Figure 11.17*. The steps are as follows:

1. Press the Windows key + *R*, enter `gpmc.msc`, and press *Enter*.
2. Expand **User Configuration** | **Policies** | **Windows Settings** | **Folder Redirection**.
3. Right-click on **Documents** and select **Properties**.
4. Select the **Basic - Redirect everyone's folder to the same location** setting.
5. In the **Target folder location** section, choose **Redirect to the following location**.
6. Specify the root path to your redirected folder.
7. Click on **OK** to close the **Document Properties** window:

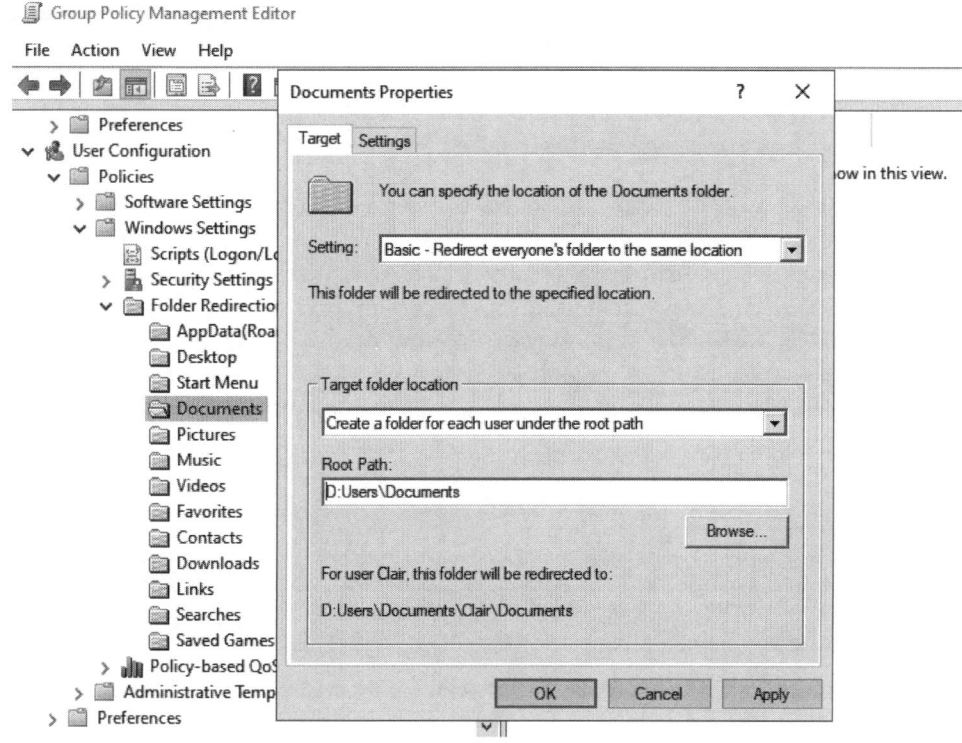

Figure 11.17 – Creating a GPO for folder redirection

Losing your data may hamper continuity. To prevent this, you can back up your data. We'll learn more about this next.

Backup and restore

A fundamental requirement when working with servers is that *data on a server must be protected from being lost*. In line with that, backups are usually used to copy the data if it's lost. However, unlike a backup, a restore is the process of data recovery whenever data on a server is lost or corrupted. The following are the different types of backups:

- A **full backup** makes a copy of all of the data. Therefore, you only require the last set of full backups to restore your data.

- An **incremental backup** copies the data that has changed since the last backup, regardless of the type. Usually, incremental backups are done from Monday to Thursday, and the full backup takes place on Friday. Therefore, to restore your data, you need the last set of full backups and incremental backups between the full backup and the day you want to restore the data. Because of this, it takes less time to do the backup but more time to restore the data.

- A **differential backup** copies the data that has changed since the last full backup. In the same way as an incremental backup, a differential backup is done from Monday to Thursday, and on Friday, the full backup takes place. You need the last set of full backups and incremental backups to restore your data. Because of this, it takes more time to do a backup and less time to restore the data.

When it comes to choosing a backup media, usually, it depends on the importance of the data and its quantity. Storage technologies such as CDs, DVDs, removable HDDs, backup tapes, **network-attached storage** (**NAS**), and **storage area networks** (**SANs**) are potential storage technologies for backing up. These days, organizations use online backup services too. Convenience, security, and cost are the decisive factors in choosing online backup services. Last but not least, the most common backup rotation scheme, **Grandfather-Father-Son** (**GFS**), is worth mentioning. The son backup is done daily, the father backup is done weekly, and the grandfather backup is done monthly.

In Windows Server 2022, Windows Server Backup is a feature that can be added using Server Manager. To add Windows Server Backup, follow these steps:

1. Press the Windows key + R, enter `servermanager.exe`, and press *Enter*.
2. From the **Server Manager** console, select **Add Roles and Features**.
3. In the **Before You Begin** option, click **Next**.
4. In the **Installation Type** step, ensure that **Role-based** or **feature-based installation** is selected, and click **Next**.
5. Select a server from the server pool under **Server Selection** and click **Next**.
6. There is no need to add roles; therefore, click **Next**.
7. In the **Features** step, scroll down the list of features and select **Windows Server Backup**, as shown in the following screenshot. Then, click **Next**:

Understanding business continuity

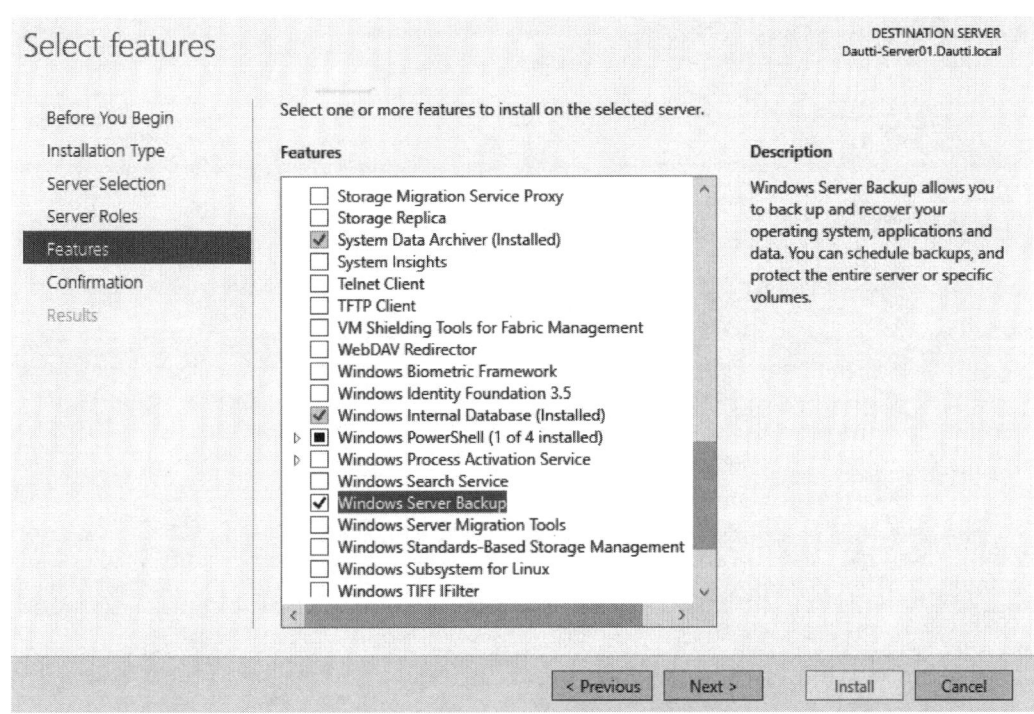

Figure 11.18 – Installing the Windows Server Backup feature

8. In the **Confirmation** area, click **Install**.
9. When the installation is complete, click **Close** to close **Add Roles and Features Wizard**.

Once we have created a backup, we can restore the data. Let's learn how to do this.

Active Directory (AD) restore

As shown in *Chapter 5, Directory Services in Windows Server 2022*, while adding the AD DS role in one of the Active Directory Domain Services Configuration Wizard steps, the **Directory Services Restore Mode** (**DSRM**) password was required (see *Figure 11.19*). This password is essential for an AD restore, so you must be careful. DSRM is to AD what Safe Mode is to the OS. It is a way of restoring AD when the latter has failed or needs to be fixed:

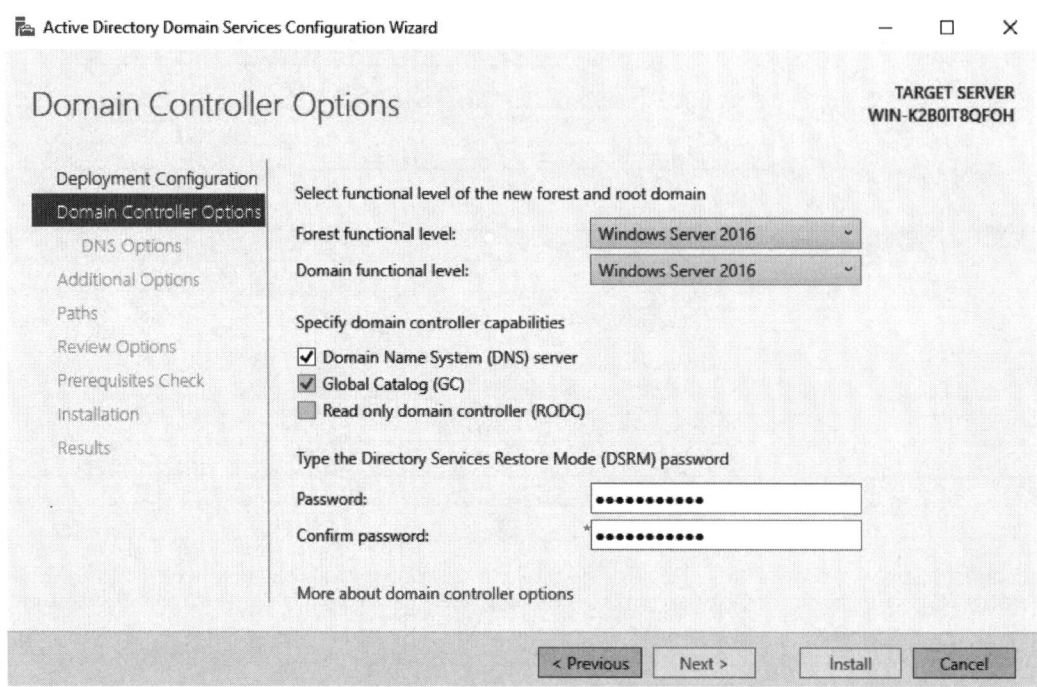

Figure 11.19 – Setting up DSRM

Usually, there are two methods for restoring data replicated on a **domain controller** (**DC**). The first method involves reinstalling the OS, reconfiguring the DC, and then, through normal replication, it will get populated from the second DC on a network. The second method considers the backup to restore the DC's replicated data. From that, the replicated data from a backup medium can be restored in the following two ways:

- **Non-authoritative restore**: This is applied when a DC has failed due to hardware or software-related problems. The AD structure is restored from a backup medium, and then it will be populated from the second DC on a network through normal replication.

- **Authoritative restore**: This takes place after a non-authoritative restore, thus helping to restore the entire system to a state before the AD objects were deleted. It uses the `Ntdsutil` command, which enables an authoritative restore of AD.

We cannot keep our server running without a proper power supply. So, let's learn how to overcome this problem.

Power redundancy

Regardless of the processor's power, memory capacity, data storage capacity, and the number of available network interfaces the server can have, all of this is useless if there is no constant power supply. Because the continuous power supply is crucial for a server's overall functionality, the **uninterruptible power supply** (**UPS**) device (see *Figure 11.20*) has an important place in the server's world. UPS is a battery-driven device that supplies the server with power during a power outage. However, despite the capabilities offered by UPS, it still does not provide a solution for lengthy power outages. For that reason, electric generators represent an alternative solution to overcome such issues:

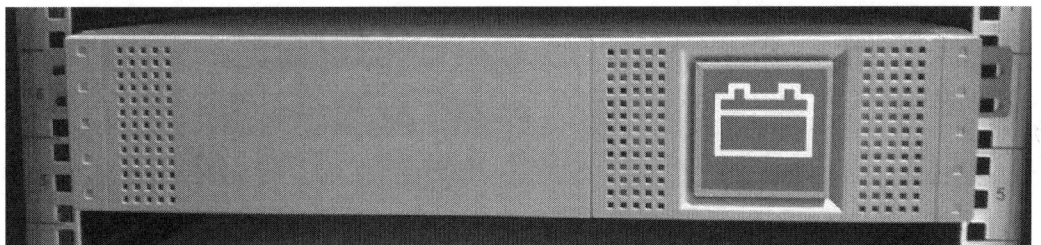

Figure 11.20 – A rack-mountable UPS

This section has helped you learn about various redundancy technologies. Next, you will complete an exercise where you will monitor and manage logs using Event Viewer.

Chapter exercise – using Event Viewer to monitor and manage logs

This exercise will teach you how to set up centralized monitoring, filter Event Viewer logs, and change the default log location. Let's dive right in!

Setting up centralized monitoring

To set up centralized monitoring in Windows Server 2022, follow these steps:

1. Open the command prompt with elevated admin rights on a **Remote Server**, enter `winrm quickconfig`, and press *Enter* to configure `LocalAccountTokenFilterPolicy` to grant administrative rights remotely to local users.
2. Right-click on the **Start** button and select **Computer Management**.
3. Expand **Local Users and Groups** and click **Groups**.
4. Open the administrator's group and add the central server.

5. Open a command prompt with elevated admin rights on a **Central Server**, enter `wecutil qc`, and press *Enter*.
6. Press *Y* (for yes) when prompted to do so.
7. From the command prompt window, enter `eventvwr.exe` to open **Event Viewer**.
8. Right-click on **Subscriptions** and select **Create Subscription…**
9. Enter a **Subscription name** and its **description**.
10. Set **Forwarded Events** to **Destination log**.
11. Select **Remote Server** by clicking the **Select Computers** button, as shown in the following screenshot, and click **OK**:

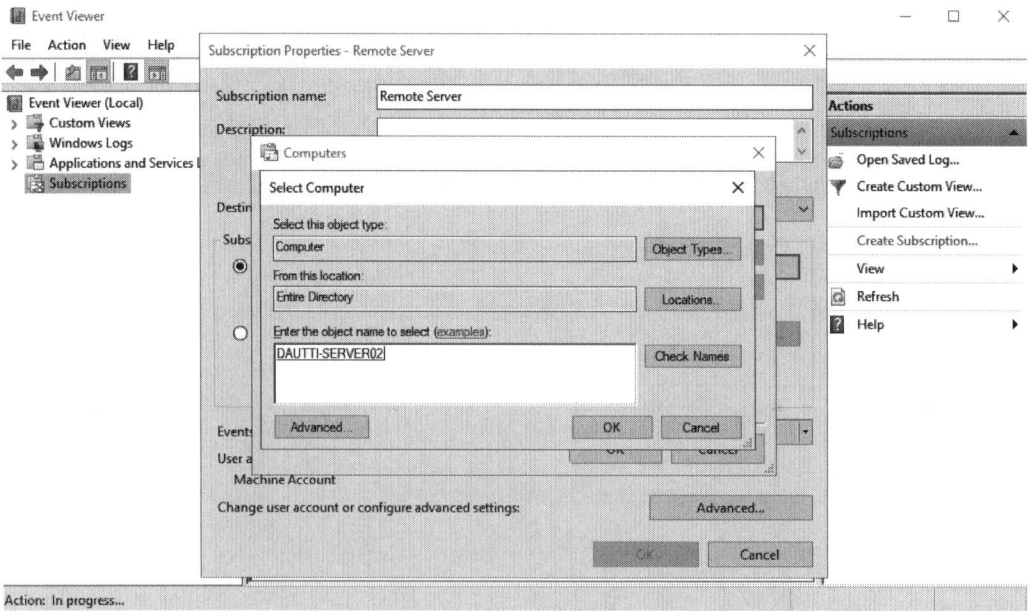

Figure 11.21 – Adding a Remote Server to collect events

12. In the **Subscription Properties** window, click on the **Select Events…** button and select **Edit**.
13. Set the event logs filtering criteria you want to collect in the **Query Filter** window and click **OK**.
14. Click on the **Advanced…** button to ensure that the machine account is the chosen option. Then, click **OK**.
15. Click **OK** to close the **Subscription Properties** window.

Now that we have centralized monitoring, let's filter Event Viewer logs.

Filtering Event Viewer logs

To filter the Event Viewer logs in Windows Server 2022, follow these steps:

1. Press the Windows key + *R*, enter `eventvwr.msc`, and press *Enter*.
2. Expand **Windows Logs** and select the log type that you want to filter.
3. In the **Actions** pane, click on **Filter Current Log...**, as shown in the following screenshot:

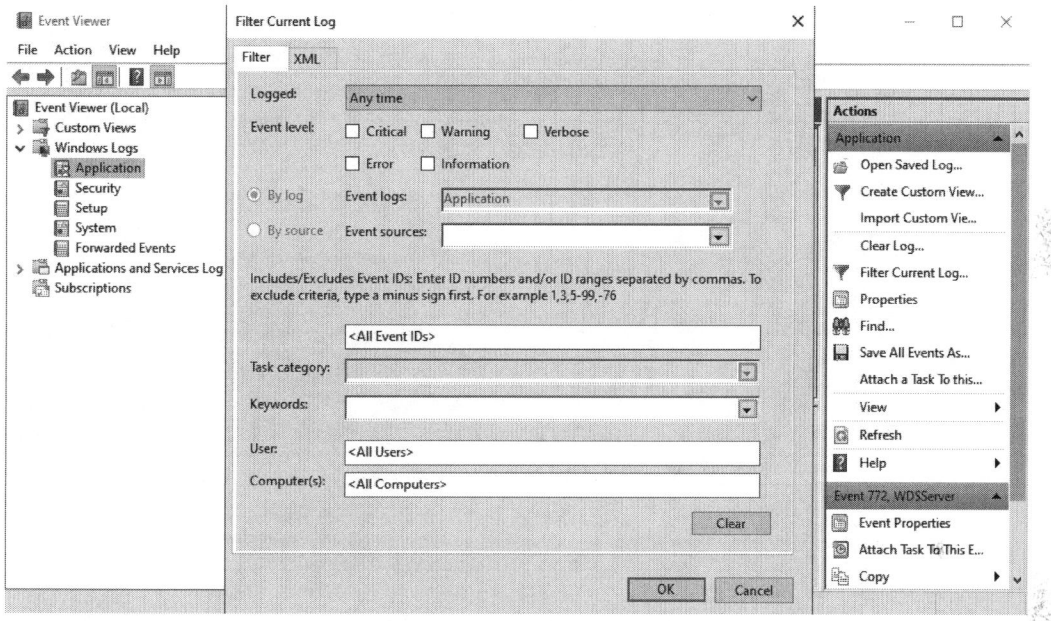

Figure 11.22 – Filtering Event Viewer logs

4. Set the filtering criteria in the **Filter Current Log** window to get the desired results.
5. Click **OK** to close the **Filter Current Log** window.

Finally, let's change the default logs location, which specifies the logging directory.

Changing the default logs location

To change the default logs location in Windows Server 2022, follow these steps:

1. Press the Windows key + *R*, enter `regedit`, and press *Enter*.
2. Locate the `HKEY_LOCAL_MACHINE\System\CurrentControlSet\Services\EventLog\System` path.

3. Open the `File` value within the `System` folder, enter the new path in the **Value data** text box, and click **OK**:

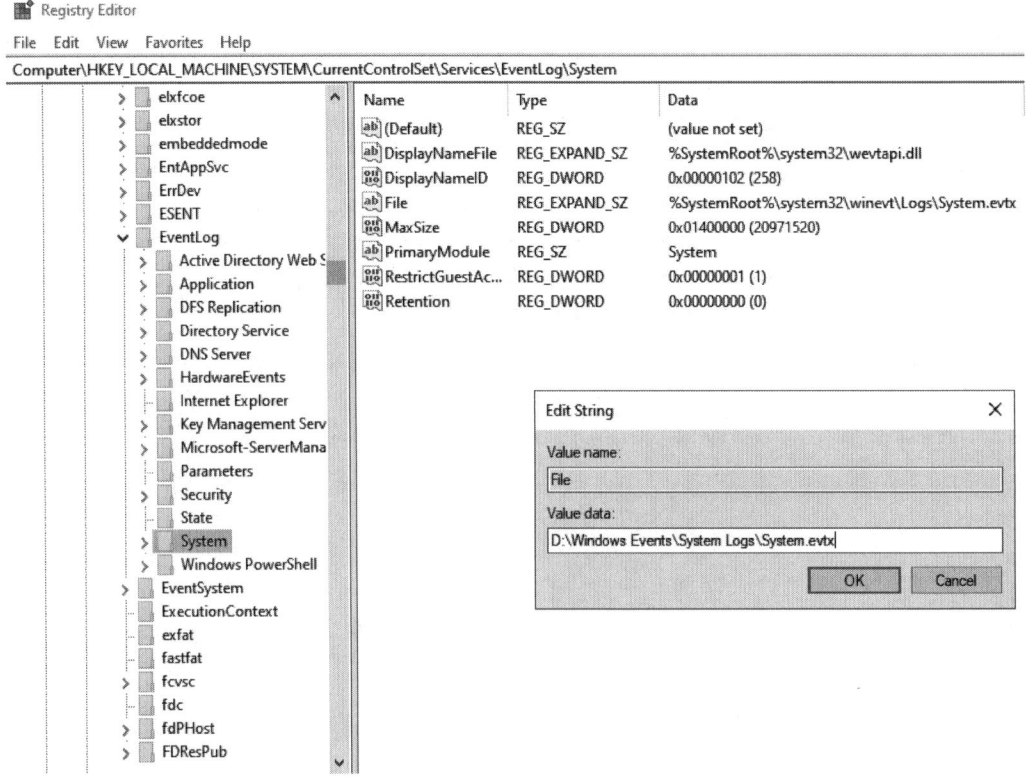

Figure 11.23 – Changing the default logs location in Windows Server 2016

4. Locate `HKEY_LOCAL_MACHINE\System\CurrentControlSet\Services\EventLog\Application` to change the default location for application logs.

5. Locate `HKEY_LOCAL_ MACHINE\System\CurrentControlSet\Services\EventLog\Security` to change the default location for security logs.

6. Close the **Registry Editor** window.

This was a helpful exercise as it explained various ways of using **Event Viewer** to manage and monitor logs in Windows Server 2022.

Summary

This chapter taught you how to update and troubleshoot Windows Server 2022. You were provided with valuable information about updating Windows Server 2022, which helps add new features to the server and increase its security. Moreover, you learned about various troubleshooting methodologies that will help you take on the challenge of resolving issues. Furthermore, you learned about the different technologies you can use to add redundancy to the server.

Finally, this chapter concluded with an exercise that provided instructions on using Event Viewer to monitor and manage logs.

In the next chapter, you will learn how to study and prepare for the Microsoft certification exam.

Questions

Answer the following questions to test your knowledge of this chapter:

1. A boot sector is a sector on a server's ROM that contains the required information so that you can boot the server. (True | False)
2. _____ is an MMC snap-in that enables system administrators to monitor events in servers.
3. Which of the following is a troubleshooting method?

    A. Rational approach

    B. Pragmatic approach

    C. Systematic approach

    D. Specific approach

4. Apple Product and Support Services engineers use a six-step troubleshooting model known as the detection method. (True | False)
5. _____ is a device with a battery that continues to supply the server with power when a power outage occurs.
6. Which of the following are Event Viewer types of logs?(Choose two)

    A. Application

    B. Security

    C. Software

    D. Driver

7. The DRP is a well-structured plan that ensures the organization will continue providing services or recovering from situations when a disaster occurs as soon as possible. (True | False)

8. _____ is a diagnostic test that verifies whether or not the server hardware is working correctly.

9. Which of the following are Windows bootloaders? (Choose two)

   A. NTLDR
   B. BOOTMGR
   C. BOOT.INI
   D. BCDEDIT.EXE

10. The Basic Input/Output System, known as the BIOS, is a program that controls the functionality of the server's hardware components. (True | False)

11. _____ refers to a group of servers that combine processor power, RAM, storage capacity, and network interfaces to achieve the high availability of services.

12. Which of the following are backup types? (Choose two)

    A. Incremental
    B. Differential
    C. Arithmetic
    D. Geometric

13. Discuss the startup process.

14. Discuss the troubleshooting process.

15. Discuss Event Viewer filtering and central logging.

Further reading

To learn more about the topics that were covered in this chapter, take a look at the following resources:

- *How Windows Update works*: https://docs.microsoft.com/en-us/windows/deployment/update/how-windows-update-works
- *Troubleshooting Windows Server components*: https://docs.microsoft.com/en-us/windows-server/troubleshoot/windows-server-troubleshooting
- *Advanced troubleshooting for Windows boot problems*: https://docs.microsoft.com/en-us/windows/client-management/advanced-troubleshooting-boot-problems
- *What is BCDR? Business continuity and disaster recovery guide*: https://www.techtarget.com/searchdisasterrecovery/definition/Business-Continuity-and-Disaster-Recovery-BCDR

Part 5: Studying and Preparing for Microsoft Certification Exams

Part 5 will focus mainly on Microsoft certification exams and how to prepare for them. In this part, you will be introduced to AZ-800 domain objectives, enabling you to prepare for the AZ-800 certification exam. By getting acquainted with these domain objectives, you will be able to identify your weak points and convert them into strengths, and thus attain the level of self-confidence needed to pass the certification exam. Finally, this part provides detailed information about the newly introduced Microsoft Role-based Certifications.

This part of the book comprises the following chapter:

- *Chapter 12, Preparing for Microsoft Certifications*

12
Preparing for Microsoft Certifications

This chapter is designed to provide an overview of Microsoft Certifications. Furthermore, this chapter contains detailed explanations about certificates and certifications. Finally, this chapter includes suggestions on preparing for certification exams, including considerations for the exam day.

Throughout the chapter, you will find valuable resources that will help you become informed about the Microsoft Certifications and what it takes to pass a certification exam successfully and launch a successful career with it. Only by doing so will you achieve the adequate skills to pass a Microsoft Certifications exam without any hurdles and become a **Microsoft Certified Professional** (**MCP**).

This chapter will cover the following topics:

- What is Microsoft Certification?
- What is Microsoft role-based certification?
- Who should take a Microsoft Certification exam?
- Which skills are measured by Microsoft Certification exams?
- What should you expect in a Microsoft Certification exam?
- How should you prepare for a Microsoft Certification exam?
- How do you register for a Microsoft Certification exam?
- On the day of the Microsoft Certification exam
- New Microsoft Certification validity period and renewal format

What is Microsoft Certification?

From identification to validation and training to certification, a Microsoft certificate is a document that shows and proves that the person with the certificate has the skills, knowledge, experience, and talent for the technology for which the certificate is issued.

Todd Thibodeaux, president and CEO at CompTIA, said, *"The certificate represents a reliable current and advanced knowledge mechanism."* But that does not mean that every Microsoft certificate expressed in the general term necessarily means that you have completed Microsoft Certification. For example, suppose you have attended a Microsoft technologies training course at one of the Microsoft Learning Partners. It may have happened that the training was not mapped to a specific Microsoft Certification exam. In that case, you ended up receiving an attendance certificate. As such, that is not considered Microsoft Certification! From there, the question then arises: what is Microsoft Certification? I think, by now, you have already learned the answer.

Microsoft Certification is a process in which the participant attends the training, which teaches the candidate the technology skills and prepares the candidate for the certification exam. Then, successfully passing the certification exam and obtaining the credential or, more precisely, the title offered by the certification exam completes the cycle of the so-called Microsoft Certification. In conclusion, a candidate who attends training and passes the Microsoft Certification exam is awarded a Microsoft certificate and considered a Microsoft Certified Professional.

> **Important note**
> You can find more information about Microsoft Certifications at `https://docs.microsoft.com/en-us/learn/certifications/`.

Since, in recent years, Microsoft has introduced role-based certifications, it would make sense to introduce you to this new type of certification.

What is Microsoft role-based certification?

In early 2019, Microsoft launched new certifications designed to offer career-focused qualifications that neatly align with existing job roles. These new certifications from Microsoft were called **role-based certifications** to show that certified candidates are keeping pace with today's technical roles and requirements. By becoming a certified candidate, you prove your expertise to employers and peers and get the recognition and opportunities you have earned.

Maybe in the back of your head, you still question why Microsoft needed to switch from product-oriented to role-based certifications. As you might know, nowadays, within any company or organization, be it tech-based or otherwise, IT roles are no longer as general as a *system administrator* or *system engineer*. Instead, these titles have subheadings such as *Azure administrator* or *Azure security engineer,* and some even further subcategorize into the high niche and specific roles. Therefore, Microsoft targeted these particular career roles with the new certification rollout to provide more focused and job-specific training instead of a general overview of a career sector.

> **Important note**
> You can find more information about Microsoft role-based certifications at https://docs.microsoft.com/en-us/learn/certifications/posts/new-role-based-certification-and-training-is-here.

If you are unsure whether to take the Microsoft Certifications exam, the following section should help you clear the fog and get a clearer view of Microsoft Certifications.

Who should take a Microsoft Certification exam?

By the time you read this chapter, you either are or are not certified in Microsoft technologies. But, if you don't hold a Microsoft Certification, don't feel embarrassed because this chapter is for you as much as it is for those with a Microsoft Certification. While certified candidates will be informed about the new format in Microsoft Certification in this chapter, you will have the opportunity to understand Microsoft Certifications and how to obtain one.

From having evidence that you have the right technical skills and knowledge to the desire for better opportunities, many other factors will make you want to obtain an IT certificate. However, it is essential to remember that organizations and businesses are interested in hiring certified candidates. While hiring certified candidates in developed economies has become standard, companies and organizations in transition and developing economies have gradually begun to practice the search and recruitment of certified candidates. Thus, while an application from a certified candidate means a less time-consuming and low-cost hiring process for businesses, having an IT certificate implies that training and certifications are vital to fuel a successful career. Therefore, Microsoft Certification makes technology professionals more likely to do the following:

- Have increased employment opportunities.
- Prove that you are an expert in the field.
- Be capable of applying theoretical knowledge to practical work.
- Quickly adapt to the work environment.
- See an increase in opportunities for a prosperous professional career.

Microsoft Certification provides a professional edge by providing globally recognized industry-endorsed evidence of skills mastery, demonstrating the ability and willingness to embrace new technologies. Hence, getting certified all begins with the desire to learn. The higher the desire to learn, the greater the chances of success in obtaining the technology skills. Note that learning underlines that the candidate must be prepared theoretically and practically in the preferred technology. Microsoft understands that everyone has different learning preferences, and provides training and certification options throughout the certification journey. The following sites offer resources to candidates interested in learning about technology and getting certified:

- **Free of cost learning paths to prepare**: Microsoft Learn (`https://learn.microsoft.com/`), where anyone can master core concepts at their desired speed and schedule. In addition, candidates can access training materials, code samples, and test-drive products at no cost.

- **Prepare with instructor-led training**: Microsoft Learning Partners (`https://partner.microsoft.com/en-US/membership/learning-partners`), which offers a breadth of solutions to suit learning needs, empowering to achieve training goals.

- **Practice before the exam**: MeasureUp (`https://www.measureup.com/`), which offers official practice tests designed to help candidates prepare for and pass certification exams.

Once you pass a certification exam and receive an email from Microsoft claiming that you are now a certified candidate, you will receive an email to claim your certification badge shortly after. All badges earned from Microsoft based on your past certification history are available through the Credly portal (`https://info.credly.com/`). So, log in with your account and claim them!

> **Important note**
> This link lets you know how Microsoft Certification exams are organized in the relevant tracks: `https://query.prod.cms.rt.microsoft.com/cms/api/am/binary/RE2PjDI`.

Now, let's look at the skills measured by the Microsoft Certification exam to help you get prepared for a certification exam.

Which skills are measured by Microsoft Certification exams?

The skills measured represent a general guideline for a particular certification exam issued by the exam client (in this case, Microsoft) of what is likely to be included in the exam. At the same time, Microsoft periodically updates the skills measured to reflect the exam's content better. So, for example, you will find that the AZ-800 exam measures the skills listed in the following subsections. The AZ-800 exam is an example because, at the time of writing this book, the MTA 98-365, the coded name for Microsoft's Windows Server Administration Fundamentals certification, is scheduled to retire on June 30, 2022, meaning that effective from July 1, 2022, the MTA exams delivered through Certiport and Pearson VUE will terminate.

The AZ-800 exam is designed to administer core Windows Server workloads using on-premises, hybrid, and cloud technologies. Therefore, candidates who intend to take the AZ-800 exam should have expertise in implementing and managing on-premises and hybrid solutions, such as identity, management, computing, networking, and storage. In addition, they must be able to use administrative tools and technologies, such as Windows Admin Center, PowerShell, Azure Arc, and **Infrastructure as a Service (IaaS)** virtual machine administration.

> **Important note**
> You can find more information about the skills measured for the AZ-800 exam at https://query.prod.cms.rt.microsoft.com/cms/api/am/binary/RWKI0r.

Next, let's look at the AZ-800 exam skills that are measured, which will help you become familiar with the topics of the questions you can expect to have on the certification exam.

Deploy and manage Active Directory Domain Services (AD DS) in on-premises and cloud environments (30–35%)

To accomplish this exam objective, you must know how to deploy and manage domain controllers and configure and manage the multi-site, domain, and forest environments. You should also be able to take care of the AD DS security principal, implement and manage hybrid identities, and finally, address the Windows Server using group policies.

Deploy and manage AD DS domain controllers

This objective may include, but is not limited to, the following:

- Deploy and manage domain controllers on-premises.
- Deploy and manage domain controllers in Azure.
- Deploy **Read-Only Domain Controllers** (**RODCs**).
- Troubleshoot **flexible single master operations** (**FSMOs**) roles.

Configure and manage multi-site, multi-domain, and multi-forest environments

This objective may include, but is not limited to, the following:

- Configure and manage forest and domain trusts.
- Configure and manage AD DS sites.
- Configure and manage AD DS replication.

Create and manage AD DS security principals

This objective may include, but is not limited to, the following:

- Create and manage AD DS users and groups.
- Manage users and groups in multi-domain and multi-forest scenarios.
- Implement **group-managed service accounts (gMSAs)**.
- Join Windows Servers to AD DS, Azure AD DS, and Azure AD.

Implement and manage hybrid identities

This objective may include, but is not limited to, the following:

- Implement Azure AD Connect.
- Manage Azure AD Connect synchronization.
- Implement Azure AD Connect cloud synchronization.
- Integrate Azure AD, AD DS, and Azure AD DS.
- Manage Azure AD DS.
- Manage Azure AD Connect Health.
- Manage authentication in on-premises and hybrid environments.
- Configure and manage AD DS passwords.

Manage Windows Server by using domain-based Group Policies

This objective may include, but is not limited to, the following:

- Implement a Group Policy in AD DS.
- Implement Group Policy preferences in AD DS.
- Implement a Group Policy in Azure AD DS.

Manage Windows Servers and workloads in a hybrid environment (10–15%)

To accomplish this exam objective, you must know how to manage Windows Servers on-premises and in the cloud. Then, you should be able to manage workloads in the Azure environment.

Manage Windows Servers in a hybrid environment

This objective may include, but is not limited to, the following:

- Deploy a Windows Admin Center gateway server.
- Configure a target machine for Windows Admin Center.
- Configure PowerShell remoting.
- Configure CredSSP or Kerberos delegation for second hop remoting.
- Configure **Just Enough Administration (JEA)** for PowerShell remoting.

Manage Windows Servers and workloads by using Azure services

This objective may include, but is not limited to, the following:

- Manage Windows Servers by using Azure Arc.
- Assign Azure Policy's Guest Configuration.
- Deploy Azure services using Azure Virtual Machine extensions on non-Azure machines.
- Manage updates for Windows machines.
- Integrate Windows Servers with Log Analytics.
- Integrate Windows Servers with Azure Security Center.
- Manage IaaS **Virtual Machines** (**VMs**) in Azure that run Windows Server.
- Implement Azure Automation for hybrid workloads.
- Create runbooks to automate tasks on target VMs.
- Implement **Data Security Center (DSC)** to prevent configuration drift in IaaS machines.

Manage virtual machines and containers (15–20%)

To accomplish this exam objective, you must know how to manage VMs using Hyper-V manager. Then, be able to set up and manage containers and work with Windows Server VMs in the Azure environment.

Manage Hyper-V and guest virtual machines

This objective may include, but is not limited to, the following:

- Enable VM-enhanced session mode.
- Manage VM using PowerShell remoting, PowerShell Direct, and HVC.exe.

- Configure nested virtualization.
- Configure VM memory.
- Configure Integration Services.
- Configure Discrete Device Assignment.
- Configure VM resource groups.
- Configure VM CPU groups.
- Configure hypervisor scheduling types.
- Manage VM checkpoints.
- Implement high availability for VMS.
- Manage VHD and VHDX files.
- Configure a Hyper-V network adapter.
- Configure **Network Interface Card (NIC)**.
- Teaming.
- Configure a Hyper-V switch.

Create and manage containers

This objective may include, but is not limited to, the following:

- Create Windows Server container images.
- Manage Windows Server container images.
- Configure container networking.
- Manage container instances.

Manage Azure Virtual Machines that run Windows Server

This objective may include, but is not limited to, the following:

- Manage data disks.
- Resize Azure VMs.
- Configure continuous delivery for Azure VMs.
- Configure connections to VMs.
- Manage Azure VMs network configuration.

Implement and manage an on-premises and hybrid networking infrastructure (15–20%)

To accomplish this exam objective, you must know how to deploy name resolutions and manage IP addressing and network connectivity both on-premises and in the cloud.

Implement on-premises and hybrid name resolution

This objective may include, but is not limited to, the following:

- Integrate DNS with AD DS.
- Create and manage zones and records.
- Configure DNS forwarding/conditional forwarding.
- Integrate Windows Server DNS with Azure DNS private zones.
- Implement **Domain Name System Security Extensions (DNSSEC)**.

Manage IP addressing in on-premises and hybrid scenarios

This objective may include, but is not limited to, the following:

- Implement and manage **Internet Protocol Address Management (IPAM)**.
- Implement and configure the **Dynamic Host Configuration Protocol (DHCP)** server role (on-premises only).
- Resolve IP address issues in hybrid environments.
- Create and manage scopes.
- Create and manage IP reservations.
- Implement DHCP high availability.

Implement on-premises and hybrid network connectivity

This objective may include, but is not limited to, the following:

- Implement and manage the Remote Access role.
- Implement and manage Azure Network Adapter.
- Implement and manage Azure Extended Network.
- Implement and manage the Network Policy Server role.
- Implement Web Application Proxy.
- Implement Azure Relay.

- Implement a site-to-site VPN.
- Implement Azure Virtual WAN.
- Implement Azure AD Application Proxy.

Manage storage and file services (15–20%)

To accomplish this exam objective, you must know how to configure and manage file synchronization in Azure and file shares and storage in Windows Server.

Configure and manage Azure File Sync

This objective may include, but is not limited to, the following:

- Create an Azure File Sync service.
- Create sync groups.
- Create cloud endpoints.
- Register servers.
- Create server endpoints.
- Configure cloud tiering.
- Monitor File Sync.
- Migrate DFS to Azure File Sync.

Configure and manage Windows Server file shares

This objective may include, but is not limited to, the following:

- Configure Windows Server file share access.
- Configure file screens.
- Configure **File Server Resource Manager** (**FSRM**) quotas.
- Configure BranchCache.
- Implement and configure DFS.

Configure Windows Server storage

This objective may include, but is not limited to, the following:

- Configure disks and volumes.
- Configure and manage Storage Spaces.

- Configure and manage Storage Replica.
- Configure Data Deduplication.
- Configure SMB Direct.
- Configure Storage **Quality of Service (QoS)**.
- Configure filesystems.

Now that you have understood the domain objectives (that is, skills measured) of the AZ-800 exam, let's look at what you should expect in the Microsoft Certification exam.

What should you expect in a Microsoft Certification exam?

As you already know, exams have questions, and in Microsoft Certification exams, you can expect to have between 40 to 60 questions. Furthermore, the exam's duration varies depending on the category of the certification exam you are taking, including additional minutes for an introduction and a survey. Usually, the passing score is 700. There is a **mark for a review** or a **flag for a review** option, which means you can utilize it if you want to review the question(s) once you have answered all of them and have not yet ended the exam's timing. However, that is available only if you have managed the exam time accordingly. You can move back and forth within questions by clicking the **Previous** and **Next** buttons. In this way, you can change your best choices (that is, answers) if you need to.

> **Important note**
>
> You can learn more about exam duration and question types by exploring the content at `https://docs.microsoft.com/en-us/learn/certifications/exam-duration-question-types`.

How should you prepare for a Microsoft Certification exam?

There is no written standard for preparing for Microsoft Certification exams. However, you can rely upon best practices for exam preparation. Therefore, part of the best practices while getting ready for Microsoft Certification exams is as follows:

- Actively working and gaining experience in the ICT industry for 6-12 months
- Attending Microsoft training at a Microsoft Learning Partner
- Reading Microsoft technologies books
- Practicing with Microsoft technologies
- Getting certified by other vendors, for example, CompTIA certification
- Attempting practice tests to become familiar with the exam questions you are about to take

- Reviewing the exam's skills measured to identify and pay closer attention to areas of improvement
- Interacting with online and offline friends who have passed a Microsoft Certification exam and learning from their experiences

> **Important note**
> Get familiar with the Microsoft Certification exams user interface by exploring the demonstration exam simulator that looks, works, and feels like a real Microsoft exam: `https://aka.ms/examdemo`.

How do you register for a Microsoft Certification exam?

As far as registering for a Microsoft Certification exam, it can be done in two ways:

- Online via `www.pearsonvue.com` (requires a web account)
- Contacting a nearby Pearson VUE-authorized testing center (find a test center at `https://wsr.pearsonvue.com/testtaker/registration/SelectTestCenterProximitySearch/MICROSOFT?conversationId=199832`)

Regarding delivering the exam, Microsoft Certification exams are offered by both Certiport and Pearson VUE. In both cases, there are two methods of exam delivery:

- Proctored exams delivered at a test center. A test center is a facility that Certiport or Pearson VUE has authorized to provide certification exams for test-takers.
- Self-administered online exams are delivered at home or office. During the Covid-19 pandemic, many vendors (client exams) including Microsoft have begun providing online examinations that can be taken in the home or office through Pearson VUE.

> **Important note**
> To schedule your Microsoft Certification exam with Pearson VUE, navigate to `https://home.pearsonvue.com/microsoft`.

On the day of the Microsoft Certification exam

Make sure that you have slept well the night before the exam. Do not stress trying to remind yourself of what you have learned while preparing for the exam. Make sure to arrive at the test center 30 minutes before the scheduled exam. Ensure that you carry the required ID. In the case of an online examination, begin with admission procedures 30 minutes in advance. Be polite with the test center administrator and carefully read the Pearson VUE Candidate Rules Agreement when you are offered that. Then, when sitting in front of the delivery workstation, relax, take a deep breath, build up your self-esteem by saying a prayer, read the exam instructions carefully, and begin the exam. Read each question attentively and do not rush to answer the questions before reading each given answer with the same amount of focus. Remember, you can mark questions for review or hit the **Previous** button to go back to questions you have already answered. So, do not waste time on questions you have doubts about.

At the same time, be rational with the exam time because even though there is enough time at your disposal, if you do not manage it properly, that may not suffice. Do not let panic get in your way. Instead, enjoy the exam fully by having fun while reading the exam questions and selecting the best choice from among the given answers. Like that, you can answer all exam questions in time.

Once you have answered all the questions, ended your exam, and completed the survey, you will be shown the result. Naturally, it is joyful to realize that you have passed the exam. However, if the exam result is not what you expected, do not despair. Instead, accept the result as it is, and as of the next day, begin preparing to retake the exam by identifying the exam objectives for which you performed insufficiently. Remember that you now have exam experience, which will help you prepare for the exam again and pass it successfully. Good luck!

> **Important note**
> You can familiarize yourself with the Pearson VUE Candidate Rules Agreement by visiting the following URL: `https://home.pearsonvue.com/candidate-rules-agreement`.

New Microsoft Certification validity period and renewal format

Technology is moving fast! Therefore, it is required to be at a pace similar to technology development. That means your skills must be current so that you do not have to play catch-up constantly. That is probably one of the main reasons Microsoft continually strives to provide new and improved certification formats to make it easier for candidates to keep their Microsoft Certifications active.

In the spring of 2021, Microsoft announced a new certification validity period and renewal format. At the time of writing this book, according to the new design, the validity of Microsoft exams, excluding the Fundamentals track, is valid for one year. Regarding the renewal of certification, this process is free of cost and is done through the Microsoft Learn portal. Of course, Microsoft will notify the candidate of the certification renewal process over email six months before the certification expires.

> **Important note**
> You can learn more about the new renewal format of Microsoft Certification here: `https://docs.microsoft.com/en-us/learn/certifications/renew-your-microsoft-certification`.

Summary

This chapter has taught you about Microsoft Certification and how to earn one. At the same time, you became familiar with the types of Microsoft Certification and the steps that will help you obtain certification, thus learning who should take a Microsoft Certification exam.

Through the example of the AZ-800 exam, this chapter provided factual information about the skills measured in the Microsoft Certification exam that will help you get an idea of how the questions in the exam will be organized. Furthermore, you have learned what to expect, how you should prepare, and, more importantly, how to register for the exam. Finally, you have been informed about Microsoft Certifications' new validity period and renewal format.

I wish you good health and success in your certification exam.

Further reading

- *Microsoft Technical Certifications*: `https://docs.microsoft.com/en-us/learn/certifications/`
- *Microsoft's Learning and Development Services*: `https://learn.microsoft.com/`
- *MTA: Windows Server Administration Fundamentals – Skills Measured*: `https://query.prod.cms.rt.microsoft.com/cms/api/am/binary/RWIw9X`

Assessments

This section contains answers to questions from all chapters.

Chapter 1 – Getting Started with Windows Server

1. True
2. Clients and servers
3. All of the above
4. True
5. Windows Server
6. True
7. Peer-to-peer (P2P) and client/server
8. True
9. Hardware and software
10. IPv4 and IPv6

Chapter 2 – Introducing Windows Server 2022

1. False
2. Docker
3. Windows Server 2022 Datacenter and Windows Server 2022 Standard
4. True
5. System Insights
6. A 1.4 GHz 64-bit processor
7. False
8. Storage Migration Services
9. True
10. Kubernetes
11. Windows Admin Center

Chapter 3 – Installing Windows Server 2022

1. GUID Partition Table (GPT)
2. False.
3. Nano server.
4. Windows Assessment and Deployment Kit (Windows ADK) and Microsoft
5. Deployment Toolkit (MDT)
6. False
7. A migration
8. Desktop Experience, Server Core, and Nano Server
9. Windows Server 2019 installation files must be on DVD media and bootable. Like DVD media, a USB flash drive is required to contain the Windows Server 2019 installation and be bootable. The network boot requires setting up a WDS server so that Windows Server 2019 is installed over the network.
10. The clean installation overwrites the existing operating system on a hard disk. Then, the WDS server enables installation over the network. An unattended installation has little or no interactivity with the operating system installation. Tools such as the Windows ADK and MDT provide a unique platform to automate desktop and server deployments. An upgrade replaces your existing OS with a new one. Migration occurs when you bring in a new machine (physical or virtual) and want to move the roles, features, apps, and settings into it.

Chapter 4 – Post-Installation Tasks in Windows Server 2022

1. True
2. Plug and play
3. Interrupt Request (IRQ) and Direct Memory Access (DMA)
4. True
5. Windows registry
6. Devices and Device Manager
7. Services Control Manager and Registry Editor
8. Service account

9. Any changes made to your server are stored in the registry. The Windows Registry is a hierarchical database that stores the hardware/software configuration and system security information. After you access the registry, you will notice that its console tree (on the left-hand side) consists of five registry keys known as hives (that is, HKEYs): HKEY_CLASSES_ROOT (HKCR), HKEY_CURRENT_USER (HKCU), HKEY_LOCAL_MACHINE (HKLM), HKEY_USERS (HKU), and HKEY_CURRENT_CONFIG (HKCC).

10. Services are background services that keep the OS alive. When accessing services through the Services Control Manager, you will notice that each service has a description that helps us understand its purpose. Each service has the following start-up types: Automatic, Automatic (Delayed Start), Manual, and Disable.

Chapter 5 – Directory Services in Windows Server 2022

1. True
2. Group nesting
3. Roaming Profile and Mandatory Profile.
4. False
5. Replication topology
6. Global group and Universal group.
7. True
8. Domain Controller
9. Active Directory Administrative Center and Active Directory Users and Computers
10. True
11. Primary Zone
12. Master schema and domain naming master
13. Active Directory (AD), a Microsoft technology, is a distributed database that stores objects in a hierarchical, structured, and secure format. AD objects represent users, computers, peripheral devices, and network services. Each object is uniquely identified by its name and attributes. DNS has a tree structure (hierarchical) where each branch represents the root zone, and each leaf has zero or more resource records. Each zone represents a root domain or multiple domains and subdomains. A domain name consists of one or more parts, called labels, and these are separated by points (for example, packtpub.com). DNS is maintained by a database that uses distributed clients/server architecture where network nodes represent the servers' names.
14. Microsoft's recommendations for effectively using group nesting when assigning permissions are both Accounts, Global, Domain Local, Permissions (AGDLP) and Accounts, Global, Universal, Domain Local, Permissions (AGUDLP).

Chapter 6 – Adding Roles to Windows Server 2022

1. True
2. File Transfer Protocol (FTP).
3. Modify, Write, and Read.
4. False
5. Software port
6. Simple Mail Transfer Protocol (SMTP) and Post Office Protocol (POP).
7. True
8. Secure Sockets Layer (SSL)
9. 3389
10. False
11. Share permissions
12. Change and Read
13. The remote Access role in Windows Server 2019 enables remote access to resources inside an organization's network. Remote Desktop Services (RDS) allows GUI remote access to computers within an organization's network and over the internet.
14. Users can be allowed or denied access to the objects, which can be said to be related to user rights. Each allowance or denial has specific permissions that determine the type of access to the objects. For example, share permissions have to do with user access to the shared folders and drives on the network.

Chapter 7 – Group Policy in Windows Server 2022

1. True
2. Group Policy Objects (GPOs)
3. Enabled and Disabled
4. True
5. Forest pane and GPOs pane
6. gpupdate /force
7. True
8. Local Group Policy Editor
9. Turned on

Chapter 8 – Virtualization with Windows Server 2022

1. True
2. Hyper-V architecture
3. Fully Virtualized mode and Paravirtualized mode
4. True
5. Virtualization Service Providers (VSP) and Virtualization Service Consumers (VSC)
6. Production Checkpoints and Standard Checkpoints.
7. True
8. Hyper-V Manager
9. Hypervisor and Root
10. Nested virtualization refers to a VM that runs inside another VM. In other words, the server's hardware can run the Hyper-V inside a VM, which also runs on a Hyper-V.
11. When virtualization becomes the primary network service driver, organizations are migrating their Active Directory Users and Computers (P2V) for cost, ease of management, and future expansion. Hence, knowing that VMs are using VHDs, Microsoft engineers have developed the Disk2vhd app to convert the Physical Disk Drive (PHD) to the Virtual Hard Disk (VHD).
12. Despite the reasons that may stand behind the decision to do Virtual to Physical (V2P) conversion, it is good to remind ourselves in the technological era that we live in, that the trend is for Physical to Virtual (P2V) conversion. Other than that, it can be said that hypervisor manufacturers, including Microsoft, will not encourage you to conduct V2P conversions.

Chapter 9 – Storing Data in Windows Server 2022

1. True
2. Storage area network (SAN)
3. Direct-Attached Storage (DAS) and network-attached storage (NAS).
4. False
5. Disk controller
6. Small Computer System Interface (SCSI) and Fiber Channel (FC)
7. True
8. High Availability (HA)
9. RAID 1 and RAID 5. True.
10. Advanced Technology Attachment (ATA), also known as Integrated
11. Drive Electronics (IDE)

12. CD-ROM and DVD-RAM

13. The idea behind data deduplication (dedup) is to provide disk space savings

14. Storage Spaces Direct (S2D) is an enhanced feature in Windows Server 2019 that enables you to group disks into storage pools, creating software-defined storage spaces

15. Distributed File Systems (DFS) enable the sharing of data from the server in an authorized and controlled way

Chapter 10 – Tuning and Maintaining Windows Server 2022

1. True
2. Server baseline
3. Cache and Cores
4. True
5. Processors and HDDs
6. Network Load Balancing (NLB) and network separation
7. False
8. Word size
9. Performance Monitor and Resource Monitor
10. True
11. Performance monitoring
12. Random Access Memory (RAM) and Read-Only Memory (ROM)
13. Performance Monitor is a Windows MMC that monitors server performance. Resource Monitor is at your disposal to view the real-time usage of both hardware and software resources.
14. Logs are useful for detailed analysis and archiving records. Alerts enable you to be vigilant about the performance and configuration of servers.

Chapter 11 – Updating and Troubleshooting Windows Server 2022

1. False
2. Event Viewer
3. Systematic approach and specific approach
4. False
5. UPS

6. Application and Security
7. True
8. POST
9. NTLDR and BOOTMGR
10. True
11. Clustering
12. Incremental and Differential
13. The Basic Input/Output System (BIOS) is a program that controls the functionality of the server hardware components. The boot sector is the sector on the server's disk containing the information to boot your server. The bootloader is a program that loads the OS kernel into RAM. The bootloader is located in MBR. There are two types of bootloaders in Windows OSes: NTLDR and BOOTMGR. MBR is also created when disk partitions are made; however, MBR resides outside disk partitions. Multiboot: every time you turn on your computer, you notice a boot menu that lists multiple OSes. Boot Configuration Data (BCD) represents a store consisting of a specific file that enables control of what should happen when an OS boots. POST is a diagnostic test that verifies that the server hardware is working correctly. Finally, safe mode is a Windows diagnostic mode that uses a minimal set of drivers and services.
14. Among the dozens of available methodologies, a six-step troubleshooting model known as the detection method is used by Microsoft Product Support Services engineers. The steps are: discover the problem, evaluate system configuration, list or track possible solutions, execute a plan, check results, and take a proactive approach.
15. The Event Viewer generates an enormous number of logs; hence, event filtering is used to find the correct information to help overcome the issues. Setting the wrong criteria will result in filtered results that will not help find the accurate information to overcome the issues. The problem with event logs is that they consume storage space. Hence, changing the default logs' locations helps to overcome the lack of storage space for storing logs. That enables writing event messages to any log files due to a lack of storage space.

Index

Symbols

32-bit architectures 278
64-bit architectures 278

A

accounts
  about 146
  computer account 150
  domain account 147
  local account 148, 149
  user profile 149
Accounts, Global, Domain Local, Permissions (AGDLP) 153
Accounts, Global, Universal, Domain Local, Permissions (AGUDLP) 153
Active Directory (AD) 126
Active Directory (AD) infrastructure
  about 126, 127
  child domain 131
  domain 128, 129
  domain, comparing with workgroup 133
  domain controller (DC) 128
  forest 130
  forest functional level (FFL) 133
  functional levels 133-135
  Microsoft Passport 136
  namespaces 135
  operations master roles 132
  replication 136
  schema 136
  sites 136
  tree domain 129
  trust relationship 133
Active Directory Administrative Center 126
Active Directory (AD) protocols
  Domain Name System (DNS) 126
  Kerberos 126
  Lightweight Directory Access Protocol (LDAP) 126
Active Directory (AD) restore 323, 324
Active Directory (AD) services
  Active Directory Administrative Center 126
  Active Directory Domains and Trusts 126
  Active Directory Module for Windows PowerShell 127
  Active Directory Sites and Services 126
Active Directory Domains and Trusts 126
Active Directory Domain Services (AD DS)
  about 211
  deploying, in cloud environments 337
  deploying, in on-premises environments 337

Index

domain-based Group Policies, used for Manage Windows Server 338
domain controllers, deploying 337
domain controllers, managing 337
hybrid identities, implementing 338
hybrid identities, managing 338
multi-domain environments, configure and manage 337
multi-forest environments, configure and manage 337
multi-site environments, configure and manage 337
role, installing 154-157
security principals, creating 338
security principals, managing 338
Active Directory Domain Services Configuration Wizard 128, 130
Active Directory Federation Services (AD FS) 178
Active Directory Module for Windows PowerShell 127
Active Directory Sites and Services 126
Active Directory Users and Computers 150
Active Server Pages (ASP) 168
Administrative Tools menu
  used, for accessing GPM console 207, 208
administrator account
  renaming 216
AD namespace 141
advanced startup options
  accessing 43-45
Advanced Technology Attachment (ATA) 247
alternating current (AC) 280
AMD Virtualization 225
application programming interface (API) 50, 223

application servers
  about 164
  collaboration server 166
  database server 166
  data protection server 167, 168
  email servers 164
  monitoring server 167
Application Virtualization (App-V) 185
authoritative DNS 142
authoritative restore 324
Azure File Sync
  configuring 342
  managing 342
Azure hybrid center 30, 31
Azure Kubernetes Service 34, 36
Azure services
  used, for hybrid workloads 339
  used, for Manage Windows Servers 339
Azure Virtual Machines
  managing, with Windows Server 340

B

backup domain controllers (BDCs) 128
backups
  about 321
  differential backup 322
  full backup 321
  incremental backup 321
basic disk
  about 263
  converting, into dynamic disk 263, 264
  MBR and GPT partition schemes based 263
Basic Input/Output System (BIOS) 42, 310, 311
beep codes
  reference link 313

Index 357

blade servers 17
block-level storage
　versus file-level storage 250
Boot Configuration Data (BCD) 314, 315
bootloader
　exploring 316
Boot Manager (BOOTMGR) 314, 316
boot menu 316, 317
boot options
　about 43
　installation media 43
　network boot 43
　USB flash drive 43
boot sector
　exploring 316
bootstrap loader 316
business continuity
　Active Directory (AD) restore 323, 324
　backup and restore 321-323
　data redundancy 319
　disaster recovery plans (DRPs) 319
　exploring 318
　folder redirection 320, 321
　power redundancy 325
business continuity and disaster
　　recovery (BCDR) 167

C

case coolers 280
case fans 280
centralized monitoring
　setting up 325, 326
central processing unit (CPU) 15, 274
Certificate Authority (CA) 177
checkpoints
　about 232
　production checkpoint 234
　standard checkpoint 234
　types 234
child domain 131
classful addressing 13
classful networks 14
clean installation
　performing 46-50
Client Access Licenses (CALs) 182
clients 9
client/server network architecture 12
clustering
　about 319
　failover clustering 319
　load-balancing clustering 320
collaboration server 166
command descriptor blocks (CDBs) 253
Command-Line Interface (CLI) 18
Common Internet File System (CIFS) 252
communication protocols, database server
　data 166
　database application 166
　Java Database Connectivity (JDBC) 166
　Object Linking and Embedding
　　Database (OLEDB) 166
　Open Database Connectivity (ODBC) 166
　users 166
communication protocols, email server
　Internet Message Access
　　Protocol (IMAP) 164
　Mail Delivery Agent (MDA) 164
　Mail Transport Agent (MTA) 164
　Mail User Agent (MUA) 164
　Post Office Protocol (POP) 164
　Simple Mail Transfer Protocol (SMTP) 164
compact disks (CDs) 262
computer account 150
computer configuration GPO settings 214, 215
computer device drivers 78

computer devices
    about 78
    external device 78
    internal device 78
    network device 78
    peripheral device 78
Computer Management 37
computer network architectures
    client/server network architecture 12
    investigating 11
    P2P network architecture 11
computer network components
    clients and servers 9
    hosts and nodes 10, 11
container deployment approach 36
containerization mode 223
containers 36, 37, 143
contiguous namespace 135
Control Panel and PC settings
    access, restricting to 218
cooling 280
Credly portal
    URL 336

D

database server 166
Data Collector Sets
    setting up 288, 289
Data Protection Manager (DPM) 167
data protection server 167, 168
data redundancy 319
Dedup (Data Deduplication)
    about 254
    enabling, on Windows Server 2022 268
    installing 254

default containers
    about 144, 145
    uses 145
default groups 151, 152
default logs location
    modifying 327, 328
defer feature updates
    reference link 300
Desktop Experience
    Server Manager, using 105
Desktop Experience installation 46
device drivers
    about 78, 79
    disabling 85
    installing 82
    rolling back 85, 86
    troubleshooting 86, 87
    uninstalling 84
    updating 82, 83
    updating, in Windows Server 2022 303, 304
    working with 80
device drivers, troubleshooting options
    disable device 86
    roll back driver 86
    uninstall device 87
    update driver 86
Device Manager
    about 80
    accessing 80
devices
    about 78
    accessing 80
    adding 82
    managing 84
    removing 83
    working with 80
differential backup 322

Index   359

digital certificate 177
Direct-Attached Storage (DAS) 248, 277
direct current (DC) 280
Direct Memory Access (DMA) 87-89
Directory Services Restore Mode
    (DSRM) 156, 323
disaster recovery plans (DRPs) 319
disk 16 260
Disk2vhd app
  about 236
  reference link 236
disk controller 251
disks 277
disk types
  basic disk 263
  dynamic disk 263
  HDDs 260
  ODDs 262
  SSDs 261
  virtual hard disk (VHD) 266
Distributed File Systems (DFS)
  about 267
  installing 267
DNS namespace 141
DNS role
  installing 154-156
DNS zone
  about 141
  primary zone 141
  secondary zone 141
  stub zone 141
Docker 36
Docker Engine 36
domain
  about 128, 129
  versus workgroup 133
domain account 147

domain-based Group Policies
  used, for Manage Windows Server 338
domain controller (DC)
  about 108, 128, 324
  server, promoting 154-157
domain functional level (DFL) 133
Domain Name System (DNS) 126
  about 137
  hostname 140, 141
  hosts files 139, 140
  lmhosts files 139, 140
  role, adding 137-139
  Universal Naming Convention (UNC) 143
  Windows Internet Name
    Service (WINS) 142
  working 137
double-device data correction (DDDC) 276
driver signing 89
Driver Store 88
dynamic disk
  about 263
  basic disk, converting into 263, 264
  mirrored volume 263
  RAID-5 volumes 263
  simple volume 263
  spanned volume 263
  striped volume 263
Dynamic Source Routing (DSR) 28

E

email servers 164, 165
Encrypting File System (EFS) 265
error-correcting code (ECC) 276
Event Viewer
  about 309, 310
  using, to manage logs 325
  using, to monitor logs 325

Event Viewer logs
  filtering 327
Exchange Server 15
Exchange Server 2022 165
Extended File Allocation Table (exFAT) 266
external device 78
external switch 231
EZ Gig IV cloning software
  reference link 237

F

failover clustering 319
Fast ID Online (FIDO) Alliance 136
FC switch 253
Fiber Channel (FC) 250
File Allocation Table (FAT) 265
file-level storage
  versus block-level storage 250
file server auditing 196
File Server Resource Manager (FSRM) 342
File Services role 186, 187
file-sharing protocols
  about 252
  File Transfer Protocol (FTP) 252
  Hypertext Transfer Protocol (HTTP) 253
  Network File System (NFS) 252
  Secure Shell (SSH) 253
  Server Message Block (SMB) 252
filesystems
  about 265
  Extended File Allocation Table (exFAT) 266
  File Allocation Table (FAT) 265
  New Technology File System (NTFS) 265
  Resilient File System (ReFS) 265
File Transfer Protocol (FTP) 171, 172, 252

flag for a review option 343
flexible single master operation
    (FSMO) 133, 337
folder redirection
  examining 320, 321
forest 130
forest functional level (FFL) 133
form factor 16
full backup 321
fully virtualized mode 222
functional levels 135

G

gigahertz (GHz) 274
Global Unique Identifier (GUID) 42
GPM console
  accessing, from Administrative
    Tools menu 207, 208
  accessing, from Run dialog box 208, 209
  accessing, from Server Manager menu 209
GPOs for system administrators, examples
  about 216
  access, denying to removable
    storage classes 218
  access, restricting to Control Panel
    and PC settings 218
  administrator account, renaming 216
  guest account, renaming 217
  Microsoft accounts, blocking 217
GP settings reference spreadsheet
  download link 211
Grandfather-Father-Son (GFS) 322
graphical processing unit (GPU) 227, 279
Graphical User Interface (GUI) 18, 46
graphics cards 279

group-managed service accounts
    (gMSAs) 28, 338
group nesting 153
Group Policy (GP) 206
Group Policy (GP), editors
  about 211
  Local Group Policy Editor 211
  Group Policy Management Editor 211
Group Policy Management (GPM) 207
Group Policy Object (GPO)
  about 144, 206, 209, 320
  computer configuration GPO
      settings 214, 215
  configuration settings 209, 213
  configuring, ways 211
  managing 207
  processing 210
  user configuration GPO settings 215
groups
  about 150
  default groups 151, 152
  group nesting 153
  group scope 152
group scope
  about 152
  global group 152
  local group 152
  universal group 152
group types
  about 150
  distribution groups 151
  security groups 151
guest account
  renaming 217
guest virtual machines 339, 340
GUID Partition Table (GPT) 42, 263, 311

H

Hard Disk Drive (HDD)
  about 16, 246, 260
  reference link 261
hardware components 274
hardware port 175
hardware RAID deployment 258
hertz (Hz) 274
hidden default containers 145
high availability (HA)
  about 259, 277
  standard 259
high-density-15 (HD-15) 281
hives (HKEYs)
  about 90
  HKEY_CLASSES_ROOT 90
  HKEY_CURRENT_CONFIG 90
  HKEY_CURRENT_USER 90
  HKEY_LOCAL_MACHINE 90
  HKEY_USERS 90
Host Bus Adapter (HBA) 250, 253
host files 139
hostname 140, 141
host OS 223
hosts 10
hybrid environment
  Manage Windows Server 339
hybrid identities
  implementing 338
  managing 338
hybrid name resolution
  implementing 341
hybrid network connectivity
  implementing 341
hybrid workloads
  Azure services, using 339

Hypertext Markup Language (HTML) 170
Hypertext Transfer Protocol (HTTP) 168, 253
Hyper-V
   architecture 224, 225
   configuration settings 227
   installation requirements 225
   installing, on Windows Server 2022 240-242
   VHD format 235
   VHDX format 235
   VM management 239
   VM settings, configuring 237, 238
Hyper-V Manager
   about 226
   checkpoint, setting up for specific VM 233
   operations 226
   user interface 226
   virtual hard disks (VHDs), creating 228
   virtual network, setting up 231
   virtual switch, setting up in Windows Server 2022 231, 232
   VM's memory, managing 229, 230
Hyper-V, virtual switches
   external switch 231
   internal switch 231
   private switch 231

I

IIS Manager 169
incremental backup 321
Industry Standard Architecture (ISA) 248
information and communications technology (ICT) 246
Information Technology Infrastructure Library (ITIL)
   examining 308, 309
   reference link 309

initiators 253
in-place upgrade
   performing 58-61
input device 79
Institute of Electrical and Electronics Engineers (IEEE) 279
Integrated Drive Electronics (IDE) 247
Intel Virtualization Technology (VT) 225
interfaces, for storage technologies
   ATA 247
   PATA 247
   SATA 247
   SCSI 247
internal device 78
internal switch 231
Internet Client Printing (ICP) 188
Internet Information Services (IIS)
   about 168-170
   components, adding to 173
Internet Protocol (IP) 10
Internet Protocol version 4 (IPv4) 13
Internet Protocol version 6 (IPv6) 13
Internet Service Provider (ISP) 137
Internet Small Computer System Interface (iSCSI) 252
Interrupt Request (IRQ) 87-89
IP address
   identifying 13
   IPv4 network addresses 13
   IPv6 network addresses 14
IP addressing
   managing, in hybrid scenarios 341
   managing, in on-premises scenarios 341
IP Address Management (IPAM) 163
IP socket 186
IPv4 network addresses
   about 13
   reference link 14

IPv4 subnetting 14
IPv6 network addresses 14
iSCSI hardware 253
isolated containers 34

K

Kerberos 126
Kubernetes 34

L

lanes 248
LAN manager hosts (lmhosts) files 139
Lightweight Directory Access
    Protocol (LDAP) 126
Line Printer Daemon (LPD) Service 188
Line Printer Remote (LPR) 188
Linux community 19
Linux Server
    overview 19
Linux subsystem, on Windows Server 2022
    reference link 19
load-balancing clustering 320
local account 148, 149
local area network (LAN) 43
local GPOs
    updating 212, 213
Local Group Policy Editor
    about 210, 211
    console, accessing 212
local printer 189
local storage 248
Logical Block Addressing (LBA) 42
logical unit number (LUN) 253
logs
    managing, with Event Viewer 325
    monitoring, with Event Viewer 325
logs and alerts 289

M

macOS Server
    overview 19, 20
    reference link 20
mail server 164
manage Hyper-V 339, 340
Manage Windows Server
    Azure services, using 339
    domain-based Group Policies, using 338
    in hybrid environment 339
mark for a review option 343
Master Boot Record (MBR) 42, 263, 311-314
MeasureUp
    URL 336
memory 15, 275, 276
Microsoft accounts
    blocking 217
Microsoft Azure
    about 64
    URL 64
Microsoft certification
    about 334
    reference link 334
Microsoft certification exam
    expectations 343
    need for 335, 336
    preparations 343, 344
    registering 344
    renewal format 345
    requirement 344, 345
    skills, measuring 336, 337
    validity period 345

Microsoft Desktop Optimization
    Pack (MDOP)
  about 185
  reference link 185
Microsoft Edge Chromium 30
Microsoft Learn
  URL 336
Microsoft Learning Partners
  reference link 336
Microsoft Management Console
    (MMC) 126, 211, 284
Microsoft (MS) 166
Microsoft Passport 136
Microsoft programs
  updating, in Windows Server 2022 301
Microsoft role-based certification
  about 334, 335
  reference link 335
Microsoft technical documentation
  reference link 307
monitoring server 167
mount point 265
multi-booting 316

N

namespaces 135
Nano Server installation 46
nested virtualization
  about 225
  setting up, in Windows Server 2022 225
NetBIOS name resolution 142
network access technologies, RA
  DirectAccess 178
  Routing and Remote Access
    Service (RRAS) 178
  Web Application Proxy 178
network-attached storage (NAS) 78, 249, 322

network device 78
Network File System (NFS) 252
network installation
  performing 51-53
network interface 16, 277, 278
network interface cards (NICs)
  about 29, 277
  benefits 277
Network Load Balancing (NLB) 277
Network Operating System (NOS)
  about 17, 186
  Linux Server 19
  macOS Server 19, 20
  Windows Server 18
network printer 189
network separation 277
network services
  migrating 61-64
network storage
  about 249
  Network Attached Storage (NAS) 249
  Storage Area Network (SAN) 249
New Technology File System (NTFS) 18, 265
New Technology File System
    (NTFS) permissions
  versus share permissions 194-196
NIC teaming 277
nodes 10, 35
non-authoritative DNS 142
non-authoritative restore 324
NT Loader (NTLDR) 314, 316

O

on-premises network connectivity
  implementing 341
on-premises resolution
  implementing 341

Index   365

operations master roles 132
Optical Disk Drive (ODD)
  about 262
  reference link 263
optical disks (ODs)
  about 246, 262
  recording types 262
organizational units (OUs)
  about 143, 144
  control, delegating 146
output device 79

P

P2P network architecture 11
Parallel ATA (PATA) 247
paravirtualized mode 223
partition schemes
  about 42
    GUID Partition Table (GPT) 42
    Master Boot Record (MBR) 42
Patch Tuesday 298
PCI Express (PCIe) 248
PDS role
  about 187-189
  installing 198, 199
  services, installing 188
Peer-to-Peer (P2P) 133
performance counter alerts
  setting up 293, 294
performance data logs
  creating 292, 293
Performance Logs & Alerts service
  working with 290, 291
Performance Monitor 284, 285
performance monitoring
  about 282
  counters 288

methodology 282
  procedures 283
  server baseline 283, 284
  Task Manager 287
Performance Monitor logs folder
  accessing 291
Peripheral Component Interconnect
    Express (PCIe) ports 281
Peripheral Component
    Interconnect (PCI) 248
peripheral device 78
Personal System/2 (PS/2) ports 281
physical port 281
physical to virtual (P2V) conversion 236
Plug and Play (PnP) 79, 87
Pods 35
Power-On Self-Test (POST)
  exploring 313
  reference link 313
power redundancy 325
PowerShell Gallery
  reference link 127
power supply 280, 281
power supply units (PSUs) 280
Preboot Execution Environment (PXE) 51
primary domain controller (PDC) 128, 132
primary zone 141
printer driver deployment 192, 193
printer pooling 190, 191
private switch 231
processor 274, 275
public key infrastructure (PKI)
  about 177
  reference link 178

Q

Quality of Service (QoS) 343
Quick UDP Internet Connections (QUIC) 28

R

rack-mountable servers 16
RAID
  about 257
  hardware RAID deployment 258
  software RAID deployment 259
  RAID types
    about 258
    RAID 0 258
    RAID 1 258
    RAID 01 or 10 258
    RAID 5 258
RDS Licensing 182, 183
Read-Only Domain Controllers (RODCs) 337
read-only memory (ROM) 275
redundant array of independent disk (RAID) 280
Redundant Array of Independent Disks. *See* RAID
Redundant Array of Inexpensive Disks. *See* RAID
registry key
  about 101, 102
  adding 102
relative identifier (RID) 132
Remote Access (RA) 178
remote access server
  setting up 178
Remote Access Service (RAS) 178
remote access VPN 184
Remote Assistance 179, 180
Remote Desktop Connection (RDC) 183
Remote Desktop Gateway (RDG) 183
Remote Desktop Services (RDS) 181, 182
Remote Desktop Session Host (RDSH) 182
Remote Server Administration Tools (RSAT) 27, 37, 180, 181
removable drive 278
removable storage classes
  access, denying to 218
replication 136
Requests for Comments (RFC) 137
Resilient File System (ReFS)
  about 18, 265
  reference link 18
Resource Monitor 285, 286
restore 321
restore points 232
role-based certifications 334
role services
  adding 162
ROM chip 310
Rotations Per Minute (RPM) 260
Routing and Remote Access Service (RRAS) 178
Run dialog box
  GPM console, accessing from 208, 209

S

Safe Mode 317, 318
schema 136
Script Center
  reference link 127
secondary zone 141
Secured-core server 33
Secure Shell (SSH)
  about 253
  reference link 253

Secure Sockets Layer (SSL) 176, 177
Security Account Manager (SAM) 133
security baseline
　download link 284
Security Identifiers (SIDs) 132, 145
Serial ATA (SATA) 247
serial bus technologies 251
server
　about 9
　exploring 15
　promoting, to domain controller
　　(DC) 154-157
server baseline 283, 284
Server Configuration
　used, for performing Windows Server
　　initial configuration 114
　using, in Server Core 106
Server Core
　installation 46
　Server Configuration, using 106
server features 162
server, hardware and software
　about 15
　Central Processing Unit (CPU) 15
　disk 16
　memory 15
　network interface 16
server hardware components
　32-bit and 64-bit architectures 278
　about 274
　cooling 280
　disk 277
　graphics cards 279
　memory 275, 276
　network interface 277, 278
　physical ports 281
　power supply 280, 281

processor 274, 275
removable drive 278
Server Manager
　about 37, 163
　used, for performing Windows Server
　　initial configuration 107
　using, in Desktop Experience 105
Server Manager menu
　GPM console, accessing from 209
Server Message Block (SMB) 28 252
server roles 162
server size 16
server virtualization 222
service accounts
　about 101, 102
　adding 103, 104
service accounts, in Windows Server 2022
　local system 101
　NT Authority\LocalService 101
　NT Authority\NetworkService 101
service dependencies
　about 101, 102
　adding 104, 105
Services Control Manager 90
shape 16
share permissions
　about 195, 196
　versus NTFS permissions 194
SharePoint Server 15
SharePoint Server 2022 167
Simple Network Management
　　Protocols (SNMPs) 250
single-device data correction (SDDC) 276
single point of failure (SPOF) 280
sites 136
site-to-site VPN 184
site (website) 174, 175

Small Computer System Interface (SCSI)
  about 246, 247, 252
  reference link 252
  SCSI Parallel Interface (SPI) 247
  Serial-Attached SCSI (SAS) 247
Small Office/Home Office (SOHO) 17
Software-Defined Storage (SDS) 259
software-defined storage spaces 253
software port (application port) 175
software RAID deployment 259
Solid State Drive (SSD)
  about 16, 246, 261
  reference link 261
specific approach 307
SQL Server 15
SQL Server 2022 166
startup process
  about 310
  Basic Input/Output System (BIOS) 310, 311
  Boot Configuration Data (BCD) 314, 315
  bootloader, exploring 316
  boot menu 316, 317
  boot sector, exploring 316
  Master Boot Record (MBR) 313, 314
  Power-On Self-Test (POST) 313
  Safe Mode 317, 318
  Trusted Platform Module (TPM) 312
  Unified Extensible Firmware
    Interface (UEFI) 311
storage area network (SAN) 78, 249, 250, 322
storage management
  with Server Manager 256
  with Windows PowerShell 256, 257
Storage Migration Service 31, 32
storage protocols
  about 252
  Fibre Channel (FC) 252

Internet Small Computer System
  Interface (iSCSI) 252
Small Computer System
  Interface (SCSI) 252
Storage Replica 32
Storage Spaces Direct (S2D)
  about 253
  used, for resiliency 259
storage technologies
  about 246
  adapters 251
  block-level storage, versus
    file-level storage 250
  controllers 251
  Dedup 254
  exploring 246
  FC switch 253
  file-sharing protocols 252
  HBA 253
  HDDs 246
  interfaces, used for connecting 247
  iSCSI hardware 253
  local storage 248
  network storage 249
  optical disks 246
  PCI 248
  PCIe 248
  S2D 253
  serial bus technologies 251
  SSDs 246
  storage management 256, 257
  storage protocols 252
  storage tiering 255
storage tiering 255
stretch cluster 33
stub zone 141

subnetting
    about 13, 14
    IPv4 subnetting 14
superuser 101
systematic approach 307
System Center 2022 167
System Center Operations
        Manager (SCOM) 167

T

targets 253
TechNet
    reference link 282
Terminal Services (TS) 181
third-party programs
    updating, in Windows Server 2022 301, 302
three-dimensional (3D) 279
tower servers 17
traditional deployment approach 36
Transport Layer Security (TLS) 177
tree domain 129
troubleshooting methodology
    about 306
    best practices 306
    Event Viewer 309, 310
    guidelines 306
    Information Technology Infrastructure
        Library (ITIL), examining 308, 309
    procedures 306
    procedures, examining 307, 308
    process 306
    systematic versus specific approach 307
Trusted Platform Module (TPM) 34 312
trust relationship 133

U

unattended installation 53-57
Unified Extensible Firmware
        Interface (UEFI) 42, 311
Uniform Resource Locator (URL) 135
uninterruptible power supply (UPS) 325
Universal Naming Convention (UNC) 143
Universal Serial Bus (USB) 278
USB flash drive
    reference link 43
user configuration GPO settings 215
user profile 149
user rights 193

V

virtual hard disk (VHD)
    creating, with Hyper-V Manager 228
    mounting 266
virtualization modes
    about 222
    containerization mode 223
    fully virtualized mode 222
    paravirtualized mode 223
virtualization service consumer (VSC) 224
virtualization service provider (VSP) 224
virtualized deployment approach 36
virtual machine bus (VMBus) 224
Virtual Machine Connection
        (VMConnect) 227
Virtual Machines (VMs) 35, 339
Virtual Private Network (VPN) 178, 184, 185
virtual to physical (V2P) conversions 237
Volume Shadow Copy Service (VSS) 265

W

web management 192
web printing 191, 192
Web Server (IIS) role
  installing 197, 198
web services
  about 168
  certificates 177, 178
  FTP 171, 172
  IIS 168, 170
  ports 175, 176
  separate worker processes 172
  sites (websites) 174, 175
  SSL 176, 177
  WWW 170, 171
Windows Admin Center
  about 37
  downloading 37, 38
Windows as a service 300
Windows Assessment and Deployment
    Kit (Windows ADK)
  reference link 57
Windows Deployment Services (WDS)
  installing 67-69
  setting up 67-74
Windows Internet Name Service (WINS) 142
Windows PowerShell
  used, for setting up nested virtualization
    in Windows Server 2022 225
Windows Preinstallation Environment
    (Windows PE) 54
Windows Registry
  about 90
  working with 91
Windows Registry, with Registry Editor
  accessing 91
  managing 91

registry value, deleting 93, 94
registry value, modifying 92
registry value, renaming 92, 93
service recovery options, setting up 95, 96
service, restarting 100
service, starting 98
service, stopping 99
settings for a service, running 97, 98
start of a service, delaying 96, 97
Windows services, accessing 94, 95
Windows services, managing 94, 95
Windows Server
  about 20
  Azure Virtual Machines,
    managing that run 340
  overview 18
Windows Server 2016 135
Windows Server 2019
  Secured-core server 33
  versus Windows Server 2022 28
Windows Server 2022
  Azure hybrid center 30, 31
  Azure Kubernetes Service 34, 36
  containers 36, 37
  Dedup, enabling 268
  editions 27
  hardware requirements 29
  Hyper-V, installing on 240-242
  Microsoft Edge Chromium 30
  minimum system requirements 29
  overview 25-27
  Storage Migration Service 31, 32
  Storage Replica 32
  updating 298
  versus Windows Server 2019 28
  Windows Admin Center 37

Windows Server 2022 installation
  about 42
  advanced startup options, accessing 43-45
  boot options 43
  partition schemes 42
Windows Server 2022 installation methods
  about 45
  clean installation, performing 46-50
  Desktop Experience installation, selecting 46
  in Microsoft Azure 64-67
  in-place upgrade, performing 58-61
  Nano Server installation, selecting 46
  network installation, performing 51-53
  network services, migrating 61-64
  Server Core installation, selecting 46
  unattended installation 53-57
Windows Server Backup
  installing 322, 323
Windows Server container images
  creating 340
  managing 340
Windows Server file shares
  configuring 342
  managing 342
Windows Server initial configuration
  about 105
  performing 107
  performing, with Server Configuration 114
  performing, with Server Manager 107
  Server Configuration, using in Server Core 106
  Server Manager, using in Desktop Experience 105
Windows Server initial configuration, with Server Configuration
  about 114
  IP address, setting up 116, 117
  Remote Desktop, enabling 116

server, joining to domain 115, 116
server name, modifying 114, 115
time zone, modifying 118
updates, checking 118
Windows Server, activating 119, 120
Windows Server initial configuration, with Server Manager
  about 107
  IE enhanced security, turning off 112
  IP address, setting up 110, 111
  Remote Desktop, enabling 109, 110
  server, joining to domain 108, 109
  server name, modifying 107, 108
  time zone, modifying 112, 113
  updates, checking 111
  Windows Server, activating 113, 114
Windows Server Migration Tools (WSMT) 61
Windows Server Registry 90
Windows Server services
  about 90
  startup types 90, 91
Windows Server storage
  configuring 342
Windows Server timeline 21
Windows Server Update Services (WSUS)
  about 304
  installing 304, 305
Windows services
  about 90
  working with 91
Windows Update
  about 298-300
  URL 298
WINS server 142
workgroup
  versus domain 133
World Wide Web (WWW) 170, 171, 253

Packt.com

Subscribe to our online digital library for full access to over 7,000 books and videos, as well as industry leading tools to help you plan your personal development and advance your career. For more information, please visit our website.

Why subscribe?

- Spend less time learning and more time coding with practical eBooks and Videos from over 4,000 industry professionals
- Improve your learning with Skill Plans built especially for you
- Get a free eBook or video every month
- Fully searchable for easy access to vital information
- Copy and paste, print, and bookmark content

Did you know that Packt offers eBook versions of every book published, with PDF and ePub files available? You can upgrade to the eBook version at packt.com and as a print book customer, you are entitled to a discount on the eBook copy. Get in touch with us at customercare@packtpub.com for more details.

At www.packt.com, you can also read a collection of free technical articles, sign up for a range of free newsletters, and receive exclusive discounts and offers on Packt books and eBooks.

Other Books You May Enjoy

If you enjoyed this book, you may be interested in these other books by Packt:

Mastering Windows Security and Hardening - Second Edition

Mark Dunkerley, Matt Tumbarello

ISBN: 9781803236544

- Build a multi-layered security approach using zero-trust concepts
- Explore best practices to implement security baselines successfully
- Get to grips with virtualization and networking to harden your devices
- Discover the importance of identity and access management
- Explore Windows device administration and remote management
- Become an expert in hardening your Windows infrastructure
- Audit, assess, and test to ensure controls are successfully applied and enforced
- Monitor and report activities to stay on top of vulnerabilities

Windows Server Automation with PowerShell Cookbook - Fourth Edition

Thomas Lee

ISBN: 9781800568457

- Perform key admin tasks on Windows Server 2022/2019
- Keep your organization secure with JEA, group policies, logs, and Windows Defender
- Use the .NET Framework for administrative scripting
- Manage data and storage on Windows, including disks, volumes, and filesystems
- Create and configure Hyper-V VMs, implementing storage replication and checkpoints
- Set up virtual machines, websites, and shared files on Azure
- Report system performance using built-in cmdlets and WMI to obtain single measurements
- Apply the right tools and modules to troubleshoot and debug Windows Server

Packt is searching for authors like you

If you're interested in becoming an author for Packt, please visit `authors.packtpub.com` and apply today. We have worked with thousands of developers and tech professionals, just like you, to help them share their insight with the global tech community. You can make a general application, apply for a specific hot topic that we are recruiting an author for, or submit your own idea.

Share Your Thoughts

Now you've finished *Windows Server 2022 Administration Fundamentals*, we'd love to hear your thoughts! Scan the QR code below to go straight to the Amazon review page for this book and share your feedback or leave a review on the site that you purchased it from.

https://packt.link/r/1803232153

Your review is important to us and the tech community and will help us make sure we're delivering excellent quality content.

Made in the USA
Monee, IL
28 April 2026